<u>Front Cover Photograph</u>:
Copyright 1988 Marc Heesters
All rights reserved
Reprinted by permission

The photograph depicts a mid-air collision involving aircraft of the Italian Air Force flight demonstration team during an airshow at Ramstein Air Base, Germany, on 28 August 1988.

Military and Aviation
books by Marion Sturkey

BONNIE-SUE: A Marine Corps Helicopter Squadron in Vietnam

Murphy's Laws of Combat

Warrior Culture of the U.S. Marines (first edition)

Warrior Culture of the U.S. Marines (second edition)

MAYDAY: Accident Reports and Voice Transcripts from Airline Crash Investigations

MID-AIR: Accident Reports and Voice Transcripts from Military and Airline Mid-Air Collisions

Regional-Interest-Only
books by Marion Sturkey

GONE, BUT NOT FORGOTTEN: An Introduction

GONE, BUT NOT FORGOTTEN: Cemetery Plots and Verbatim Transcripts of Inscriptions and Epitaphs from Cemeteries of European Settlers, and their Descendants, in McCormick County, South Carolina

Flying is not inherently dangerous but, like the sea, it is unforgiving of human error.

[words on a sign at the entrance to the flight line at NAS North Whiting Field, Milton, Florida]

Rear Cover Photograph: photograph courtesy of U.S. DOD Visual Information Center

MID-AIR

Accident Reports and Voice Transcripts
from Military and Airline
Mid-Air Collisions

Marion F. Sturkey

Heritage Press International

MID-AIR: Accident Reports and Voice Transcripts from Military and Airline Mid-Air Collisions

First Edition

Library of Congress Control Number: 2007930944

ISBN: 978-0-9650814-7-4

Heritage Press International
204 Jefferson Street
P.O. Box 333
Plum Branch, SC 29845 USA

Manufactured in the United States of America

for
Abigail

Acknowledgments

<u>Grateful Acknowledgment</u> is extended to the U.S. National Transportation Safety Board and the U.S. Department of Defense for making their Aircraft Accident Reports available to the public.

<u>Grateful Acknowledgment</u> is extended to the Federative Republic of Brazil, the Kingdom of Spain, the Republic of India, the Federal Republic of Germany, the Republic of Croatia, and the United Kingdom for making their Aircraft Accident Reports (by whatever name) available to the public.

<u>Grateful Acknowledgment</u> is extended to Marc Heesters, private pilot and renowned aviation photographer, for authorizing reprinting of his photograph of a mid-air collision in Germany on 28 August 1988.

<u>Grateful Acknowledgment</u> is extended to Adrian R. Pingstone, renowned Senior Aerodynamicist, for authorizing reprinting of 12 aviation photographs from his collection of thousands of photographs.

<u>Grateful Acknowledgment</u> is extended to Hans Wendt, renowned professional photographer, for authorizing reprinting of his post-collision photograph of PSA Flight 182 prior to ground impact.

<u>Grateful Acknowledgment</u> is extended to The Boeing Company for authorizing reprinting of its illustration of a Boeing PW-9 biplane.

<u>Grateful Acknowledgment</u> is extended to U.S. Government entities, listed below, for making their photographs available to the public:

    U.S. Library of Congress
    U.S. National Weather Service
    U.S. Department of Defense Visual Information Center
    U.S. Air Force (includes the former U.S. Army Air Force)
    U.S. Navy
    U.S. Army
    U.S. Marine Corps
    U.S. Federal Aviation Administration
    U.S. National Aeronautics and Space Administration

Purpose of this Book

Military and civil Aircraft Accident Reports are created by the United States and other countries identified herein. These reports are published for *educational purposes only*.

This book consists of summaries of Aircraft Accident Reports which concern mid-air collisions. Like the lengthy government and technical reports upon which it is based, this book is published for *educational purposes only*. Experience is a cruel teacher. Yet, by examining past mid-air collisions we can enhance military and civil aviation safety in the future.

The goal of aircraft accident investigation is to (1) determine the circumstances and causes of an accident in order to (2) prevent future accidents and loss of life. The aim is not to apportion blame or liability. Consequently, this book does not assign or imply fault or misconduct on the part of any government, any business entity, or any person living or dead.

Table of Contents

For Part 121 and Part 135 flights listed in the Table of Contents, the name of the airline or operator is enclosed in parentheses. Abbreviations are used for other flights as follows:

In aviation you start with (1) a bag full of luck and (2)

an empty bag of experience. The trick is to fill your

bag of experience before you empty your bag of luck.

[Marion F. Sturkey, *MAYDAY*, 2005]

Introduction

Aircraft Accident Investigation: All modes of surface transportation are subject to collisions. Ships, trains, and automobiles sometimes collide. Aviation is more safe per mile than surface travel, but it is not immune to the perils of runway or mid-air collisions.

Skilled analysts from government and industry investigate each military and civil aviation collision. The object is not to apportion blame or liability. Instead, the investigative goal is to (1) identify the circumstances and causes of the accident in order to (2) prevent future accidents and loss of life.

Aircraft Accident Reports: Modern-day military and civil Aircraft Accident Reports are usually hundreds of pages in length and contain results of detailed analyses, research, tests, and other investigative findings. These technical reports are the primary source documents for each chapter in this book. A person desiring more detail should refer to the source document. Unless otherwise specified, each such document is in the public domain. Many are available through the U.S. Freedom of Information Act. The source document and other sources of information are identified in each chapter.

An Aircraft Accident Report is not always the "final word" on an accident. The so-called final report may be followed by updated or revised reports over a period of many years. Therefore, a report used herein as a source document may not be the "final" report.

Recommendations: Aircraft Accident Reports entail recommendations for corrective action. These recommendations are intended to prevent similar accidents in the future. They cover a host of issues which may include crew resource management, training, air traffic control procedures, navigation facilities, aircraft systems, etc. These proposals are crucial to aviation safety. Yet, they are _**another story**_ and are _**not included in this book**_.

Legal Standing: After an accident swarms of civil attorneys usually descend upon survivors and relatives of those who were killed or injured. In *theory* their goal is to ensure that justice is served. In practice they have two objectives: (1) convince a jury that government, an aircraft manufacturer, or some entity with "deep pockets" was negligent and caused the accident, and (2) convince a jury to award actual and punitive monetary damages to them and their clients. Toward these ends they are not legally bound by the content of an Aircraft Accident Report.

Voice Transcripts: Air Traffic Control (ATC) radio transcripts have been included in many military and civil Aircraft Accident Reports for over 60 years. Cockpit Voice Recorder (CVR) transcripts have been common for at least a quarter-century. Nonetheless, readers are cautioned that creation of a written transcript from an audio recording is not a precise science. Instead it is the best product possible from a collective technical effort. Further, written transcripts or parts thereof can be misleading if taken out of context. In the cautionary words of the U.S. National Transportation Safety Board:

> Transcripts should be viewed as an accident investigation tool to be used in conjunction with other evidence gathered during the investigation. Conclusions or interpretations should not be made using the transcripts as the sole source of information.

ATC and CVR transcripts included in Aircraft Accident Reports often are extremely lengthy. Therefore, this book usually does not include complete transcripts. Non-pertinent noises and conversation generally are excluded in the interest of clarity. A person wishing to see the complete transcript should refer to the source document.

Name of Aircraft Manufacturer: The two manufacturing giants, Douglas and McDonnell, merged in 1967 and became McDonnell Douglas, which later merged with Boeing. Raytheon acquired Beechcraft. The European-built Fokker F-27 became known as the Fairchild F-27 when later built in the United States. Confusing!

For continuity this book uses the name of the manufacturer when the aircraft in question was built. For example, readers will find no reference herein to a "McDonnell Douglas DC-3."

Quotations: This book uses quotations in the interest of specificity. Quotations are usually *displayed (meaning, indented without use of quotation marks)* to enhance readability. Unless otherwise stated the source is always the Aircraft Accident Report, which is identified in the Probable Cause section of each chapter.

Interpolations: In the interest of clarity, essential interpolations are inserted into quoted text and quoted dialogue. Each interpolation is enclosed in brackets.

Explanatory Text: In subsequent chapters in this book all explanatory text is presented in *italic* type.

Chapter Format: Subsequent chapters summarize Aircraft Accident Reports and contain the following sub-headings:

The Flights: .... The planes, the crews, the routes, etc.
Trouble Ahead: .... Events which led to the collision.
The Collision: ..... The collision and its aftermath.
The Investigation: .. Analysis and technical findings.
Probable Cause: .... Direct and contributing causes.

In addition to the sub-headings above, several chapters conclude with a Postscript, a look at unique circumstances or events which followed the accident and investigation.

Chapter Sequence: Based upon the date of the accident, each chapter is sequenced in chronological order.

Errata: On a project of this magnitude and complexity it is possible that errors may have crept in. The author will appreciate notification (*with credible documentation*) of any such errors so that they may be corrected in future printings or editions.

The First Mid-Air Collisions

For millennia the concept of human flight remained mired in the realm of escapist fantasy. Yet, in France in 1783 two brothers rose into the sky in a tethered hot air balloon – the first viable human flight technology. Ballooning quickly became "the rage" in Europe. Two aerial pioneers perished two years later when their ruptured balloon plunged to earth. They became the world's first aviation fatalities.

Experimentation with fixed-wing gliders marked the next century. By 1853 manned and controlled glider flights evolved from theory into reality. The first (1) powered, (2) manned, (3) controlled, and (4) sustained flight in a heavier-than-air machine took place near Kitty Hawk, North Carolina, in 1903. The first powered airplane fatality occurred in 1908 when a Wright Flyer crashed at Fort Myer, Virginia, and crushed the skull of a passenger.

At the time, aviation theorists considered crashing to earth to be the primary aeronautical peril. Little thought had been given to the remote possibility that two aircraft might collide in the sky. However, in 1910 an Antoinette monoplane and a Farman biplane rammed into each other in flight over Milan, Italy. The fledgling aviation community had experienced the ***world's first mid-air collision***.

During World War I, thousands of warplanes swirled through the sky in aerial battles over Europe, and many collided. One of the most famous aces in the German Air Service (54 victories) lost his life in 1918 when he rammed into his wingman in flight.

Financiers soon inaugurated scheduled air passenger service in both Europe and America. In 1922 a Daimler Airways plane (de Havilland DH-18) and a Cie des Cgrands Express plane (Farman Goliath) collided head-on in the sky over France. The seven people aboard the two aircraft all perished in what is believed to be the first mid-air collision involving a commercial airliner.

The first documented mid-air collision between a military plane and a commercial airliner would take place in the United States seven years later in 1929 . . . .

Those who cannot remember the past are condemned
to repeat it.

[George Santayana, *The Life of Reason*, 1906]

MID-AIR

Accident Reports and Voice Transcripts
from Military and Airline
Mid-Air Collisions

Give the Passengers a Thrill
Mid-Air Collision over San Diego, California
21 April 1929

Boeing PW-9 Pursuit Biplane, Serial No. 28-037
U.S. Army Air Force post-maintenance test flight
1 aboard, 1 killed

Ford 5-AT Tri-Motor, Registration NC9636
Maddux Airlines scheduled passenger flight
5 aboard, 5 killed

Synopsis: The military pilot in a Boeing PW-9 pursuit biplane tried to dive in front of a Ford 5-AT Tri-Motor airliner. He misjudged his dive, and the biplane slammed into the cockpit of the passenger plane. The mid-air collision and subsequent crashes killed all six people aboard the two aircraft.

The Boeing PW-9 Pursuit Biplane Flight: Fifteen minutes before noon on Sunday morning a U.S. Army Air Force pilot walked out to the flight line at Rockwell Field near San Diego, California. The young pilot, age 30, climbed into the cockpit of a then-state-of-the-art flying machine, a Boeing PW-9 pursuit biplane. He fired up the in-line gasoline engine and took off on a post-maintenance test flight:

In the decade after the end of World War I the term "fighter" plane had not yet come into vogue. Instead, airplanes designed for air-to-air offensive combat were called "pursuit" planes.

Boeing designed the PW-9 (PW means "pursuit, water-cooled") based upon studies of the German Fokker, and deliveries to the U.S. Army Air Force had begun in 1925. The PW-9 sported a 435 horsepower liquid-cooled Curtiss D-12 engine which gave the warplane a top speed of 160 miles per hour. Its internal fuselage structure utilized welded metal tubing instead of the wood and wire bracing found in older aircraft designs. The wings featured wooden spars and ribs covered by fabric.

Boeing would deliver 113 new PW-9 pursuit biplanes to the U.S. Army Air Force over seven years. A variant, the FB-2, featured

Boeing PW-9 Pursuit Biplane (copyright The Boeing Company, all rights reserved, reprinted by permission)

a tailhook and a 525 horsepower Packard engine. It went to the Navy for the fledgling aircraft carrier fleet. With a gross weight of 3,120 pounds the biplane could be armed with two .30 caliber machineguns and could carry two 122 pound bombs.

The Ford 5-AT Tri-Motor Flight: As the military biplane climbed skyward from Rockwell Field, a commercial Ford 5-AT Tri-Motor prepared to take off from nearby Ryan Airport. The Tri-Motor airliner, owned and flown by Maddux Airlines, would carry two pilots and three revenue passengers on the regularly scheduled flight to Imperial Valley, California, and Phoenix, Arizona. Two passengers, a wealthy attorney and his teenage daughter, planned to deplane at Imperial Valley. The remaining passenger, a female "high society" newspaper employee, planned to fly all the way to Phoenix.

The attorney's daughter, age 19, had never before ridden in an airplane, and perhaps she had a premonition of the tragedy to come. As she boarded the Ford Tri-Motor she waved goodbye to her sister and lamented: "You'll be sorry if I get my neck broken."

Henry Ford, the automobile entrepreneur, had branched out into commercial aviation in 1926. That year he rolled out his first airplane, the Ford 5-AT (AT meant "air transport") Tri-Motor. This rugged and dependable aircraft would earn a permanent place in commercial aviation history. Ford engineers designed the plane so that it could fly well on two engines and maintain level flight – under optimum conditions – on only one.

The new airliner featured corrugated aluminum construction which led to its slang moniker, the Tin Goose. Three Pratt & Whitney 420 horsepower Wasp air-cooled radial engines gave the Tri-Motor a cruising speed of almost 120 miles per hour. In commercial service the plane would accommodate two pilots and over 10 passengers, plus a stewardess on some flights. During a production run that ended in 1933, Ford sold the new aircraft to airlines around the globe. The reliable Tri-Motor is generally considered the world's first "modern" airliner.

Trouble Ahead: In the era before air traffic control, radar data, and documentation from CVRs and FDRs, witnesses on the ground played crucial roles in reconstructing aircraft accidents. In this instance many people watched the two planes as they flew over San Diego at about 2,000 feet that sunny Sunday. All agreed the Army biplane was "stunting" around the Tri-Motor, and one explained:

The Army plane, after maneuvering around the [Tri-Motor] for a few minutes, did a "wing-over" below the [Tri-Motor] and then shot up in front of it to do another wing-over. Instead, the [Army] plane . . . crashed down on the nose of the [Tri-Motor] plane.

The Collision: As the biplane pilot tried to dive earthward in front of the Tri-Motor, he flew too close. At an estimated dive speed of 130 to 150 miles per hour the right wing of his biplane sliced down through the cockpit of the airliner. The impact tore off the front of the Tri-Motor cockpit and the No. 2 engine, which was mounted in the nose. Mutilated remains of the two Tri-Motor pilots fell out of their plane through the gaping hole where the front of the cockpit had been. The fuselage of the fatally wounded airliner pitched up, rolled inverted, and then dove toward the ground.

The Tri-Motor crashed upside-down in the Wabash Canyon district of San Diego. The severed No. 2 engine and part of the cockpit fell to earth one-quarter mile away. Bodies of the two pilots, one of whom had been decapitated, were found nearby.

The attorney in the Tri-Motor cabin died upon impact with the ground, but the two female passengers still showed signs of life when potential rescuers arrived. An ambulance raced the two women toward Mercy Hospital. The teenager died in the ambulance, and the older woman succumbed to her injuries in the hospital. The next day the *San Diego Union* newspaper headline would blare the news:

ARMY CRAFT DIVES ON BIG MADDUX SHIP 1800 FEET ABOVE CITY

In graphic detail the newspaper article described the collision and crashes. The narrative included many statements from eyewitnesses to the tragedy. A sub-headline summed up the fate of those aboard the two doomed aircraft:

Three Passengers, Two Pilots, and Service Flier Victims as Huge Air Liner Plunges to Earth and Pursuit Machine Drops like Bullet; Parachute's Failure to Open Proves Fatal to 'Stunter'

Ironically, the Army pilot had not been seriously injured in the collision. Minus its right wing, his biplane had tumbled earthward out of control. He had managed to bail out and pull his D-ring, but then his luck ran out. Although his parachute had begun to deploy, it became tangled in the falling wreckage of the PW-9. Thus snared, the pilot had plummeted downward with the remains of his biplane. Impact with the ground had killed him instantly.

Ford 5-AT Tri-Motor (official U.S. Air Force photo)

The Investigation: Several investigative entities began delving into the circumstances which led to the accident. A coroner's jury began a criminal investigation. The Army convened a Board of Inquiry at Rockwell Field. The San Diego Air Control Board began searching for answers. In addition the federal Department of Commerce sent aviation experts to San Diego to reconstruct the tragedy.

Teams of investigators found that paint smears on the Tri-Motor wreckage matched paint from the biplane. A strut and wires from the Army craft were found wrapped around the No. 1 propeller on the passenger plane's left wing. Witnesses each testified that they had watched the biplane flying – or looping – over and around the Tri-Motor. The San Diego Air Control Board soon announced:

> The stunting of Lt. [pilot's name] was an open and wanton violation of the San Diego air traffic rules as well as the federal Department of Commerce rules . . . . It was just a plain case of [the Army pilot] misjudging his distance and hitting the Maddux [Tri-Motor] with his own [plane].

This was Maddux Airlines' first accident in their 19 months of scheduled passenger flights. Their now-deceased chief pilot, age 34, had logged over 7,000 hours of accident-free flight time. The president of the airline wired a protest to the Department of Commerce. He complained that both Army and Navy pilots had been "stunting" near his airliners "to give passengers a thrill." The airline vice president examined the crumpled wreckage of his Tri-Motor, and he subsequently testified:

> The seats in the cockpit were covered with large quantities of blood, indicating that the [two pilots] were killed instantly.

The investigative teams found a bloody handwritten note in the Tri-Motor wreckage. Analysis revealed it had been written in flight before the collision by the female newspaper employee. In the note she described various things she saw through the cabin window, and she ended the note with the words: "Army plane stunting." The supervising inspector from the Department of Commerce released a statement to the newspapers on 23 April:

> No structural failure of any kind may fairly be charged to either of the aircraft involved. On the contrary, the entire blame probably will be charged to failure of the human element to function properly.

Probable Cause: The coroner's jury finished its criminal investigation and made its findings public. Physical evidence confirmed that the two planes had collided. The aforementioned bloody note was accepted as evidence. Numerous witnesses testified that the Army biplane had been "stunting" around the Ford Tri-Motor airliner. The jury members concluded:

> The Army plane struck the [air]liner from above as it neared the completion of what was described as a combination wing-over and loop, crashing head-on into the cockpit of the Maddux plane.

The reports from the (1) Army Board of Inquiry and the (2) U.S. Department of Commerce are now missing. After nearly a century perhaps they were lost, burned, purged, misfiled, or merely thrown away. In any event they are not to be found. However, newspaper reports which chronicle results of the investigations are available. Documentation of the coroner's jury action still exists.

On 24 April 1929, three days after the accident, the jury charged the deceased Army lieutenant with "criminal negligence." In part the jury verdict stated the Maddux Airlines plane was:

> ... rammed by a United States Army plane piloted by Lt. [pilot's name] and that said [pilot's name] was violating the rules of aviation when his plane rammed said Maddux plane and [he] was responsible for the accident and deaths of all persons in the said Maddux plane; and we, the jury, further find that said Lt. [pilot's name] was criminally negligent in following and approaching said Maddux Airlines airplane.

Postscript: The collision of the airliner and biplane over San Diego in 1929 was not the first aerial collision in recorded history. On many prior occasions planes had collided in flight. In an amazing coincidence, the most recent of these mid-air collisions had taken place (1) near San Diego (2) only two days before the Tri-Motor and biplane collided. On that Friday two U.S. Navy planes had rammed into each other over the beach, killing the four military crewmen.

Commercial passenger planes in Europe had been involved in mid-air collisions. Yet, in the United States there had been no *known* instance wherein a passenger airliner had been involved in a mid-air collision before 21 April 1929. A technical publication, *Notable California Aviation Disasters*, offers the following explanation:

> The Maddux Airlines Tri-Motor crash was the first mid-air collision involving a United States commercial airline plane in aviation history.

We Hit the Airliner!
Mid-Air Collision near Palm Springs, California
23 October 1942

Douglas DC-3, Registration NC16017
American Airlines Flight 28
12 aboard, 12 killed

Lockheed B-34 Ventura, Serial No. not documented
U.S. Army Air Force ferry flight
2 aboard, 0 killed

Synopsis: The pilot of a Lockheed B-34 Ventura bomber flew close to a Douglas DC-3 airliner to wave at his friend, the first officer. The planes collided. The military B-34 landed safely. The DC-3 airliner, with its empennage torn away, crashed and killed all 12 people on board.

The Douglas DC-3 Flight: Late in the afternoon at 1636 Hours (Pacific War Time) a Douglas DC-3 took off from the Lockheed Air Terminal at Burbank, California. American Airlines Flight 28 began climbing toward its cruising altitude of 9,000 feet. On this long transcontinental flight to New York the first stop was scheduled for Phoenix, Arizona:

The famous DC-3 had first flown in 1935, and it revolutionized air travel. It quickly replaced trains as the favored means of coast-to-coast transit, for with two or three refueling stops a transcontinental flight had become possible. Several types of radial air-cooled engines were used over the years, the most popular of which were the Pratt & Whitney R-1830 Twin Wasp and the Wright R-1820 Cyclone.

Douglas turned out over 10,000 of the twin engine DC-3 planes as civil airliners or as C-47 and R4D military workhorses. After World War II thousands of war-surplus C-47s and R4Ds were converted to civil transports. They became standard equipment on most of the world's airlines. As late at 1998 over 400 reliable DC-3s remained in service, primarily in developing countries.

Paratroopers board a Douglas C-47, a military variant of the civil Douglas DC-3 airliner (official U.S. Air Force photo)

The DC-3 captain, age 42, had logged 17,155 flight hours and had been flying for American Airlines for 14 years. He held the requisite ATP certificate. His first officer, age 24, had earned a Commercial Pilot certificate and had 863 hours of flying experience. A lone stewardess in the cabin took charge of the nine paying passengers.

The Lockheed B-34 Ventura Flight: Ten minutes before the DC-3 airliner took off, a Lockheed B-34 Ventura had climbed into the CAVU sky from Long Beach, California. The U.S. Army Air Force bomber was flown by two military pilots. This "ferry" flight had been scheduled to deliver the plane to the Army airfield at nearby Palm Springs, California:

The Lockheed factory rolled out over 3,000 B-34 bombers for World War II. Based upon the earlier Model 18, the B-34 entered military service with Great Britain, Canada, New Zealand, and the United States. The medium bomber could cruise at 200 knots and had a combat radius of around 950 miles.

Twin 2,000 horsepower Pratt & Whitney R-2800 Double Wasp radial engines powered the B-34. The warplane had a maximum takeoff weight of 27,750 pounds. Two .50 caliber machineguns bristled from a dorsal turret, and two fixed forward-firing .50

Lockheed B-34 Ventura (photo courtesy of U.S. Library of Congress)

caliber weapons were mounted in the nose. Lighter .30 caliber machineguns were positioned on the sides of the fuselage behind the wings. The B-34 could carry up to 3,000 pounds of ordnance in an internal bomb bay.

The B-34 bomber pilot, an Army lieutenant, was 25 years of age and had logged 1,500 flight hours. In addition to military qualifications he had earned a Commercial Pilot certificate. He was stationed at Long Beach and assigned to the Air Transport Command. The lieutenant had logged only nine hours as pilot-in-command in B-34 aircraft. His "acting" copilot was a staff sergeant who had logged about 100 hours in small Army training planes.

Trouble Ahead: The DC-3 climbed to 9,000 feet and cruised along Green Airway No. 5. At 1702 Hours the captain checked-in by radio with his base at Burbank and reported his position over Riverside, California. The next check-point would be at Indie Intersection, and the captain estimated he would arrive there at 1722 Hours.

Meanwhile, the B-34 pilot harbored a secret. The DC-3 first officer was his friend. Consequently, on a "contact" flight plan the lieutenant flew in circles and waited on the airliner. Finally spotting it, he flew along the left side of the civilian craft and rocked his wings. No response. He throttled back and banked to the right, passing behind the airliner to try his luck on the other side.

There were no known radio messages to or from the B-34. The radio transcript, below, has been abstracted from the handwritten American Airlines radio log at Burbank, so some verbiage may not be verbatim. The transcript begins before takeoff as the DC-3 airline captain makes a routine radio check:

> American 28: . . American Airlines Flight 28
> Burbank: . . . . . . American Airlines dispatch office at Burbank

American 28: Burbank, Flight twenty-eight, radio check. (1631)

Burbank: American twenty-eight, [you are] loud and clear. (1631)

[5 minutes later the DC-3 takes off at 1636]

American 28: Burbank, American twenty-eight. (1702)

Burbank: American twenty-eight, go ahead. (1702)

American 28: Burbank, we're over Riverside at nine-thousand [feet], estimating Indie at two-two. (1702)

Burbank: American twenty-eight, roger. (1702)

[the DC-3 and the B-34 __collide__ at 9,000 feet at apx. 1715]

[Although his DC-3 is spinning out of control the captain is able to transmit a single frantic radio message]

American 28: Flight twenty-eight from Burbank! Ah! Correction! Burbank from Flight twenty-eight! (1715)

Burbank: Go ahead, Flight twenty-eight. (1715)

Burbank: Go ahead, Flight twenty-eight. (1716)

Burbank: Flight twenty-eight, this is Burbank, go ahead. (1716)

[there is no response]

The Collision: From a position off the left wing of the airliner, the B-34 bomber pilot had banked right and crossed behind the commercial flight. From the DC-3 five o'clock position the military pilot had decided to ease in closer to attract the attention of his friend:

[The Army B-34 bomber pilot], feeling he was still too far from the airliner to recognize his friend, turned his plane to the left to approach closer.

The bomber slipped in closer – closer – too close! Realizing the danger the Army lieutenant tried to throw his craft to the right, but it was too late. The B-34 overran the DC-3. The No. 2 propeller on the right wing of the bomber chewed into the rudder and vertical stabilizer of the airliner. The damaged DC-3 rolled left and disappeared from the bomber pilots' field of view. The military copilot would later acknowledge that he shouted:

Aaahhh! We hit the airliner!

The mid-air impact had torn off part of the de-icing boot on the B-34 right wing and bashed in the oil radiator air scoop. All three propeller blades on the No. 2 engine, as well as the engine nacelle, had been bent and scored. Despite the damage the B-34 remained flyable, and the shaken pilot landed safely on the Army airfield at Palm Springs. He ran to the office of the commanding officer and reported his accident.

After the collision two witnesses, a housewife and a volunteer airplane spotter, had watched the DC-3 airliner as it spiraled down toward the ground. They both watched the empennage break away:

During the descent the entire tail assembly was torn loose from the fuselage.

The following day the *Los Angeles Times* newspaper headline would explain the result of the mid-air collision:

Transport Crash
Fatal to 12

In vivid detail the newspaper chronicled the spiraling descent and crash of the mortally wounded DC-3:

Like a great wounded bird [the DC-3] spun earthward to crash in an almost inaccessible spot northwest of the desert resort near what is known as the Old Rodeo Ground.

The aerial impact had not caused any damage forward of the empennage and had caused no injury to the passengers and crew. The 12 uninjured people aboard, aware of their imminent fate, rode the doomed and spinning DC-3 down into the ground. The sudden impact on a rugged ledge of San Jacinto Peak near Palm Springs, plus a raging ground fire, killed them all. A Palm Springs police officer reported that the bodies had been "burned beyond recognition."

The Investigation: The CAB office in Santa Monica, California, got word of the accident at 1930 Hours, and investigators arrived at the DC-3 crash site before midnight. Over the next few days they found almost no useful evidence in the burned rubble:

Impact with the ground and [the] subsequent fire resulted in such extensive damage to the [DC-3] that little could be learned from an inspection.

At the Army airfield, analysis of score marks on the damaged bomber revealed the precise mechanics of the collision. In addition investigators were favorably impressed with the "great frankness" exhibited by the Army lieutenant prior to his court martial:

He diverged from his military mission to fly the bomber in close proximity to the airliner for the express purpose of signaling a friend in the copilot's seat of the latter plane.

The lieutenant explained his ill-fated plan to investigators. He and the now-deceased DC-3 first officer had been close friends. The night before the accident they had talked about their upcoming flights. They discovered they were scheduled to take off at about the same time in the afternoon, so they agreed to an aerial rendezvous:

They had trained together . . . in small type aircraft and thought it would be pleasant to "see each other in the air."

The radio in the military B-34 and the radio in the civil DC-3 operated on different frequencies, so the two pilots had discussed various means of communicating with each other by rocking their wings. They also had planned to "salute" each other by waving their hands if they got close enough.

Probable Cause: Four months after the accident the CAB adopted the Report of the Civil Aeronautics Board, File No. 2362-42, on 23 January 1943. Board members found that the conduct of the Army pilot had been "wholly without justification," and the official CAB report elaborated:

> We are driven to the conclusion that this collision resulted from the reckless and irresponsible act of the bomber pilot, and that the captain of the airliner was without fault.

The board could find no indication that the DC-3 captain had been aware of plans for the surreptitious aerial rendezvous. The board found the following ***Probable Cause*** of this accident:

> Reckless and irresponsible conduct of [the Army B-34 pilot] in deliberately maneuvering a bomber in dangerous proximity to an airliner in an unjustifiable attempt to attract the attention of the first officer of the latter plane.

The Fatal "Training" Dogfight
Mid-Air Collision near Adel, Georgia
22 June 1943

Waco UPF-7, Registration NC29335
War Training Service basic airwork training flight
2 aboard, 0 killed

Beechcraft AT-10 Wichita, Serial No. not documented
U.S. Army Air Force formation training flight
2 aboard, 2 killed

Synopsis: Two military Beechcraft AT-10 Wichita planes and a civil Waco UPF-7, flown by the War Training Service, engaged in a simulated dogfight. The AT-10 wingman collided with the Waco. Both aircraft tumbled out of control and crashed. Impact with the ground killed both pilots in the AT-10. The Waco pilots bailed out and survived.

The Waco UPF-7 Flight: A few minutes after 1500 Hours a Waco UPF-7 biplane took off from the civil airport near the town of Adel in southern Georgia. Aboard were a War Training Service (WTS) instructor and his WTS student. They climbed to 2,500 feet for basic airwork, and after a half hour they began practicing power-on and power-off stalls:

During World War II the United States established the civil War Training Service. Civilian flying schools and flight instructors became government contractors. They cranked out thousands of neophyte pilots. These new basically-trained aviators then could join the armed forces and immediately begin advanced military flight training.

The Waco UPF-7 biplane, designed for flight training, had an open-air tandem cockpit. A Continental W-670 radial engine produced 220 horsepower. The Waco became the standard advanced trainer for fledgling pilots during World War II.

The young WTS instructor, age 22, held a Commercial Pilot certificate and had logged 675 flying hours, 266 of which were in the

Waco YMF-5, a biplane aesthetically similar to the Waco UPF-7 (photo courtesy of Adrian R. Pingstone)

Waco. His WTS student recently had received his Student Pilot certificate and had only 22 hours of flight experience. Their Waco was assigned to Carson Chalk Flying Service for use in the WTS flight training program at Adel Airport.

The Beechcraft AT-10 Wichita Flight: A half hour after the Waco took off, two U.S. Army Air Force trainers climbed skyward from nearby Moody Field (later to become Moody AFB) near Valdosta, Georgia. The two Beechcraft AT-10 Wichita planes were based at the U.S. Army Air Force flying school at Moody, which had geared-up to train hundreds of Army pilots:

Beechcraft built the AT-10 primarily out of wood, for scarce metal was reserved for combat airplanes during World War II. Even the fuel tanks, sealed with neoprene, were made of wood. Only the engine cowlings and cockpit enclosures were constructed of aluminum. Beechcraft would roll out 1,771 of these planes during the war, and 600 more would be built under license in Texas. Two 295 horsepower Lycoming R-680 radial engines powered the advanced-training military aircraft.

An experienced instructor pilot, an Army captain, led the flight. In the lead AT-10 cockpit with him was a student pilot, an air cadet. Another air cadet and a lieutenant, both student pilots, manned the second AT-10 in this military flight-of-two. The instructor and his three students planned to practice formation flying.

In the second AT-10 the air cadet and the lieutenant concentrated on maintaining their position off the right wing of their flight leader. Both student pilots were members of Flight Training Class 43-G at Moody Field. The air cadet had logged 163 hours of flight time. The lieutenant had equal flight experience under his belt. He had been an experienced U.S. Army infantry officer, but he had volunteered for Army Air Force flight training.

Trouble Ahead: As the lead Army AT-10 flew at 2,500 feet the instructor spotted the civil Waco. Later the survivors – two from the Waco and two from the lead AT-10 (which was not involved in the collision) – would offer contradictory versions of the aerial melee:

> None of [the four survivors were] in accord as to the direction, manner of approach, or their relative positions when the two flights encountered one another. It is apparent, as well, that none of them agree as to the maneuvers just preceding the collision.

On the ground four farmers witnessed the aerial dogfight. Some of them later would testify that the three aircraft completed only two or three circles; others would say they completed as many as six:

> The Army formation was circling to the right, and the Waco was also circling to the right *inside [emphasis added]* of the circular course being followed by the Army formation.

The Collision: The speed of the AT-10 was about 30 knots faster than the speed of the Waco. To counter that airspeed disadvantage the Waco instructor stayed inside the turn radius of the faster Army planes. However, as the Waco rolled out of a turn:

> The No. 2, or [wingman], of the Army formation overtook the Waco. The left wing of the AT-10 struck the lower right wing of the Waco from the rear, and both planes fell separately out of control to the ground.

Beechcraft AT-11, a military training aircraft similar to the Beechcraft AT-10 (photo courtesy of U.S. Air Force Historical Research Agency)

The aerial impact ripped off the lower right wing of the Waco. The front and rear spars tore away only one foot outboard from the fuselage attachment fittings, throwing the plane into a violent left spin. The WTS instructor and his student bailed out and parachuted to the ground without injury five miles south of Adel.

The mid-air collision had torn off most of the left wing of the Army AT-10 wingman. The aircraft spun to the right, out of control. No one knows exactly what happened inside the cockpit. The two student pilots wore parachutes, but they never escaped from their violently spinning aircraft. They rode the crippled AT-10 to the ground, and the impact instantly killed both young men. Two days later the *Valdosta Daily Times* newspaper would capsule the accident and loss of life:

> Two student fliers were killed about five miles south of Adel late Tuesday afternoon in a mid-air collision between a primary training ship and a twin-engine advanced trainer from Moody Field.

The Investigation: The CAB investigated this accident because the Waco was of civil registry. The accident location had been free of clouds with unlimited visibility, so weather had not been a factor. There was no known pre-collision mechanical problem with either aircraft. Investigators determined that the three planes had engaged in tight "circling maneuvers" around one another until the aerial impact occurred. Subsequent impact with the ground had completely destroyed both planes.

The four pilots who survived separately gave detailed accounts of events that led to the accident. However, the four versions did not agree. Consequently, investigators turned to other WTS and Army students and instructors for assistance, and they discovered:

> Such [simulated dogfights] between aircraft of the Army and the WTS were apparently not a rare sight in this vicinity.

Army flight instructors, all commissioned officers, claimed that WTS planes routinely dived at their formations. The Army officers denied that they ever responded by "stunting" or "dogfighting" with WTS aircraft. However, WTS flight instructors told investigators a vastly different story:

> [WTS] flight instructors from Adel Field testified that they had been "attacked" many times by Army planes, and [they] freely admitted that they had engaged in *playful maneuvers [emphasis added]* in the air with Army pilots.

Investigators realized the Army students who died had been innocent victims. The two students had followed orders; they had followed their flight instructor. They had concentrated on maintaining their position 45 degrees behind the right wing of their leader. The CAB pointed out that the Army instructor had led the two students "into the circumstances which brought about the collision" and their deaths.

Probable Cause: The CAB adopted the Report of the Civil Aeronautics Board, File No. 2791-43, a week before Christmas in 1943. Board members made it clear that aerial combat training involves "violent maneuvers, simulated surprise attacks," and entails "special hazards." The board cautioned that these maneuvers should not be forced upon non-military WTS instructors and WTS students. Nonetheless, the board explained:

The fact remains that [Army and WTS instructor] pilots . . . were aware [that they were] maneuvering with, or around, each other.

The board found that evidence as to who started the dogfight was contradictory. Based upon the totality of the information available the board noted:

Such unnecessary and unauthorized maneuvers between Army and WTS ships had frequently occurred around these two flight training centers, and it is apparent that the responsibility for this reckless type of flying is fairly equally divided between the pilots of both groups.

The board found the ***Probable Cause*** of this accident to be:

The action of both instructors in [conducting] hazardous and uncalled-for maneuvers.

Crash in a Cotton Field
Mid-Air Collision near Darlington, South Carolina
12 July 1945

Douglas A-26 Invader, Serial No. 44-35553
U.S. Army Air Force local training flight
3 aboard, 2 killed

Douglas DC-3, Registration NC25647
Eastern Airlines Flight 45
21 aboard, 1 killed

Synopsis: A military Douglas A-26 Invader pilot practiced a radio range instrument approach procedure while a civil Douglas DC-3 descended to land. In clear weather the two aircraft collided. The A-26 pilot bailed out and survived, but his two crewmen were killed. The DC-3 crash-landed in a cotton field, killing one passenger.

The Douglas A-26 Invader Flight: At 1315 Hours a Douglas A-26 Invader took off from Florence Army Airfield in South Carolina. The U.S. Army Air Force pilot flew to a restricted military operations area 40 miles from the base, and there he practiced various flight maneuvers. He then headed for home. About 15 or 20 miles from the field he tuned to the Florence radio range transmitter and began practicing aural null approach procedures:

The A-26, designed as an attack bomber, had entered military service in 1944 in Europe and the Pacific Theater. It could be armed with up to 14 heavy machineguns plus bombs and rockets. Two Pratt & Whitney R-2800 radial engines gave the warplane power to spare. In view of its primary bombing role, in 1948 the aircraft would be re-designated as the B-26.

On this flight the A-26 carried only one pilot, a first lieutenant 24 years of age who had logged 1,400 flight hours. In addition to his military qualifications he had earned a Commercial Pilot certificate. There was no copilot aboard, but two unlucky gunners had tagged along for the ride.

Douglas A-26 Invader (official U.S. Air Force photo)

The Douglas DC-3 Flight: Almost an hour before the A-26 took off, Eastern Airlines Flight 45 had left Washington, D.C., at 1222 Hours. The Douglas DC-3 headed south toward an intermediate stop at Columbia, South Carolina, on the way to its final destination at Miami, Florida. As he neared the military airfield at Florence the captain veered to the west of his assigned airway, Amber 7, to avoid heavy air traffic at the military training base.

The DC-3 on this flight had accumulated 21,154 flying hours since Eastern Airlines had purchased it new from the factory in 1940. It carried a crew of four: the captain, age 35, and a first officer plus two flight attendants. Seventeen passengers relaxed in the cabin. Several planned to deplane in Columbia, but most intended to fly all the way to Miami.

Trouble Ahead: At 3,000 feet the A-26 pilot banked 15 degrees to the left at 220 miles per hour, and he listened to the aural tone from the radio range station:

He was listening for the change in signal which would indicate the relative bearing of the station from the aircraft.

The A-26 pilot concentrated on the radio range tone, then chanced to look up. Horrified, he saw the DC-3 dead ahead. He jammed his control column forward.

Douglas C-47, a military variant of the famed Douglas DC-3 commercial airliner (official U.S. Air Force photo)

Meanwhile, the DC-3 captain had been flying toward the airport at Columbia, 62 miles away. The weather was clear with a forward visibility of 15 miles, and the only clouds were high above him. He expected to land in Columbia in about 20 minutes, so he began a cruise descent of 200 feet per minute. Suddenly the captain saw the A-26 through the left side of his windshield. It was flying straight toward him. In desperation he pulled back on his control column.

The Collision: The A-26 fuselage skimmed under the DC-3, but the tall vertical stabilizer on the attack bomber ripped into the airliner:

> The vertical fin of the A-26 [sliced into] the leading edge of the DC-3 [left] wing [and] progressed along the leading edge until it struck the left engine nacelle, tearing loose that engine. The left engine then moved sufficiently to the right to allow its still rotating propeller to cut into the [DC-3] fuselage.

As the A-26 whipped under the DC-3, it failed to clear the No. 2 propeller on the airliner's right wing. The whirling DC-3 propeller sawed off the entire tail of the Army plane. With the empennage and its control surfaces torn away the attack bomber rolled inverted and dove earthward. At about 900 feet the pilot managed to bail out successfully. His pilotless A-26 screamed down into the ground near the rural cotton-farming community of Darlington, South Carolina, instantly killing the two hapless gunners.

Like the A-26, the DC-3 also headed for the ground. The No. 1 engine and propeller assembly had been torn completely off the left wing. The engine and its propeller blades tumbled down into a small vegetable garden, narrowly missing several nearby houses.

On the DC-3 right wing all No. 2 engine propeller blades had been sheared off in the process of cutting through the A-26 rear fuselage. With one engine totally ripped off and with no propeller blades left on the other engine, the DC-3 had become an expensive glider. Worse yet, it had two heavily damaged wings plus major structural damage where its No. 1 propeller had sliced into the fuselage behind the cockpit.

The DC-3 captain pulled his craft out of a steep dive. As the wounded airliner glided down toward the ground a long cotton field fortuitously chanced to lie straight ahead:

> The aircraft continued a sharp glide toward an open area and was landed with flaps and wheels up in a cotton field . . . . It ground looped . . . and the right engine fell from the aircraft.

Several passengers were battered, broken, and bleeding. An ambulance arrived and rushed them toward McLeod Infirmary in Florence, 20 miles away. One of the injured, a two year old boy, died on the way to the infirmary. The mother of the dead child and three others were hospitalized with critical injuries. The next day *The State* newspaper headline in Columbia, South Carolina, would read:

Airliner, Army Plane in Mid-Air Collision

The newspaper explained how uninjured passengers were transported from the crash site so that they could resume their trip:

Passengers who were unhurt were taken to Florence by bus, and Eastern sent a relief plane to carry them on to their destinations.

The Investigation: The CAB dispatched investigators to delve into the circumstances that led to the accident. The two planes had been on headings roughly perpendicular to each other. Thirty seconds before impact they had been about two miles apart. Each plane would have been visible from the cockpit of the other craft and would have remained visible until the collision. Investigators realized, however, that neither pilot had seen the other aircraft until about one second prior to the mid-air impact:

For the last thirty seconds of the A-26 turn the DC-3 was almost directly ahead of the Army pilot, and the only possible reason for failing to see the [DC-3] was the preoccupation of the A-26 pilot within the cockpit and his lack of attentiveness to other traffic.

The DC-3 captain had maintained a straight course. Normally he would have had right-of-way because his aircraft was on the right of the A-26, and both planes were on a "contact" clearance (later to be called a *VFR clearance*). Yet, where neither pilot is aware of the presence of the other aircraft the rule of right-of-way does not apply. In any event, right-of-way does not relieve a pilot of responsibility for safe conduct of the flight:

Both pilots were operating under contact clearances and [they] understood that the only traffic separation possible was provided by their own vigilance and flying technique.

Probable Cause: The CAB adopted the Accident Investigation Report, File No. 2773-45, on 12 June 1946, exactly 11 months after the mid-air collision. The DC-3 had been eight miles off its assigned airway. This deviation had been well within the "safety measure" limits authorized under Civil Air Regulations, but unfortunately:

In order to reach the restricted area northwest of Florence it was necessary for Army aircraft to fly through the region which [the DC-3 captain] chose in deviating from the airway.

The A-26 had been practicing procedures in an area "not designated or reserved" for such military practice. Still, there was no specific regulation that prohibited military practice in that area. The board members emphasized that – simply stated – both pilots should have seen the other aircraft in time to take necessary evasive action. Consequently, the board found the *Probable Cause* of this accident to be:

> The lack of vigilance on the part of the pilots of both aircraft, resulting in the failure of each pilot to see the other aircraft in time to avoid [the] collision.

Skyscraper Dead Ahead
Mid-Air Collision over Manhattan, New York
28 July 1945

North American B-25 Mitchell, Serial No. 43-0577
U.S. Army Air Force navigation training flight
3 aboard, 3 killed

Empire State Building
Masonry and steel structure, 350 Fifth Avenue, New York
1,500 occupants, 11 killed

Synopsis: A military North American B-25 Mitchell bomber flew low over New York City through solid clouds. The aircraft slammed head-on into the 79th floor of the Empire State Building in downtown Manhattan. The mid-air impact killed all three occupants of the B-25 plus 11 people inside the building.

The North American B-25 Mitchell Flight: Two days before the accident a North American B-25 Mitchell bomber took off from its military base at Sioux Falls, South Dakota. The U.S. Army Air Force navigation training flight proceeded to Bedford, Massachusetts, near the pilot's hometown. While the pilot visited his family the bomber stayed at the Bedford airport for two nights:

North American built the B-25 as a twin engine medium bomber. Its "Mitchell" moniker had been bestowed in honor of Gen. Billy Mitchell, an early advocate of airpower. The bomber earned its greatest fame for the "Doolittle Raid" on Tokyo, Japan, in 1942. Most B-25 variants were powered by twin Wright R-2600 radial engines. Weapons included bombs, six .50 caliber machineguns, and a huge 75mm cannon.

The morning of the accident dawned cloudy and rainy along the eastern seaboard of the United States. At 0855 Hours the pilot took off from Bedford and headed south. There were two additional men aboard, a copilot and a U.S. Navy machinist's mate headed for home on emergency leave. The pilot wanted to fly to the Army airfield at Newark, New Jersey, where he planned to pick up his commanding officer for the return flight to Sioux Falls:

North American B-25 Mitchell (official U.S. Air Force photo)

[The pilot] had planned to file a flight plan [to] Newark Army Airfield, but filed for La Guardia Airport, New York, because of existing traffic at Newark and the weather, which was *below visual flight minimums [emphasis added]*.

The Empire State Building: The Empire State Building had been completed in 1931 during the Great Depression. Towering 1,250 feet over Manhattan at 350 Fifth Avenue, the contemporary art deco structure was the tallest building in the world. An observatory atop the structure offered a breathtaking view of the city far below. Constructed of masonry and steel, the building weighed a staggering 300,000 metric tons and contained 2,200,000 square feet of office space for roughly 15,000 workers. Standing 102 stories tall, it was considered one of the modern-day Seven Wonders of the World.

Fortunately this was a Saturday, and most of the building was vacant. However, volunteers from the National Catholic War Relief Service were working on the 79th floor to arrange economic aid and emergency supplies for war-torn Europe.

Trouble Ahead: The pilot flew to the vicinity of La Guardia Airport and radioed the control tower. He said he was 15 miles south of the airport, but moments later the tower controller spotted him circling northeast of the field. The pilot told the controller that he was flying at 700 feet in visual flight conditions. He requested the weather at Newark, and the La Guardia controller radioed:

> Newark weather, *six-hundred feet overcast [emphasis added]*, visibility two-and-a-half to three miles.

La Guardia Tower cleared the pilot to fly toward Newark if he could maintain a forward visibility of three miles. If not, he was required to return to La Guardia. One technical report would note:

> New York City [was] bathed in a thick fog. Visibility on the ground was bad enough, but . . . the fog [the pilot] found himself in was more like soup.

The pilot had never flown over downtown New York before, and now he could not see the city below. Flying through thick clouds, he lowered his landing gear, apparently thinking he was approaching the airfield at Newark. Unfortunately he was over Manhattan. The Empire State Building towered dead ahead, shrouded and hidden in the opaque white mist.

The Collision: The instant before impact the pilot apparently saw the masonry and steel monster materialize out of the mist in front of him. In desperation he pulled back on his control column. At over 200 knots the B-25 angled upward, but it was too late:

> The aircraft struck the Empire State Building while attempting to fly *visual [emphasis added]* under instrument conditions.

A news media report would later claim: "The force of the impact was such that many of the bodies were blown to bits." That report turned out to be an exaggeration, for most deaths in the building were caused by burns. A technical analysis written long after the accident would more accurately explain the collision:

> The majority of the plane hit the 79th floor, creating a hole in the building 18 feet wide and 20 feet high. The plane's high octane fuel exploded, hurtling flames down the side of the building and

inside through hallways and stairwells all the way down to the 75th floor.

The three men aboard the B-25 perished instantly. The impact caused a thunderous noise, leading many people in nearby buildings to believe that New York City was being bombed. One witness in the Empire State Building was quoted by United Press International:

> There was an awful explosion. The building quaked, and I saw a sheet of flame [go] 100 feet down the hallway . . . . Burning pools of high octane gasoline seemed to be all around. Screaming men and women were running everywhere.

Inertia had torn both heavy engines off the wings and had turned them into missiles. Engine No. 1 blasted completely through the building and out the other side, and it tumbled down atop the roof of a shorter skyscraper. The No. 2 engine ripped through an elevator shaft, severing the elevator cables. Most of the 79th floor burst into flames, and one survivor later would lament:

> One man was standing in the flames. I could see him. It was a coworker [who failed to survive]. His whole body was on fire, and I kept calling out to him: "Come on, Joe! Come on, Joe!"

The sensational aerial collision with the tallest building in the world made front page news in Europe and North America. One newspaper headline proclaimed:

Empire State Building Becomes Pillar of Horror

Eleven people in the building had perished in the inferno. Dozens more had suffered horrible burns but managed to escape alive. The most miraculous survival involved an elevator operator on the 75th floor, four floors below the point of impact. A written report from the Otis Elevator Company would explain:

> All the governor and safety cables were sheared, so the car went into free-fall. A few seconds later the car was in the basement.

From the 75th floor the elevator plunged all the way to the bottom of the elevator shaft in the sub-basement. Fortunately for the elevator operator, the close-fitting lower portion of the shaft created a partial compressed-air cushion below the free-falling elevator car. Although severely injured, the woman lived to tell about it (she still holds the Guinness World Record for the longest recorded elevator fall).

The Empire State Building survived with relatively little structural damage. Unlike tall skyscrapers built in later years, it had been reinforced with limestone blocks backed with bricks. The floors were steel-reinforced concrete designed to prevent the spread of fire. The building would be repaired and fully operational within three months. It would remain the world's tallest building until 1970.

The Investigation: The military Accident Investigation Board noted that the pilot, a lieutenant colonel, was a decorated war hero who had flown the heavier Consolidated B-24 Liberator bomber in combat in Europe. However, the Army response to a Congressional Inquiry explained that he had:

> ... occupied the left seat and received a pilot's check-out en route [to Bedford]. [He] had not previously flown [a] B-25.

The board realized the pilot had believed he was 15 miles south of La Guardia Airport when he first radioed the tower. Yet, in reality he had been northeast of the airfield.

Because the B-25 had been flying on a "contact" clearance, and because of the bad weather, the controller had initially told the pilot to land at La Guardia. Nonetheless, the pilot had *insisted* on flying to Newark. Investigators knew the pilot's commanding officer had been waiting to be picked up there. That fact undoubtedly influenced the pilot's desire to forge ahead despite the foul weather.

Ultimately the controller had given in and had cleared the flight to Newark Army Airfield. However, the clearance contained an admonition to return to La Guardia if forward visibility fell below three miles. At the time of the collision the forward visibility at the location and altitude at which the B-25 was flying was much less, something on the order of a mere 200 to 300 feet. One aviation sage explained that the pilot elected to descend:

> . . . out of the thickest part of the fog and take a peek at the ground, hoping to get his bearings straight. What he found was a forest of skyscrapers, the tops of which were all around him.

Probable Cause: The Army issued an exhaustive report on this accident. Because of intense media and political interest that report is now officially "out-of-file." However, the Army's (1) internal correspondence, (2) *Summary of Circumstances,* (3) responses to a Congressional Inquiry, and (4) letters to a U.S. Senator and private citizens have been retained. In addition, many independent technical reports are available.

The above sources reveal that the board found the ***Primary Cause*** of the accident to be the pilot's insistence on continuing the flight to Newark. The weather obviously had been below prescribed minimums for such a "contact" (later to be called, *VFR*) flight.

A ***Contributing Cause*** was the clearance given by the controller at La Guardia. It was deemed improper in view of existing weather conditions.

The verbiage of the tower controller's final radio message before the accident had been released to the news media. In a tragic twist of irony the controller had radioed to the pilot:

> From where I'm sitting [in the control tower], at the present time I can't see the Empire State Building.

Acrobatics on the Airway
Mid-Air Collision near Chesterfield, New Jersey
30 July 1949

Grumman F6F Hellcat, Bureau No. 72877
U.S. Navy cross-country proficiency flight
1 aboard, 1 killed

Douglas DC-3, Registration NC19963
Eastern Airlines Flight 557
15 aboard, 15 killed

Synopsis: The military pilot of a Grumman F6F Hellcat began "buzzing" a single engine civil aircraft. Concentrating on the small plane, the Hellcat pilot failed to see a big Douglas DC-3 airliner and blindly rammed into it. The Hellcat and the airliner crashed, killing all 16 people on board both planes.

The Grumman F6F Hellcat Flight: On a warm and clear morning at 0937 Hours a U.S. Navy pilot walked out to his aircraft for a cross-country proficiency flight. He fired up a Grumman F6F Hellcat fighter plane and roared skyward from NAS Anacostia in the District of Columbia. His VFR flight plan called for a trip to NAS Quonset Point, Rhode Island.

Because of the colorful history of the F6F Hellcat a detailed look at its military background is warranted:

At the start of the United States' involvement in World War II in 1941 the premier Japanese fighter, the Mitsubishi A6M Zero, totally outclassed its American rivals. The United States needed a fast, nimble, rugged, lethal, and heavily armored fighter to control the sky over the Pacific.

Grumman designed the F6F Hellcat and equipped it with a 2,000 horsepower Pratt & Whitney R-2800 Double Wasp radial engine. The new Hellcat could fly faster than the Zero, climb faster, dive faster, and it carried more fuel and ammunition. Six .50 caliber machineguns in the wings would later be supplemented with a 20mm cannon and five-inch rockets. The new fighter was heavily

Grumman F6F Hellcat (official U.S. Navy photo)

armored, and in a dogfight it could turn with the famed Zero. It would soon earn its nickname, the Ace-Maker.

Hellcats first saw combat on 31 August 1943 and began cutting a swath through ranks of the outmatched Zeros. U.S. Navy and Marine Corps pilots in Hellcats would eventually destroy 5,156 Japanese planes. The most spectacular shoot-downs took place on 19 June 1944, when Hellcats downed over 160 enemy aircraft. Officially this aerial shootout was known as the Battle of the Philippine Sea. Most historians would simply call it "The Great Marianas Turkey Shoot."

Grumman turned out an astounding 12,272 Hellcats for World War II. Most later versions sported a water-injection engine that increased output to 2,200 horsepower, and a radar equipped variant became a potent night fighter.

The flight plan to Rhode Island called for a time en route of two hours. The military pilot, a lieutenant, was a properly qualified Naval Aviator. Records revealed that he had never been reprimanded or disciplined for any infraction of flight rules.

The Douglas DC-3 Flight: Twenty-three minutes after the Hellcat took off heading northeast, a Douglas DC-3 airliner left La Guardia Airport at New York and flew southwest. The DC-3, Eastern Airlines Flight 557, was bound for Wilmington, Delaware, and Washington, D.C. The pilots had filed a VFR flight plan, and their airliner cruised toward Wilmington at 2,000 feet:

Douglas C-47, a military variant of the civil Douglas DC-3 airliner (official U.S. Air Force photo)

No greater accolade for the DC-3 exists than the realization that, over seven decades after it first flew in 1935, many still remain in service worldwide. The aircraft had become a commercial success in the years before World War II. When the United States went to war in 1941 the Douglas production line started turning out thousands of C-47s, the military variant of the civil DC-3.

The DC-3 on this flight had been bought by Eastern in 1940 and had flown 37,840 hours. Its two 1,200 horsepower Wright R-1820 Cyclone radial engines powered Hamilton Standard propellers. The captain, age 35, had earned an ATP certificate and had 10,013 hours of flight experience. His young first officer, age 27, had logged 1,397 flight hours. In the cabin a male flight attendant catered to the needs of his 12 revenue passengers.

Trouble Ahead: The airliner leisurely motored southwest on Red Airway 3. At 1017 Hours the pilots radioed their position over Freehold, New Jersey, and they estimated arrival over Philadelphia, Pennsylvania, at 1045 Hours. This would prove to be the last radio contact with Flight 557. Those aboard the DC-3 had no way to know they all would be dead in 13 minutes.

Meanwhile, a single engine Piper Super Cruiser flew eastward at 1,000 feet, far below the altitude of the airliner. The pilot of the Piper did not know the Hellcat was flying below him. However, without warning the Hellcat fighter suddenly zoomed vertically upward right in front of the startled Piper pilot:

> While [the Piper pilot] was in level flight, the fighter passed an estimated 100 feet directly in front of him and headed *vertically [emphasis added]* upwards.

The mesmerized Piper pilot watched the Hellcat loop over him and then disappear behind him. Moments later -- here it came again! The powerful Hellcat again pulled up vertically, screaming skyward in front of the small Piper Super Cruiser.

The Collision: The alarmed Piper pilot watched the fighter arc high overhead in an apparent chandelle:

> This maneuver continued until [the Piper pilot] saw the fighter and a DC-3 collide an estimated 700 to 1,000 feet above him.

At 1030 Hours the Hellcat left wing tore through the left wing of the airliner. The head-on impact tore the left wings off both aircraft. All aboard, if not killed instantly, were doomed:

> Following the collision both aircraft continued in their respective directions, losing parts as they fell, the F6F [Hellcat] striking the ground about one-half mile to the northeast, and the DC-3 about one mile to the southeast.

The violent tumbling fall to the ground slung the Hellcat pilot through his canopy and out of his fighter. His body fell to earth 200 feet away from the fuselage of his plane. Post-mortem examination showed that he had been killed by the mid-air collision.

The DC-3 airliner burned upon impact with the ground near Chesterfield, New Jersey. Everyone aboard had been killed by the aerial impact, the collision with the ground, the post-crash fire, or a combination of these events:

> The collision resulted in the destruction of both aircraft and death to all occupants.

By midday the news of the tragic collision and crashes reached the news media. *The Evening Bulletin*, a Philadelphia newspaper, postponed its afternoon edition to allow inclusion of information about the crashes. The banner headline read:

----

16 Die as Planes Collide in Air

Eastern Airliner Falls in Pieces and Explodes, Comes to Earth on Farm, Navy Hellcat Hits Two Miles Away

----

Quoting witnesses on the ground, the newspaper reporter wrote:

> The body of the [DC-3] plane seemed to split in two . . . and the whole disintegrating mass plunged to the ground. Farmers and others ran toward the scene, but almost immediately there was a terrific explosion, and the mass of the wreckage was enveloped in flames.

The Investigation: Physical analysis of the wreckage verified the mechanics of the aerial impact. It also corroborated the statement of the Piper pilot, the only known witness to the actual collision. The accident had taken place in crowded airspace:

> The airways in the New York area have a high traffic density. Furthermore, they pass over areas where the density of non-scheduled and itinerant flying is extremely high.

CAB investigators uncovered no possible causative factor other than the actions of the Hellcat pilot. The DC-3 captain and first officer had been in level flight roughly 2,000 feet above the ground, about 1,500 to 2,000 feet from the centerline of Red Airway 3. The CAB found no indication of improper conduct or flying technique on the part of the two Eastern pilots. The Piper pilot also had been flying straight and level. Investigators found nothing to indicate that his conduct contributed to the accident.

The CAB found a map in the Hellcat wreckage, and on the map the Navy pilot had plotted a direct course to NAS Anacostia. The accident site was right on the charted course. Investigators theorized the Hellcat pilot had sighted the Piper by happenstance:

> The [Hellcat] pilot must have adhered quite closely to [his] course until he reached the vicinity of the accident, where he saw the [Piper] aircraft and "buzzed" it.

By adoption, Civil Air Regulations had become part of U.S. Navy flight rules. These regulations prohibited "flying in such proximity to other aircraft as to create a collision hazard." In addition they prohibited acrobatic maneuvers on airways:

> CAR 60.16 (b): No person shall engage in acrobatic flight within any civil airway or control zone.

Probable Cause: The CAB adopted the Accident Investigation Report, File No. 1-0067, on 30 November 1949 four months after the mid-air collision. At the time of the accident the weather had been clear. Horizontal visibility had been estimated to be 10 miles with scattered clouds high above at 12,000 feet. The board concluded that the conduct of the Hellcat pilot led to the accident:

> The Navy pilot flew into the DC-3 following a pull-up in front of the small civil aircraft. It is highly probable that the F6F [Hellcat] was either not seen from the DC-3, or seen so late that evasion was impossible. *[and also]* Neither aircraft was seen by the crew of the other in time to avoid [a] collision.

The board also concluded that the Hellcat pilot had been performing acrobatics immediately prior to the collision. The board outlined the following *Probable Cause* of the accident:

> The reckless conduct of the Navy [Hellcat] pilot in performing acrobatic maneuvers on a civil airway and his failure to notice the presence of an air carrier aircraft, with which he collided.

Look Out for the P-38!
Mid-Air Collision at Washington, D.C.
1 November 1949

<u>Lockheed P-38 Lightning, Registration NX26927</u>
Test flight prior to acceptance by the Bolivian Air Force
1 aboard, 0 killed

<u>Douglas DC-4 Skymaster, Registration N88727</u>
Eastern Airlines Flight 537
55 aboard, 55 killed

<u>Synopsis</u>: A commercial Douglas DC-4 Skymaster approached the airport to land. A military Lockheed P-38 Lightning fighter with engine trouble approached the same runway, and the two aircraft collided. The subsequent crashes killed all 55 people aboard the DC-4 and critically injured the P-38 pilot.

The Lockheed P-38 Lightning Flight: A Lockheed P-38 Lightning fighter took off from Runway 3 at Washington National Airport at Washington, D.C., at 1137 Hours. The pilot had planned a test flight before acceptance of the plane by the Bolivian Air Force. After takeoff he turned left and climbed west of The Pentagon, the military nerve center of the U.S. Armed Forces:

The P-38 represented a radical change in fighter design. It had tricycle landing gear, a twin-boom empennage, two liquid cooled Allison V-1710 engines, self-sealing fuel tanks, and heavy armor protection for the pilot. The high altitude interceptor had a top speed of almost 400 knots and could climb to 20,000 feet in six minutes. It was armed with four .50 caliber machineguns plus a 20mm cannon.

With external fuel drop-tanks the P-38 had a 2,200 mile range. It was the only American fighter with enough range to conduct the famous "Yamamoto Raid" in early 1943. The Lockheed factory turned out 9,200 of these potent warbirds.

The P-38 in question had been delivered to the U.S. Army Air Force in April 1945, and shortly thereafter it had been declared war

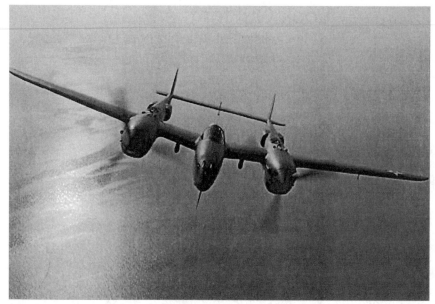

Lockheed P-38 Lightning (official U.S. Air Force photo)

surplus. At that time it had flown only 15 hours. Universal Air Marine & Supply Company, a dealer in war surplus airplanes, bought the P-38 in 1949 for later delivery to the Bolivian Government. The CAA gave the aircraft a civil registration number. At Washington National Airport, technicians installed external fuel tanks and worked on the fuel system, engines, and radios.

The P-38 pilot, with 3,600 flight hours, had trained with the U.S. Army Air Force during World War II and was properly licenced to fly the fighter. He was a Bolivian national, and in Bolivia he was the Director of Civil Aviation.

The Douglas DC-4 Skymaster Flight: In VMC weather with a visibility of 15 miles, Eastern Airlines Flight 537, a Douglas DC-4 Skymaster, flew inbound for landing at Washington National Airport. The plane carried 51 passengers, including two babies-in-arms, plus a crew of four. At 1144 Hours the Washington tower controller cleared the airliner to land on Runway 3:

The civil DC-4 had gone into production immediately before the start of World War II. The U.S. government commandeered the assembly line after war broke out, and 1,163 military variants

Douglas C-54, the military variant of the civil Douglas DC-4 Skymaster airliner (official U.S. Air Force photo)

soon rolled out of the factory. The transport became known as the C-54 in the U.S. Army Air Force and as the R5D in the Navy and Marine Corps.

In the postwar era, DC-4s and converted C-54s reigned as the premier commercial airliner in the world. Four Pratt & Whitney R-2000 Twin Wasp radial engines and huge fuel tanks allowed it to fly from New York to California with only one fuel stop. With a maximum takeoff weight of 73,000 pounds it could carry up to 86 revenue passengers. After being sidelined by new passenger jet transports, DC-4s served as "freighters" into the 1990s.

The young captain, age 33, had logged 9,033 flight hours and had earned an ATP certificate. His even younger first officer, with 4,396 hours in his logbook, had begun his flying career in the U.S. Armed Forces and held a Commercial Pilot certificate. Two stewardesses in the cabin rounded out the Eastern Airlines crew.

Trouble Ahead: Shortly after takeoff while the P-38 pilot climbed west of the airport, his No. 2 manifold pressure and engine speed began fluctuating. He did not declare an emergency. He flew south, turned 180 degrees to the left, and began a five mile, high, straight-in approach to Runway 3. Meanwhile:

> At 1144 [Hours], Eastern Flight 537 was cleared to land number one on Runway 3. This clearance was given when Eastern was on its downwind leg west of the field . . . . [At that time] the P-38 was south of Runway 3 on a high straight-in approach.

The controller, then unaware of the P-38 engine problems, would later *claim* he radioed the P-38 pilot and told him to enter the traffic pattern downwind. These alleged instructions were not acknowledged, and the CAB later admitted:

> Conceivably a good part of the tower's instructions to the P-38 may not have been received.

As the big DC-4 airliner lined up on final approach the P-38 was flying behind it and above it, heading for the same runway. Naturally the DC-4 pilots could not see the fighter plane behind them. Also, the fighter pilot could not see the DC-4, which was hidden from view below the nose of his warbird:

> There were no recordings of the [radio messages to and from] the P-38, [which] was on a frequency of 126.18 megacycles, while [the] Eastern [DC-4] was on a frequency of 119.1 megacycles and, consequently, neither plane could hear [radio] transmissions between the tower and the other plane.

Because radio messages were not recorded the ATC transcript, below, has been *reconstructed* from sworn testimony during post-accident CAB hearings. Readers are free to form their own opinions with regard to the resulting *credibility* issue:

Eastern 537: . . . Eastern Airlines Flight 537
P-38: . . . . Lockheed P-38 Lightning
Tower: . . . . Washington National Airport Control Tower

Tower: Eastern five-thirty-seven, you are cleared to land, Runway three, winds . . . . (1144)

P-38: Washington Tower, this Bolivian P-thirty-eight, I got engine trouble, request landing instructions. (1144)

Tower: Bolivian P-thirty-eight, you asking [for] landing instructions?

P-38: Yes, I have engine trouble. I am in a hurry. (1144)

Tower: Bolivian P-thirty-eight, you are cleared to land *number two [emphasis added]* on Runway three. (1144)

P-38: Tower, this is Bolivian P-thirty-eight on approach. (1145)

[the controller later will claim that he twice told the P-38 to enter a left traffic pattern; the P-38 pilot will dispute this claim]

[as the P-38 descends and overtakes the slower DC-4 on final approach, the controller sees the imminent collision; he switches to the DC-4 frequency and transmits in desperation]

Tower: Eastern! Turn left! Turn left! (1146)

Tower: Look out for the P-thirty-eight! (1146)

[the two aircraft collide on short final approach at 1146]

The Collision: In response to the "turn left" warning the DC-4 pilots had aborted their approach. Unable to see the P-38 above them, they began a climbing left turn. The two aircraft slammed into each other 300 feet above the ground and one-half mile short of the runway. The whirling No. 1 propeller on the P-38 sawed through the DC-4 fuselage behind the wing:

> [The P-38 propeller] cuts extended on both sides of the [DC-4] fuselage, from the top down to the emergency escape hatches, [and] for the remainder of the way around the fuselage the structure [tore] into two [pieces].

The rear half of the DC-4 fuselage fell like a brick and plunged upside-down into mud flats on the bank of the Potomac River. The front of the plane dove into the river and sank immediately. All 55 people aboard the airliner perished.

The P-38 pilot fared a little better. His fighter had plunged into the Potomac and had sunk, but *somehow* (neither he nor anyone else knows how) he escaped from the submerged wreckage. Severely injured and drowning, he was not able to swim to the shore. The crew of a boat from nearby Bolling Air Force Base spotted him as he sank beneath the surface, and the rescuer later explained:

> His head slipped below the surface. When he went down I knew
> it was then or never, so I plunged in after him.

The boat occupants fished the comatose pilot out of the river, and medical technicians rushed him to nearby Alexandria Hospital. For many days he would lie in critical condition with a broken neck, multiple skull fractures, multiple body trauma, and pneumonia. He eventually would recover and survive.

At the time, in terms of loss of life, this was the worst aviation tragedy in United States history. The next day *The Washington Post* newspaper headline would exclaim:

Banks of Potomac Scene of Worst Air Disaster

The accompanying newspaper article detailed the total carnage at the DC-4 crash site on the mud flats:

> The impact of the crash was so great that bodies were hurled more
> than 150 feet. Debris was scattered over the shoreline, with
> clothing hanging from trees and bushes . . . . Many of the bodies
> were missing arms and legs, and at least two were headless.

The Investigation: *What* had happened was clear. Yet, (1) *why* the pilots did not see each other and (2) *why* the tower did not transmit an early warning to both planes remained murky. The absence of an ATC recording thwarted investigators. Tower controllers admitted they could not recall all their instructions to the two aircraft. In fact, the two controllers could not agree on several things they had radioed to the pilots. The hospitalized P-38 pilot could not recall all his contacts with the tower. The DC-4 pilots were no longer alive to give their versions of events. In any event, it soon became clear that a third aircraft likely played a role in the accident:

> A United States Air Force B-25 [bomber had] made a practice instrument approach to Runway 36. This aircraft . . . passed over the field without landing at 1143 [Hours].

At the time of the B-25 low pass the P-38 had been on a long, high, final approach. During his recovery in the hospital the P-38 pilot confirmed hearing the tower initially tell him that he was cleared to land "number two." He said he saw an aircraft flying toward the runway ahead of him, a "twin engine, twin tailed" plane (an accurate description of a B-25). Investigators *theorized* that the P-38 pilot saw the B-25 and (1) *assumed* it was the aircraft that was cleared to land ahead of him and (2) *assumed* he was cleared to land "number two" behind it.

The above scenario presumes the P-38 pilot never saw the DC-4. This was deemed likely, for the DC-4 had then been on its downwind leg and out of the P-38 pilot's line of sight. The P-38 pilot would have been concentrating on the runway and the B-25 a couple of miles ahead. He could not hear radio messages to or from the DC-4. Also, he said that radio messages to him never mentioned the DC-4 (a fact that both controllers confirmed despite their other disagreements). The CAB opined:

> It remains to be determined whether those in the control tower discharged their duties as prescribed by Civil Air Regulations.

Investigators imputed "poor judgement" to ATC. The CAB felt tower controllers should have made an "earlier attempt" to warn the DC-4 pilots about the presence of the P-38.

Ironically, controllers *could* have made simultaneous transmissions on both radio frequencies. They merely needed to place both frequency toggle switches in the "on" position.

Probable Cause: The CAB adopted its Accident Investigation Report, File No. 1-0138, on 22 September 1950. The cause of the accident boiled down to one simple issue – whom the CAB elected to believe. Contradictory statements of the controllers and the P-38 pilot were mostly self-serving. The controllers *claimed* they never gave the P-38 pilot final clearance to land number one. Conversely, the P-38 pilot *claimed* that on final approach the tower had radioed:

> Bolivian P-thirty-eight, cleared to land, Runway three.

Without question the DC-4 pilots had been cleared to land and had no reason to expect any other aircraft to be making a final approach. They had been unable to hear radio transmissions to or from the P-38. The tower never alerted them to a potential conflict until the frantic "turn left" warning a few seconds before the collision. With respect to that tardy ATC warning the board noted:

> The tower did not act with the requisite alertness and promptness in communicating to Eastern the position of the P-38.

Despite ATC shortcomings the CAB more-or-less dismissed the "cleared to land" claim of the hapless bedridden P-38 pilot. Notwithstanding his engine problems, under the *assumption* that he had not been given final clearance to land, the board stated that he should not have made the approach. In addition – for reasons which it never explained – the board placed little credence in:

1. The significance of the B-25 low pass.

2. The tower's acknowledged failure to alert the P-38 pilot to the presence of the DC-4.

3. The tower's failure to alert the DC-4 pilots to the presence of the P-38 until mere seconds before the collision.

4. The tower's acknowledged failure to transmit simultaneously on both frequencies to both aircraft.

5. The fact that neither the P-38 pilot nor the DC-4 pilots, on different frequencies, could hear radio messages to and from the other aircraft.

Despite these crucial factors the board found the ***Probable Cause*** of this accident to be elementary:

> The execution of a straight-in final approach by the P-38 pilot without obtaining proper clearance to land and without exercising necessary vigilance.

The Miami-to-Havana Tragedy
Mid-Air Collision over Key West, Florida
25 April 1951

<u>Douglas DC-4 Skymaster, Registration CU-T188</u>
Compania Cubana de Aviacion Flight 493
39 aboard, 39 killed

<u>Beechcraft SNB-5, Bureau No. 39939</u>
U.S. Navy instrument training flight
4 aboard, 4 killed

Synopsis: A commercial Douglas DC-4 Skymaster rammed into a military Beechcraft SNB-5 in VMC weather. Out of control, both planes tumbled down into the sea. The accident killed all 39 people on the DC-4 and all four men aboard the SNB-5.

The Douglas DC-4 Skymaster Flight: At 1109 Hours, Compania Cubana de Aviacion (hereafter, *Cubana*) Flight 493, a Douglas DC-4 Skymaster, departed from Miami, Florida. The CAB had granted Cubana a "Permit to Foreign Air Carrier" for flights between Miami and Havana, Cuba. Pan American Airways (hereafter, *Pan Am*) was the domestic agent for Cubana in Miami. The Cuban airliner, with a crew of five plus 34 passengers, was flying on an IFR flight plan. However, the pilots had been instructed to "maintain visual flight rules" for separation from other aircraft.

The DC-4 captain, age 50, and first officer, age 29, each held an "Airline Pilot License" issued by the Republic of Cuba. By chance the Chief Pilot at Cubana elected to tag along as a cockpit observer. A purser and a steward catered to the 34 passengers. At 1119 Hours, ten minutes after takeoff, the captain reported leaving 2,000 feet and climbing to 4,000 feet.

The Beechcraft SNB-5 Flight: A Beechcraft SNB-5 climbed skyward from NAS Key West, Florida. The U.S. Navy training plane carried three pilots and a radioman. On a VFR flight plan, the crew planned to use the Key West radio range station to practice instrument approaches to the Boca Chica Airport. Instrument approach procedures stated in part:

Beechcraft AT-7, one of the many variants of the famed "Twin Beech" (photo courtesy of U.S. Air Force Historical Research Agency)

. . . the pilot proceeds outbound on the west leg of the [radio] range on a heading of 279 [degrees], reduces speed to 105 knots, lowers the landing gear, and descends to an altitude of 1,300 feet. A procedure turn is then made . . . [for an] instrument approach to Boca Chica Airport.

The SNB on this flight had logged 993 flight hours and was assigned to the Atlantic Fleet All-Weather Training Unit. A Navy instructor pilot, a lieutenant, manned the left seat in the cockpit, and his two students would alternate in the right seat:

SNB aircraft flown by the U.S. Navy were among several military variants of the famous "Twin Beech." The twin engine utility plane had first flown in 1937. It served during and after World War II in a host of roles. The U.S. Army Air Force usually flew C-45 Expediter and AT-11 Kansan variants, and the U.S. Navy flew the venerable SNB.

Two Pratt & Whitney 450 horsepower radial engines drove two-blade Hamilton Standard propellers. Navy pilots joked that the plodding and underpowered "SNB" moniker was an abbreviation for "Secret Navy Bomber." The plane would serve in a utility and training role well into the 1960s.

Trouble Ahead: As the DC-4 flew south toward Key West on the way to Havana, the SNB crew simulated an instrument approach to the airport. Listening to the tone from the radio range station, the pilots tracked outbound and contemplated the "procedure turn" back toward the airport.

The two planes slowly converged at about 4,000 feet. The SNB angled in from the left side of the DC-4. Like two ants scurrying for the same breadcrumb, the two aircraft headed straight for the same spot in the sky.

The CAB would not include a *complete* ATC radio transcript in the report dealing with this accident. However, the CAB would include *excerpts* from the ATC recordings:

> Cubana 493: . . . . Cubana Flight 493
> Center: . . . . Miami Center (ARTCC)
> Pan Am: . . . . Pan Am Flight Watch

[Cubana Flight 493 takes off from Miami at 1109]

Center: *[by interphone to Pan Am]* ATC clears Cubana four-nine-three, cruise and maintain four-thousand [feet]. (1114)

[Pan Am relays the clearance to Cubana Flight 493]

Pan Am: Cubana four-nine-three, ATC clears Cubana four-nine-three, cruise and maintain four-thousand. (1119)

Cubana 493: Roger, estimating Key West at eleven-forty-eight, leaving two-thousand, climbing to four-thousand. (1119)

[the Navy SNB takes off from Key West at 1126]

[the SNB pilots make radio contact with military Ground Control Intercept 2 minutes later at 1128]

*[the two aircraft **collide** 21 minutes later at 1149]*

The Collision: The DC-4 flying south toward Havana, and the SNB on its westerly heading, rammed into each other in level flight:

The Navy aircraft crashed into the water just west of the Naval Station. The Cubana aircraft, however, continued on for some distance before entering a left bank which became progressively steeper until . . . it crashed into the ocean approximately 1.7 miles southeast of the point of collision.

The accident took place in view of hundreds of sunbathers on the Key West beaches. Nearby vehicular traffic screeched to a sudden halt as scores of motorists abandoned their cars and raced down to the shore. They peered at the doomed remnants of the SNB protruding out of the water nearby. They could see parts of the DC-4 wreckage, although most of it was under 25 feet of water. *The Key West Citizen* newspaper later would report:

> Forty-three persons lost their lives when the two planes struck [each other] in the air off the southern point of Key West . . . . Both aircraft crashed into the sea and no survivors were found. *[and also]* Great clouds of smoke hung in the air for as long as twenty minutes afterwards.

The Navy Duty Officer got word of "an accident" at 1150 Hours. Within 10 minutes the magnitude of the tragedy became apparent, and Navy search and rescue operations began at once. For seven days Navy and Coast Guard vessels, with 20 divers, labored to retrieve bodies and pieces of the two aircraft.

Human remains were taken to the Naval Hospital for identification. Relatives could claim personal effects of deceased passengers at the Monroe County Courthouse. The waterlogged clothes, luggage, watches, shoes, handbags, wallets, and other items were piled on the courthouse floor. One grieving relative, a Cuban professor, told the coroner through an interpreter:

> That's my sister's ring. Here is her name and the name of her husband. They were on their way back from their honeymoon in Miami. They were not supposed to be on a death flight.

Navy divers recovered complete or partial remains from all 39 people aboard the DC-4 and two of those aboard the SNB. However, they never found any trace of the other two Navy men. They had been ejected from the wreckage and swept away by the sea.

Douglas C-54 Skymaster, the military variant of the commercial Douglas DC-4 airliner (official U.S. Air Force photo)

The Investigation: The DC-4 flight plan had called for a magnetic heading of 223 degrees southwest of Miami. Witnesses confirmed the airliner had been flying southwest. Passing Key West, the DC-4 had gently banked left and turned to the new heading of 179 degrees, almost due south.

The SNB flight, according to witnesses, had passed over the Key West radio range station and was flying west, approximating the required heading of 279 degrees. Wreckage analysis confirmed that the two aircraft were on their assigned headings:

> The cut on the DC-4 left wing at station 588½, which was made by one blade of the right propeller of the SNB, was the first contact between the two aircraft.

The DC-4 left wing had ripped through the SNB fuselage, totally destroying the smaller aircraft. The mid-air impact had torn off most of the DC-4 left wing outboard of the No. 1 engine. The airliner's sheared-off outer left wing fell down into the sea, entangled with the twisted SNB wreckage:

Sufficient evidence was found to establish the position of the aircraft relative to each other at the moment of initial impact.

For instrument training the SNB cockpit had been equipped with detachable orange plexiglass panels that covered the windshield. A student wearing colored goggles would be unable to see through the windshield but would retain normal vision within the cockpit. The instructor pilot, in addition to tutorial duties, had sole responsibility for heads-up vigilance outside the cockpit.

Ironically, in this instance the DC-4 had been approaching the SNB from the right. The SNB instructor had occupied the left seat. His field of vision toward the approaching airliner was much more limited than his field of vision straight ahead and to the left.

Probable Cause: CAB investigators completed their work and adopted an Accident Investigation Report on 17 October 1951. They found that both airplanes and flight crews were properly certified. Both flights had been routine and had gone according to plan up to the time of the collision:

> The weather in the Key West area at the time of the accident was clear. *[and also]* Neither the weather nor the sun's position is considered a contributing factor in this accident.

A host of people had seen the planes in flight before the collision, but no witnesses reported any evasive maneuvers. This absence of evasive action had been confirmed by wreckage analysis. The board concluded that none of the pilots saw the other aircraft flying toward them. Therefore, the board found that the *Probable Cause* of this accident was basic:

> Failure of crews of both aircraft to maintain sufficient vigilance under VFR conditions to prevent a collision.

Tragedy over the Grand Canyon
Mid-Air Collision over the Grand Canyon, Arizona
30 June 1956

<u>Lockheed L-1049 Super Constellation, Registration N6902C</u>
Trans World Airlines Flight 2
70 aboard, 70 killed

<u>Douglas DC-7 Seven Seas, Registration N6324C</u>
United Airlines Flight 718
58 aboard, 58 killed

<u>Synopsis</u>: A Lockheed L-1049 Super Constellation took off and headed east. Three minutes later a Douglas DC-7 Seven Seas took off from the same runway and flew in the same direction. On IFR flight plans the two airliners collided at 21,000 feet. Both aircraft crashed out of control, killing all 128 people on board.

The Lockheed L-1049 Super Constellation Flight: Sixty-four passengers walked out of the terminal at Los Angeles International Airport in California. They climbed up the mobile stairway into the cabin of a big Lockheed L-1049 Super Constellation. Trans World Airlines (TWA) Flight 2 was bound for Kansas City, Missouri. The aircraft took off from Runway 25, and Los Angeles Departure Control vectored the flight up through a low overcast:

The Constellation had been designed to the specifications of the eccentric entrepreneur, Howard Hughes. With its distinctive triple-tail and artfully curved fuselage, the "Super-Connie" had reigned as the undisputed queen of the sky in the early 1950s. Powered by four Wright R-3350 twin-row 18 cylinder radial engines, she could cruise at almost 300 knots. With a pressurized cabin the sleek Constellation could fly coast-to-coast "above the weather" at 20,000 feet in slightly over seven hours.

The captain broke out "on top" of the clouds at 2,400 feet and radioed Los Angeles Center to confirm his en route clearance. He had logged almost 15,000 hours of flight time. Backed up by his first officer, two flight engineers, plus two "hostesses" in the cabin, the captain climbed toward his assigned cruising altitude of 19,000 feet.

Lockheed L-1049 Super Constellation (official U.S. Air Force photo)

The Douglas DC-7 Seven Seas Flight: Only three minutes after the TWA flight climbed skyward, United Airlines Flight 718 began rumbling down the same runway. The Douglas DC-7 Seven Seas climbed up through the same overcast into the clear blue sky above. ATC cleared the pilots to climb in VFR conditions to 21,000 feet, the planned cruising altitude:

> *The DC-7 was the last of the Douglas radial engine airliners, and it entered service in 1953. Like the Constellation, it was powered by four Wright R-3350 engines. It was the first airliner that could fly nonstop, westbound, across the United States despite the prevailing headwinds. Douglas called it the Seven Seas because it could fly over 100 passengers to any country in the world. After new jet transports entered airline service most DC-7s continued to serve for decades as freighters and tankers.*

The DC-7 on this flight was painted red, white, and blue and had rolled out of the Douglas factory the previous year. The captain had 17,000 hours of flight time in his logbook. A first officer and flight engineer assisted him in the cockpit. Two young "stewardesses" would cater to their passengers on the trip to Chicago, Illinois.

Douglas C-74 Globemaster, a military variant of the civil Douglas DC-7 Seven Seas airliner (official U.S. Air Force photo)

Trouble Ahead: TWA Flight 2 droned eastward through sunny skies at 19,000 feet. Meanwhile, United Flight 718 motored in the same direction at 21,000 feet. Neither aircraft was equipped with a CVR, for this was the 1950s. However, radio messages between airliners, ATC facilities, and "aeronautical ground stations" were routinely recorded. The portion of the ATC transcript, below, begins after TWA Flight 2 requests an altitude change:

> Capt-United 718: . . . . Captain, United Airlines Flight 718
> FO-United 718: . . . . . First Officer, United Airlines Flight 718
> TWA-Ground: . . . . . . TWA aeronautical ground station
> LA-Center: . . . . Los Angeles Center (ARTCC)
> SL-Center: . . . . Salt Lake Center (ARTCC)

TWA-Ground: [Center], TWA two is coming up on Daggett [and is] requesting twenty-one-thousand feet. (0921)

[Los Angeles Center calls Salt Lake Center]

LA-Center: TWA two is requesting two-one-thousand, how does it look? I see he is [at] Daggett, direct [to] Trinidad, I see you have

United [Flight] seven-eighteen crossing his altitude, in his way at two-one-thousand.

SL-Center: Yes, their courses cross, and they are right together.

[Los Angeles Center calls the TWA ground station]

LA-Center: Advisory [for] TWA flight two, unable [to] approve two-one-thousand.

TWA-Ground: Just a minute – aaahhh – I think he wants a thousand on top – yes, a thousand on top until he can get [21,000 feet].

LA-Center: ATC clears TWA two, maintain at least one-thousand on top, advise TWA two his traffic is United seven-eighteen, direct Durango, estimating Needles at zero-nine-five-seven.

[the ground station radios the clearance to TWA Flight 2]

[At 0958 the DC-7 pilots report passing over Needles at 21,000 feet and estimate they will reach Painted Desert at 1031]

[1 minute later the Constellation pilots report passing over Lake Mohave "1,000 feet on top" at 21,000 feet and estimate they will reach Painted Desert at 1031]

[both aircraft have estimated they will reach the same place, at the same time, at the same altitude]

*[the two airliners **collide** at 1030:30]*

FO-United 718: *[a scream]* Aaahhh! Salt Lake! United seven-eighteen! Aaahhh! We're going in! (1030:53)

Capt-United 718: *[a scream]* Arrhhh! Pull! Up! (1030:53)

The Collision: Aeronautical station radiomen at Salt Lake City and San Francisco recorded desperate screams from the two doomed United DC-7 pilots at 1030:53 Hours. However, the radiomen could not decipher the words or message.

Later the tape recording would be subjected to spectrographic analysis, speech stretching, and a technique called "visible speech." These tests would reveal that the first officer began screaming first. During pauses his microphone remained keyed, and analysts detected the voice of a second person. He was identified as the captain, and the pitch of his screams was even higher. The analysts dryly noted that the two pilots were under "great emotional stress."

The two airliners had flown eastward over the Grand Canyon in Arizona. Passengers lucky enough to have window seats likely had enjoyed the scenic view. Possibly, so had the pilots. Meanwhile, the two aircraft slowly converged:

> The initial impact occurred with the [United] DC-7 moving from right to left relative to the [TWA] L-1049. . . . The lower surface of the DC-7 left wing struck the upper aft fuselage of the Constellation with disintegrating force. The collision ripped open the fuselage of the Constellation. . . . Most of the left outer wing [of the DC-7] separated during the collision.

The Constellation cabin ripped open and flung passengers out into the sky. The triple-tail empennage broke away. What was left of the airliner tumbled down into the canyon four miles below. Most of the wreckage fell into a deep draw near Temple Butte.

Meanwhile, the United DC-7, minus much of its left wing, entered a steep spiral from which recovery was impossible. The mortally wounded airliner slammed into the face of a cliff a mile from the Constellation. Much of the debris and some dismembered bodies then rained down into a deep and inaccessible "chimney" below the main impact site:

> A relatively short wreckage distribution path showed that the aircraft contacted the ground at a steep angle.

Neither airliner had reported reaching the Painted Desert fix, so ATC issued a "missing aircraft alert" at 1151 Hours. Later that day searchers sighted smoke from the two crash sites deep down inside the Grand Canyon. There had been a total of 128 people on the two aircraft, and none of them survived.

Eager to peddle misfortune and scandal for profit, the news media enjoyed a brief field day. ATC controllers had known the airliners were (1) flying at the *same altitude* (2) toward the *same place* in the sky, and (3) planned to get there at the *same time*. Insanity! The heads of inept dolts at ATC would roll, the media crowed.

This rush to judgment proved premature. Those in the aviation community knew the truth. They knew the presumed control by ATC was more fantasy than fact.

The Investigation: A brief review is necessary. In the technical aftermath of World War II, modern aircraft instrumentation had made flight in IMC weather practical. Advances in electronics led to a national network of ADF and VOR stations for aerial navigation. In all types of weather, planes could fly on "airways" in the sky. Still, a means of preventing aircraft from colliding on airways in the clouds was needed, and ATC was born:

> *ATC in the United States was in its infancy in 1956. Within en route "centers," civil servants in white shirts and bow ties became "controllers." Armed with microphones and headsets, they sat at desks facing rows of slanting metal trays. Each tray represented an airway segment. The trays held handwritten "flight progress strips," each of which represented an aircraft that ATC had "cleared" on an IFR flight.*
>
> *Each flight progress strip fit between bars on the tray, and each bar represented 1,000 feet of altitude. Therefore, in theory, if a controller could keep his flight progress strips separated, no aircraft could collide. Crude, but it worked.*
>
> *En route radar coverage did not exist. Pilots radioed "position reports" to aeronautical ground stations. The ground stations used a telephone or teletype to relay the reports to the center. They relayed clearances and pilot requests in the same manner.*
>
> *This system worked well for aircraft (1) on an IFR flight plan, (2) in IFR weather conditions, (3) in controlled airspace. Under other conditions the ATC system had severe limitations. These limitations became obvious after the Grand Canyon collision.*

The CAB conducted a meticulous review of the accident. After leaving Los Angeles both flights initially had flown on airways, and ATC separated them from each other and from all other IFR traffic.

Both flights later flew off of airways into uncontrolled airspace. Both radioed position reports to ATC. Consequently, ATC knew both aircraft were cruising at the (1) *same altitude* and estimating arrival at the (2) *same electronic fix* at the (3) *same time.*

How could ATC have allowed the flights to progress in this manner, the public wondered? The answers lay in regulations well known only to ATC, military pilots, and airline pilots:

First, in "uncontrolled airspace" off of airways, ATC was not required to provide, and in most instances could not provide, control services or advisories to aircraft on IFR flight plans. ATC procedures clearly stated:

[ATC] clearances authorize flight within control zones and control areas only; no responsibility for separation of aircraft outside these areas is accepted.

Second, although the TWA flight was at 21,000 feet, ATC had not cleared it to fly at that altitude. Instead, ATC had cleared the pilots to fly "1,000 feet on top," something totally different. The Flight Information Manual best explained the concept:

"1,000 feet on top" may be filed in an IFR flight plan, or assigned by ATC in an IFR clearance, in lieu of a cruising altitude. [This] permits the flight to be conducted at any altitude . . . 1,000 feet or more above the cloud layer.

In other words, the TWA pilots could have been flying at any altitude 1,000 feet or more above the cloud tops. By chance the captain had elected *to fly at 21,000 feet.*

Third, although both aircraft were at 21,000 feet, they were flying in clear weather. Under such conditions, Civil Air Regulations required the pilots to maintain *visual* separation:

During the time an IFR flight is operating in VFR weather conditions, it is the direct responsibility of the pilot to avoid other aircraft.

Probable Cause: On 15 April 1957, roughly 10 months after the accident, the CAB adopted the Accident Investigation Report, File No. 1-0090. The board determined that both aircraft were flying in VMC weather in uncontrolled airspace. ATC was not required to enforce separation between them:

[Air] traffic control services are not provided in uncontrolled airspace. *[and also]* Under visual flight rules . . . it is the pilot's responsibility to maintain separation from other aircraft.

The board found that ATC had issued proper clearances. After learning that both aircraft were flying in the same area at the same altitude, ATC was not required to issue an amended clearance or a traffic advisory. Both aircraft were flying in VFR conditions, and the see-and-be-seen adage applied. Therefore, the board detailed the *Probable Cause* of this accident as follows:

The Probable Cause of this mid-air collision was that the pilots did not see each other in time to avoid a collision. It is not possible to determine why the pilots did not see each other.

Postscript: The Grand Canyon collision blared a wake-up call to the aviation industry. As an interim step, U.S. military radar sites began monitoring commercial airline flights. Congress created a new agency, which would evolve into the FAA. Within a few years new radar systems and electronic wizardry enabled controllers to separate IFR aircraft from takeoff until landing. Were the days of mid-air collisions long past? Wishful thinking!

Say Goodbye to Everybody!
Mid-air collision near Sunland, California
31 January 1957

<u>Douglas DC-7 Seven Seas, Registration N8210H</u>
Douglas pre-delivery factory test flight
4 aboard, 4 killed (plus 3 killed on the ground)

<u>Northrop F-89 Scorpion, Serial No. 52-1870A</u>
U.S. Air Force fire control system factory flight check
2 aboard, 1 killed

Synopsis: A military Northrop F-89 Scorpion took off for a factory test of its updated radar fire control system. At 25,000 feet it collided with a civil Douglas DC-7 Seven Seas that was on a pre-delivery factory test flight. Both aircraft crashed. The radar operator ejected from the F-89, but the five others aboard the two aircraft were killed.

The Douglas DC-7 Seven Seas Flight: A brand new Douglas DC-7 Seven Seas accelerated down Runway 3 at Santa Monica Municipal Airport in California and climbed skyward. This was its first test flight before anticipated delivery to Continental Airlines. The crew of four (pilot, copilot, flight engineer, and radio operator) were all Douglas employees, and no one else was aboard. The tanks had been filled with enough fuel for six hours of flight, although the proposed time in the air was only 2 hours and 15 minutes. The flight engineer had calculated the takeoff weight to be 88,000 pounds, far below the maximum allowable weight of 143,000 pounds.

The Douglas pilot, age 36, held an ATP certificate and had logged 11,757 hours in the air. He enjoyed a second career as a movie actor in Hollywood and had appeared in over 70 films including *I Shot Billy the Kid* and *Sunset Boulevard*. His copilot, age 50, also held an ATP certificate and had 7,115 hours in his logbook.

The Northrop F-89 Scorpion Flight: The U.S. Air Force owned the Northrop F-89 Scorpion. Under contract, the plane had been returned to the manufacturer for an overhaul and modernization project. For the coming test flight a Northrop pilot would rendezvous with another

Northrop-piloted F-89 and check the operation of the onboard radar
fire control systems:

> *The F-89 was a twin engine fighter-interceptor. It had been
> designed to track, intercept, and destroy enemy aircraft in all
> weather conditions, day or night. A radar operator seated behind
> the front cockpit would guide the pilot into attack position. The
> pilot then would destroy the enemy plane with either air-to-air
> missiles or 20mm cannons. Two Allison afterburning turbojets
> powered the interceptor, and Northrop eventually would build
> 1,050 of these Cold War era warplanes.*

Afterburners glowing, the two F-89s took off from Runway 25 at
Palmdale, California, at 1050 Hours. The jets climbed to 25,000 feet
and began a series of scissoring maneuvers to get into position for
simulated all-weather attacks on each other. They flew over the San
Fernando Valley, an area where other air traffic was common.

Trouble Ahead: Both aircraft flew on VFR flight plans at 25,000
feet in uncontrolled airspace. Neither plane was in contact with ATC,
nor were they required to be. Each maintained contact with a "radio
room" at its company base.

Although the DC-7 and F-89 pilots did not know it, they were
flying straight toward each other. In this pre-CVR era the cockpit
conversations were not documented. However, the Douglas radio
room recorded all radio traffic to and from the DC-7. All radio
messages were routine until 1118 Hours. Thereafter the DC-7 would
spin out of control. However, its long spiral down to the ground
would allow the crew to radio their plight to their company:

> Pilot: . . . . . . Pilot, Douglas DC-7 Seven Seas
> Copilot: . . . Copilot, Douglas DC-7 Seven Seas
> ROper: . . . . Radio operator, Douglas DC-7 Seven Seas
>
> *[the DC-7 and F-89 **collide** at 1118; the F-89 disintegrates, and
> the DC-7 spins out of control]*

Pilot: Uncontrollable! (1118)

Copilot: We're – Mid-air! – Mid-air collision! – Ten how! *[aircraft
identification, using the phonetic "how" for "H"]* (1118)

Copilot: We're going in! Uncontrollable! Uncontrollable! (1118)

Copilot: We are – we've had it, boy – poor jet too – [I] told you we should take chutes. *[meaning, parachutes]* (1118)

[knowing he will die, the DC-7 copilot makes a final request]

Copilot: Say goodbye to everybody! (1119)

ROper: We're spinning in the valley! (1119)

[the DC-7 crashes, killing all aboard]

The Collision: In clear weather the two aircraft had collided almost head-on. At a closure speed of about 710 knots the F-89 had flown into the outboard left wing of the airliner:

> The left wing of the DC-7 and [the] nose section of the F-89 progressively penetrated one another until the [DC-7] left wing outboard of station 530 was sheared off . . . .

In less than one-hundredth of a second the collision had crushed the nose, front fuselage, and forward cockpit of the F-89. This impact had torn off the outboard left wing of the DC-7, and the airliner had begun a slow roll to the left. Despite the efforts of the pilots the roll gradually steepened. The doomed Douglas transport spiraled earthward. As airspeed increased, aerodynamic loads on the airframe began to take their toll:

> General breakup of the DC-7 before ground impact was the result of airloads beyond the design or required strength of the airframe. *[and also]* The DC-7 crashed on the grounds of the Pacoima Junior High School and an adjoining church.

The DC-7 fuselage screamed down onto the school playground in Pacoima and exploded, sending flaming debris flying in all directions. The impact seemed like "an earthquake" in the city. Jet-black smoke and orange flames boiled hundreds of feet skyward.

The majority of the 1,300 students were inside the schoolhouse, but about 200 boys were playing outside. Flying debris killed three of them, and 74 others suffered injuries and burns ranging from minor to critical. Chaos reigned within minutes as scores of ambulances and medical personnel flocked to the school. Hundreds of parents arrived and frantically searched for their children. One shocked bystander at the schoolyard told a newspaper reporter:

> I counted eight kids mangled. One of them had his leg torn off.

Meanwhile, after the mid-air collision the remains of the F-89 had burst into flame. The impact had crushed the forward fuselage and cockpit, instantly killing the pilot. The radar operator ejected and, although seriously burned, parachuted to safety in Burbank, California. What was left of the burning F-89 crashed in La Tuna Canyon in the rugged Verdugo Mountains:

> The impact and accompanying explosion caused extensive disintegration of the aircraft. An intense ground fire completely consumed many of the wreckage pieces.

The Investigation: The CAB interviewed 115 witnesses. Most had seen the cloud of smoke at the moment of impact, followed by what sounded like a "clap of thunder." By chance a motion picture crew accidentally filmed the explosive cloud in the process of filming a western movie scene. In addition a surveyor who saw the collision repositioned his transit and took bearings on the explosive cloud location. From other witnesses and these technical sources the CAB pinpointed the point of collision 5,000 feet northeast of Hansen Dam spillway between Pacoima and Sunland.

Score marks on metal wreckage confirmed the two aircraft had hit each other. Before the collision both crews had radioed their companies and had reported that they were flying at 25,000 feet. According to the National Weather Service the visibility at that altitude had been estimated at 50 miles:

> Both flights were operated under VFR flight plans. Accordingly, the avoidance of other aircraft was a direct responsibility of the pilots of both aircraft.

Analysis of the test flight procedures indicated these functions should not have had an adverse effect on the pilots' visual vigilance. According to the surviving F-89 radar operator his airspeed had been 380 knots. The CAB estimated the speed of the DC-7 at 330 knots:

This collision occurred nearly head-on while the DC-7 was flying straight and level . . . . It occurred while [the F-89 pilot] was in a level left turn from 135 degrees toward . . . 45 degrees.

Probable Cause: The CAB released its Accident Investigation Report, File No. 2-0020, on 22 November 1957. The board noted that the most difficult airplane to see is one that is approaching head-on because (1) the visible frontal profile is small, (2) there is no relative motion, and (3) the closure speed is greatest:

Only the minimum time opportunity existed for the pilots to have carried out the basic elements of collision avoidance.

At a closure speed of about 710 knots there had been scant time for the pilots to see each other and take evasive action. Yet, the "poor jet too" radio transmission from the DC-7 copilot revealed he had seen the F-89 a split-second prior to the collision. It was not known whether or not the F-89 pilot had seen the DC-7 before impact. The board found the *Probable Cause* of this accident to be:

The high rate of near head-on closure at high altitude which, together with physiological limitations, resulted in a minimum avoidance opportunity during which the pilots [in both planes] did not see the other aircraft.

Northrop F-89 Scorpion (official U.S. Air Force photo)

A Simple Night VFR Flight
Mid-Air Collision over Norwalk, California
1 February 1958

Douglas C-118 Liftmaster, Serial No. 53-3277
U.S. Air Force MATS transcontinental logistics flight
41 aboard, 41 killed (plus 1 killed on the ground)

Lockheed P2V Neptune, Bureau No. 127723
U.S. Navy instrument training flight
8 aboard, 6 killed

Synopsis: Two military planes, a Douglas C-118 Liftmaster and a Lockheed P2V Neptune, took off from adjacent airfields. In clear weather at night the two aircraft collided. Both crashed, killing 47 of the 49 people aboard.

The Douglas C-118 Liftmaster Flight: A big four engine Douglas C-118 Liftmaster took off from Runway 30 at Long Beach Municipal Airport in California at 1908 Hours. The U.S. Air Force transport headed for McGuire AFB in New Jersey, with an intermediate stop scheduled at Wright Patterson AFB in Ohio. Although on an IFR flight plan, the ATC clearance called for a VFR climb to the planned cruising altitude of 17,000 feet:

The C-118 (known as the R6D in the U.S. Navy) is the military variant of the Douglas DC-6 commercial airliner. In 1947 the Air Force had ordered 166 of them for the Military Air Transport Service (MATS). The C-118 had intercontinental range and could carry a 50,000 pound payload or 74 combat equipped troops. Its four 18-cylinder Pratt & Whitney R-2800 Double Wasp radial engines each pumped out 2,400 horsepower.

The aircraft commander, an Air Force major, had logged 7,819 hours of flight time since starting his flying career in the waning years of World War II. His copilot, a second lieutenant, had 701 flying hours in his logbook. The aircraft and its crew of six airmen were based at McGuire AFB. In addition to the crew the transport carried 35 military passengers, five of whom were women.

Douglas R6D Liftmaster, known as the C-118 in the U.S. Air Force (photo courtesy of U.S. DOD Visual Information Center)

The Lockheed P2V Neptune Flight: As the C-118 took off from the municipal airport a Lockheed P2V Neptune accelerated down Runway 22 Left at nearby NAS Los Alamitos. After takeoff the U.S. Navy plane turned left and clawed its way skyward on a VFR flight plan. The crew planned to practice instrument flying procedures. They and their aircraft were assigned to VP-773, a Navy long range patrol squadron based at Los Alamitos:

> *Lockheed had designed the P2V as a land-based long range patrol bomber. Two 3,500 horsepower Wright R-3350 Cyclone turbo-compound radial engines powered the warplane. Most models had been retrofitted with two Westinghouse J-34 auxiliary jet engines. The lethal P2V could carry bombs, torpedoes, depth charges, and rockets. Two .50 caliber machine guns bristled from a dorsal turret. Later models gradually removed the air-to-air defensive armament in favor of more anti-submarine electronics. One aviation historian noted the popularity of the aircraft:*
>
>> *"The P2V was the last radial-engine-powered bomber accepted for delivery by the United States. From May 12, 1945, to the end of the production run in 1962, 1,036 Neptunes were produced in seven major variants."*

Lockheed P2V Neptune (official U.S. Navy photo)

*The most famous P2V had been named the "Truculent Turtle."
In 1947 it completed an amazing (1) nonstop (2) non-refueled
flight from Perth, Australia, to Columbus, Ohio, a distance of
11,235 miles. This piston-engine world distance record would
stand for 39 years until the flight of Burt Rutan's experimental
around-the-world "Voyager" in 1986.*

The P2V pilot-in-command, a Navy lieutenant commander, had
3,508 hours of flight experience. His copilot, a lieutenant, had
logged 1,087 hours. Four additional men rounded out the crew, and
two non-crew military passengers had tagged along for the ride.

Trouble Ahead: In the clear night sky the four Pratt & Whitney
radials powered the C-118 toward its cruising altitude. Climbing
through 500 feet, the pilot made his last radio contact with ATC and
requested a right turn out of the airport traffic pattern. The controller
approved this request, and the C-118 continued its VFR climb.

Meanwhile, the P2V turned left after takeoff and climbed inland
toward the instrument practice area. With both piston engines
pounding away and both jet engines at full power, the aircraft
enjoyed excellent climb capability in the cool night air. With a
reported visibility of 20 miles no further contact with ATC was
expected until completion of the training mission.

The Collision: Five minutes after takeoff at between 2,500 and
3,000 feet the C-118 was flying east and turning right. The P2V,
flying north in the dark, rammed completely through the big MATS

transport. The impact sliced the big C-118 into two sections and mortally wounded the P2V. The next day the *Los Angeles Times* newspaper headline would report the inevitable casualties:

---

47 Killed as Planes Collide in Air

The rear of the C-118 fuselage tumbled down onto a gasoline service station in Norwalk, California, rupturing the gasoline tanks and setting them ablaze. The rest of the fuselage plummeted into the parking lot at the Norwalk sheriff's sub-station. The fiery impact, explosions, and fuel-fed infernos that followed demolished a garage and 30 vehicles. The conflagration terrified nearby jail inmates, who were unable to flee. Authorities evacuated all 20 panicky prisoners to get them away from searing heat and flames.

All 41 people aboard the C-118 died in the collision and crash. Fortunately only one person on the ground was killed, a housewife whose body was completely cut in half by flying debris. The next day a news media report would note:

> The [collision], which littered an area of more than five square miles with bits of wreckage, some of it ablaze and starting fires, attracted sight-seers who saw the [aerial] fireball halfway across the county.

After the aerial impact the P2V had remained more or less intact. The crippled patrol bomber struggled northwest for three miles. It finally crashed into a rock quarry across the street from the fire department in the city of Santa Fe Springs. The collision, crash, and raging fire on the ground killed six of the eight men aboard. One, an ordnanceman, had been seated in the bow observer station. The collision had ripped him out of the plane into the night sky. All other crewmen had remained in the doomed P2V until it slammed into the ground.

Thousands of people witnessed the collision and crashes. They had been outdoors to catch a glimpse of America's first satellite, *Explorer 1*. It had been scheduled to pass overhead at the time of the accident. The *Los Angeles Times* would explain:

> Santa Ana Freeway was jammed beyond control within 15 minutes . . . Los Angeles policemen sped to the scene to aid beleaguered deputy sheriffs, firemen, and highway patrolmen in controlling the traffic, the fires, and crowds of bystanders.

Two persons, still alive, were pulled from the P2V wreckage. One died in an ambulance on the way to a hospital. The other had sustained critical injuries, but he would survive.

In addition to the battered survivor, the P2V crash produced a *miracle-man*. The radio operator – somehow – crawled out of the tangled burning wreckage without help from anyone. He walked away totally unscathed. One firefighter explained in awe:

> There was this one survivor [just standing] outside the plane. Someone put a jacket around him, and he looked pretty good.

The Investigation: Because both the Air Force and Navy had an aircraft involved in this accident, each military service conducted its own investigation. The two investigative teams agreed upon the basic factors involved. Both flight crews had been qualified. No aircraft systems or ATC considerations played a role in the collision.

At the time of the accident it had been totally dark. Yet, the ceiling had been unlimited with a 20 mile visibility according to the Navy, 15 miles according to the Air Force. The position lights and flashing anticollision lights on both planes had been functional and turned "on." The two aircraft had converged at a 90 degree angle, with the P2V to the right of the C-118. Pilots in both cockpits should have been able to see the converging aircraft.

There was, however, a complicating factor. Both planes had been overflying the urban sprawl of the Los Angeles area. Lighting on the ground had been extensive. Street lighting, vehicle lights, neon signs, home lights, and flashing anticollision lights on towers and antennas had created a moving and blinking kaleidoscope of lights below.

As any experienced pilot knows, against such a background it is hard to discern, or identify, the lights on a nearby plane. Worse yet, with the two aircraft converging at roughly the same speed, and at right angles to each other, there would have been no relative motion. The lights on the other aircraft, as seen from either cockpit, would have appeared to remain stationary on the windshield. Thus, they would have visually blended into the sea of lights below.

Probable Cause: The Navy completed its Aircraft Accident Report, Serial No. 1-58, and the Air Force compiled a similar report. Both military boards agreed that this accident should have been avoided. Although the C-118 had been on an IFR flight plan, it had been cleared for a VFR climb. The Navy plane also had been climbing VFR. The crews of both aircraft had been operating in a see-and-be-seen environment. One military researcher noted:

> The meteorological conditions were good to excellent, with a high thin ceiling and a visibility of at least 15 miles.

Technically, when two airplanes are converging the one on the right – in this case the P2V – has right-of-way. Nonetheless, it was obvious the pilots never saw each other, or saw each other too late to take evasive action. When specifying the *Probable Cause* of this accident the investigators made a short blunt statement:

> Pilot Error, pilots of both aircraft.

Despite the sea of lights below, the pilots in both planes should have seen the converging plane in time to take evasive action. An independent technical report explained the cause in more detail:

> Both aircraft crews failed to exercise proper "see and avoid" procedures regarding other aircraft in the vicinity of their own [plane] while operating under visual flight rules.

Section ADF Approach – IFR at Night
Mid-Air Collision near Naha, Okinawa, Ryukyu Islands
7 March 1958

Fairchild R4Q Flying Boxcar, Bureau No. 128741
U.S. Marine Corps passenger and logistics flight
25 aboard, 25 killed

Douglas AD-6 Skyraider, Bureau No. 135350
U.S. Marine Corps ferry flight
1 aboard, 1 killed

Synopsis: A Fairchild R4Q Flying Boxcar and a Douglas AD-6 Skyraider flew over the Pacific Ocean from the Philippine Islands to Okinawa. During a section ADF approach in IMC weather at night the two planes collided and fell into the sea. The accident killed all 26 military men aboard the two aircraft.

The Fairchild R4Q Flying Boxcar Flight: At NAS Cubi Point, Philippine Islands, a Fairchild R4Q Flying Boxcar accelerated down the runway, climbed skyward, and headed north over the vast Pacific. Assigned to U.S. Marine Corps squadron VMR-253, the transport carried cargo, a crew of six, and 19 passengers bound for Okinawa. Although the weather was clear the pilots had filed an IFR flight plan:

The first R4Q (known as the C-119 in the U.S. Air Force) had rolled out of the factory in 1949. It was designed to haul cargo, personnel, and wheeled vehicles. The aircraft featured twin booms extending rearward from the wings on both sides of the "boxcar" fuselage. A rear ramp afforded drive-in capability for vehicles and facilitated cargo loading. Two big Wright R-3350 Cyclone twin-row radial engines, each pounding out 3,500 horsepower, drove huge four-blade propellers.

The R4Q remained in production until 1955 and would serve in the U.S. Armed Forces until 1974. It saw extensive use during the wars in Korea and Vietnam. A gunship variant, the AC-119, sported protective armor, four 7.62mm miniguns, and two 20mm cannons. Some of these rugged workhorses still fly in the armed forces of developing nations.

Fairchild R4Q Flying Boxcar (official U.S. Air Force photo)

The R4Q Transport Plane Commander, a Marine captain 29 years of age, had logged 1,960 hours of flight time. His copilot, a first lieutenant, had 955 hours of flying experience. They climbed to 9,000 feet and droned north over the ocean. The crew included a navigator for this anticipated six hour flight. The trusting passengers soon settled in for the long trip to Okinawa in the Ryukyu Islands.

The Douglas AD-6 Skyraider Flight: During the Cold War era the Marine Corps maintained infantry and aviation garrisons throughout the South Pacific. Long over-water ferry flights were common for single engine Marine planes. If a multi-engine transport plane happened to be going to the same destination the single engine aircraft usually would go along as the wingman. Otherwise the single engine plane would make the flight alone, as explained by the commanding general of the Fleet Marine Force, Pacific Theater:

> Single place aircraft [flying] extended overseas flights is necessary in developing their long range ferry capability and contributes to the combat mobility of the tactical squadrons.

Douglas AD-6 Skyraider (official U.S. Air Force photo)

A Douglas AD-6 Skyraider had to be transferred from Cubi Point to Okinawa. Assigned to Marine Corps attack squadron VMA-224, the single engine plane was capable of making the long flight alone. However, the R4Q was already scheduled to make the trip, so the Skyraider would tag along:

The Douglas AD-6 Skyraider (known as the A-1 in the U.S. Air Force) was the last of the heavy single-seat piston engine combat planes. Deliveries to the Air Force, Navy, and Marine Corps had begun in late 1945 and had continued through 1957. Douglas turned out over 3,000 of these potent warhorses. The Skyraider became the world's first combat aircraft able to carry more than its own weight in ordnance. For its ground attack role it bristled with four 20mm cannons. Its wing racks could carry an astounding load of rockets, bombs, missiles, and napalm.

A single massive Wright R-3350 Cyclone radial engine powered the Skyraider. Flown from shore bases and aircraft carriers, the

Skyraider became the ground attack platform of choice during the 1950s, 1960s, and early 1970s. It gained its greatest fame in the role of "Sandy" during "SAR North" rescue missions during the Vietnam War. Heavy armor, a huge ordnance load, and unequaled loiter time made it preferable to jet attack planes of that era. The venerable Skyraider would serve with the U.S. Armed Forces until the late 1970s.

The Skyraider pilot took off behind the R4Q and flew as wingman on the long flight over the ocean toward Okinawa. The pilot, a first lieutenant 24 years of age, had accumulated 767 hours of flight time. Alone in the Skyraider, he maintained a loose trail position behind and to the right of the R4Q as the hours slipped away.

Trouble Ahead: Four hours into the flight with nothing but empty ocean below, the weather began to turn sour. Soon the two planes had trouble staying in VFR conditions. Also, it was getting dark. The R4Q pilots radioed Okinawa Terminal Control:

At 1836 Hours, Marine 8741 *[the R4Q]* requested a change in altitude, from 9,000 feet to VFR-on-top.

Okinawa ATC approved the change, and the two aircraft climbed above the clouds and leveled off at 12,000 feet. They continued to fly north while homing on the powerful Okinawa radio beacon. Prior to reaching that navaid:

Clearance was given, and acknowledged, to proceed to the Naha radio beacon, to hold east and descend to 4,000 feet in the holding pattern. Naha weather was 3,000 broken, 7,000 overcast, 3 miles visibility . . . .

After 5 hours and 30 minutes over the Pacific Ocean the two planes finally reported "feet dry" over Okinawa. Soon thereafter the R4Q pilots lost radio contact with Okinawa Terminal Control. Twice they radioed Naha Tower in the blind and reported descending, as cleared, in the holding pattern.

The portion of the ATC radio transcript, below, contains messages from Okinawa Terminal Control (ARTCC) and the two planes. The transcript begins at 1929 Hours as the R4Q pilots get clearance to descend to 4,000 feet:

Terminal: ..... Okinawa Terminal Control (ARTCC)
Marine 8741: ... Fairchild R4Q Flying Boxcar
Marine 5350: ... Douglas AD-6 Skyraider

Terminal: [Marine] eight-seven-four-one, Okinawa Terminal, the Naha weather is three-thousand scattered, seven-thousand overcast, visibility eight [miles], winds three-six-zero at one-seven knots, altimeter three-zero-one-four. (1929)

[the R4Q pilots acknowledge the information]

Terminal: Eight-seven-four-one, Okinawa Terminal Control, cleared to descend to and maintain four-thousand [feet], over. (1929)

Marine 8741: Terminal Control, eight-seven-four-one, roger, understand [we are] cleared to descend and maintain four-thousand feet.

Terminal: Roger, eight-seven-four-one, the Naha weather . . . .

Marine 8741: . . . say again the altimeter.

Terminal: Roger, altimeter three-zero-one-four.

Terminal: Eight-seven-four-one, upon reaching Jig Nan beacon, hold east, inbound heading two-seven-zero, descend in the holding pattern until reaching four-thousand, over. (1931)

[the R4Q pilots acknowledge the clearance; 2 minutes later they report overflying the Jig Nan radio beacon]

Marine 8741: [Okinawa Terminal, this is] Marine eight-seven-four-one, descending to four-thousand feet, [passing] over the Jig Nan beacon at three-three, over. (1933)

[Terminal Control does not hear this message clearly, and hereafter the R4Q pilots lose almost all radio contact]

Terminal: Eight-seven-four-one, Terminal, are you calling, over?

Terminal: Eight-seven-four-one, Okinawa Terminal, are you calling?

Marine 8741: Okinawa, eight-seven-four-one, do you read?

Terminal: Eight-seven-four-one, Terminal Control, are you calling?

Marine 8741: Okinawa Terminal, Marine eight-seven-four-one, over.

Terminal: Eight-seven-four-one, Okinawa Terminal, go ahead.

Marine 8741: [Garbled transmission].

[Terminal Control tries a different radio frequency]

Terminal: Eight-seven-four-one, Okinawa Terminal on channel five.

Terminal: Eight-seven-four-one, Okinawa Terminal.

[unable to contact Terminal Control, the R4Q pilots radio their wingman, the Skyraider pilot]

Marine 8741: Whiskey Alpha [garbled], Marine eight-seven-four-one, do you read?

Marine 5350: Eight-seven-four-one, [I read you], go ahead.

Marine 5350: Eight-seven-four-one, did you get the altimeter?

Marine 5350: Eight-seven-four-one, did you get the altimeter?

[the Skyraider pilot gets no response from the R4Q pilots]

[at 1939 the R4Q pilots radio Naha Tower in the blind and report descending through 8,500 feet in the holding pattern]

Terminal: Eight-seven-four-one, Okinawa Terminal, do you read?

Marine 8741: [Garbled transmission].

Terminal: Eight-seven-four-one, Okinawa Terminal, over.

Marine 8741: Okinawa Terminal, Marine eight-seven-four-one.

Terminal: Eight-seven-four-one, Okinawa Terminal, go ahead.

Marine 8741: Terminal Control, eight-seven-four-one [is] presently holding over Jig Nan beacon, descending to four-thousand – *[rest of message blocked by transmissions from other aircraft]*. (1942)

Marine 8741: Terminal Control, eight-seven-four-one, did you copy?

[at 1942, unable to get a response from Terminal Control, the R4Q pilots again radio Naha Tower in the blind and report descending through 5,500 feet in the holding pattern]

*[1 minute later the R4Q and the Skyraider **collide** at 1943]*

Terminal: Eight-seven-four-one, [this is] Okinawa Terminal Control transmitting on Guard, if you read come up [garbled] frequency, over.

Terminal: Eight-seven-four-one, [this is] Okinawa Terminal Control on Guard and Delta, if you read come up on channel five or this frequency, over.

Terminal: Marine eight-seven-four-one, Okinawa Terminal Control transmitting on Guard, channel three, and channel five, over.

Terminal: Eight-seven-four-one, Okinawa Terminal Control on all [garbled] frequencies, if you read depart the Naha radio beacon, three-three-zero degrees, four-thousand feet, I say again, if you read depart the Naha radio beacon three-three-zero degrees, four-thousand feet.

[there is no response]

The Collision: At 1943 Hours while descending through about 5,000 feet in the holding pattern, the two planes had slammed together. The right wing of the R4Q had dug into the left side of the Skyraider:

> The right wing of the [R4Q] was smashed almost directly inward. . . . The severe compression of this wing is indicative of a rapid lateral closing rate.

The aerial impact had broken the Skyraider in half. The forward part of the fuselage lurched left into the No. 2 engine and propeller of the R4Q. The Skyraider's still swirling propeller spun further left, chewing its way through the R4Q cockpit and its doomed occupants. The collision destroyed both aircraft, and they tumbled down into the ocean near the Okinawa coast. Analysis later would indicate that anyone still alive died instantly upon impact with the water.

Tower controllers had seen "bright flashes" through the clouds. Unable to recontact either aircraft, controllers notified military SAR facilities. Air and sea rescue units began a search, and the next day three bodies were discovered floating near the coast. It took another 24 hours to locate the two downed planes and remaining victims:

> The main wreckage was discovered on Sunday, 9 March 1958, 3 miles south of the Naha Airport in about 60 feet of water, [and] there were no survivors.

Part of the Skyraider left wing recovered from the ocean floor (photo from
Aircraft Accident Report 1-58 courtesy of U.S. Marine Corps)

The Investigation: Based upon available information the two planes
had been in IMC weather at the time of the collision. A precision
GCA approach had been available, but the R4Q pilots had not
requested it. They had tried a section ADF approach, which involved
a much heavier pilot workload and more maneuvering:

> [It] placed the AD-6 [Skyraider] pilot in a position which required
> him to fly close formation on a transport aircraft at night under
> instrument conditions [while] making descending turns.

The ADF section (*section* means two planes flying in formation)
approach had been exceptionally demanding for the Skyraider pilot.
He had to (1) maintain his position off the R4Q right wing (2) at night
(3) in IFR conditions (4) while the R4Q maneuvered to negotiate the
non-precision approach.

The Skyraider had its own navigation radios. The pilot could
have broken off from the R4Q for an individual ADF approach. A
precision GCA would have been a cinch. Yet, the young pilot tried
to maintain formation on the wing of the R4Q with tragic results.

What caused the collision? Spatial disorientation in the clouds?
Turbulence? Mechanical problems? Sudden entry into more dense
clouds which caused the Skyraider pilot to lose sight of his leader?
Rapid maneuvering by the R4Q? No one knows:

Hypoxia and/or carbon monoxide certainly cannot be discounted. The flight was at 12,000 feet for an undetermined time, and it is impossible to ascertain whether oxygen was used.

The R4Q pilots had experienced radio problems, but "lost comm" alone would not have caused the two planes to collide. In the end investigators reasoned:

> The board *feels [emphasis added]* that the impact occurred after the R4Q rolled level after turning [left in the holding pattern] outbound from the Naha radio beacon. Location of the wreckage substantiates this theory.

Probable Cause: The board delved into the background of the flight crews, ATC concerns, communications, navaids, wreckage analysis, etc. They found no "smoking gun." Board members did agree that the R4Q commander erred by accepting a (1) section descent with (2) different types of aircraft in (3) actual instrument conditions. Instead of a Probable Cause the board set forth a *Primary Cause*:

> It is determined that the Primary Cause of the accident was poor judgement on the part of the R4Q Plane Commander.

The board listed two *Secondary Causes*:

> [First, bad] weather.

> [Second], pilot error on the part of the pilot of the AD-6.

Postscript: Marine Corps higher authority did not agree with the Secondary Cause finding of pilot error on the part of the Skyraider pilot. This finding was rejected with the following comment:

> All that is definitely known from examination of the wreckage is that there was a mid-air collision. The *assumption [emphasis added]* that the AD-6 [Skyraider pilot] failed to maintain a proper wing position is based strictly on conjecture and not on facts.

Higher authority was correct. The *specific* cause of the accident had not been determined. The following written admission by the board is an epitaph for 26 U.S. Marines who died in this accident:

> The exact cause of this mid-air collision remains unknown.

Fire in the Sky and Metal Rain
Mid-Air Collision near Las Vegas, Nevada
21 April 1958

North American F-100 Super Sabre, Serial No. 56-3755
U.S. Air Force instrument training flight
2 aboard, 2 killed

Douglas DC-7 Seven Seas, Registration N6328C
United Airlines Flight 736
47 aboard, 47 killed

Synopsis: While practicing an instrument penetration a military North American F-100 Super Sabre fighter struck a civil Douglas DC-7 Seven Seas airliner cruising at 21,000 feet. Both planes crashed and killed all 49 people on board.

The North American F-100 Super Sabre Flight: Before daybreak at Nellis AFB near Las Vegas, Nevada, a U.S. Air Force instructor began a 40 minute review of the flight to come. The instructor and his student then climbed into their supersonic North American F-100 Super Sabre fighter. Fifteen minutes later at 0745 Hours they lit the afterburner, accelerated down Runway 4, climbed, and flew toward a military practice area south of Las Vegas:

The F-100 had first flown in 1953, and North American would build 2,294 of these warplanes before production would end in 1959. The F-100 was the world's first production airplane capable of flying faster than the speed of sound in level flight. Heat-resistant titanium was used extensively in the airframe. Powered by a single afterburning Pratt & Whitney J-57 turbojet, the F-100 had a range of over 1,000 miles and a service ceiling of 55,000 feet. The swept-wing supersonic fighter was armed with four 20mm cannons and equipped to fire missiles and rockets. It had been designed to carry a nuclear bomb. However, it would make its combat debut in Vietnam in the 1960s as a fighter-bomber armed with conventional ordnance.

The pilot, a captain, had been flying in the Air Force since 1953 and had logged 1,542 flight hours. His student, a second lieutenant,

North American F-100 Super Sabre (official U.S. Air Force photo)

had 363 hours in his logbook. He was transitioning into the F-100, in which he had flown a mere four hours. Both pilots and their Super Sabre were based at Nellis.

The Douglas DC-7 Seven Seas Flight: United Airlines scheduled Flight 736 from Los Angeles to New York with the first en route stop at Denver, Colorado. The Douglas DC-7 Seven Seas had been loaded with 130 pounds of mail, 1,468 pounds of freight, 1,824 pounds of baggage, and 2,400 gallons of aviation fuel. With 42 passengers and five crewmembers aboard the aircraft weighed 98,654 pounds, well under the maximum allowable takeoff weight. The airline pilots had filed the following IFR flight plan:

> UAL Trip 736, N6328C, DC-7, TAS 305 knots, cruising level 21,000 feet, ETD 0735, ETE 2 hours plus 45 minutes, via V-16 to Ontario, V-8 to Denver, pilot will accept VFR climb.

Powered by four Wright R-3350 radial engines, the DC-7 began its takeoff roll down Runway 25 Left at Los Angeles International Airport at 0737 Hours. The airliner rose and climbed up through the smoke and haze of the Los Angeles basin into VFR-on-top conditions. The pilots flew northeast through clear skies under the control of Los Angeles Center. The captain, first officer, and flight engineer manned the flight deck. Two stewardesses in the cabin took charge of their slim load of only 42 revenue passengers.

Trouble Ahead: Thirty-seven minutes after takeoff the DC-7 pilots reported their position over Daggett at 21,000 feet at 0814 Hours. Droning routinely northeast, they estimated they would reach Las Vegas VOR at 0831 Hours.

Meanwhile, the two F-100 pilots had completed most of their basic Instrument Mission Number One procedures, described by the Air Force as follows:

> Military climb schedule to 30,000 feet . . . Normal turns . . . Steep turns . . . [Level] changes of airspeed . . . Constant airspeed climbs and descents . . . .

The F-100 instructor pilot radioed Nellis VFR Control and got clearance for a simulated instrument penetration from 28,000 feet. At 0828 Hours the student pilot chopped his power, popped his speed brakes, dropped his nose, and began a rapid descent.

The radio transcript, below, begins as the DC-7 pilots get their ATC clearance from Los Angeles Ground Control. The transcript includes military UHF and civil VHF transmissions to and from both aircraft. The F-100 and the DC-7 use different radio frequencies, so the pilots can not hear radio messages to or from the other plane:

> <u>United 736</u>: . . . . United Airlines Flight 736 (Douglas DC-7)
> <u>AF 755</u>: . . . . . . U.S. Air Force F-100 Super Sabre
> <u>Center</u>: . . . . Los Angeles Center (ARTCC)
> <u>GndControl</u>: . . . Los Angeles Ground Control
> <u>Nellis VFR</u>: . . . Nellis Air Force Base VFR Control

<u>GndControl</u>: United seven-three-six is cleared to the Los Angeles radio beacon, maintain VFR-on-top, climb magnetic heading two-five-zero degrees to [VFR]-on-top. (0730)

[The DC-7 takes off at 0737; the pilots contact Center]

<u>Center</u>: United seven-three-six is cleared to the Denver Airport, continue to climb in VFR conditions to and maintain two-one-thousand feet. (0740)

[at 0745 the F-100 takes off from Nellis AFB]

[at 0811 the DC-7 reports overflying Daggett at 21,000 feet, estimating Las Vegas VOR at 0831]

Douglas C-74 Globemaster, a military variant of the civil Douglas DC-7
Seven Seas airliner (official U.S. Army photo)

*[at 0823 the F-100 requests a simulated instrument penetration;
Nellis VFR Control clears the F-100 down to 28,000 feet]*

*[at 0827 the F-100 reports level at 28,000 feet; Nellis VFR
Control clears the fighter for an instrument penetration]*

<u>AF 755</u>: Roger, leaving two-eight-thousand. (0828)

<u>Nellis VFR</u>: Air Force seven-five-five, report completing procedure
turn. (0828)

[at 0828 the F-100 begins its descent]

*[at 0831 the F-100 and the DC-7 **collide** at 21,000 feet]*

[pilots in both planes transmit desperate "mayday" messages]

<u>AF 755</u>: Mayday! Mayday! Seven-fifty-five! Bailout! (0831)

<u>United 736</u>: United seven-thirty-six! Mayday! Mayday! Mid-air
over Las Vegas! (0831)

[there are no further transmissions from either aircraft]

The Collision: As the DC-7 airliner cruised straight and level at
21,000 feet along airway Victor 8, the F-100 fighter was diving
earthward. The two aircraft ripped into each other about nine miles

southwest of Las Vegas VOR. An independent technical report would describe the impact:

Nearly head-on, both aircraft clipped their outer right wings, slicing off more than 10 feet from the DC-7 and ripping both the right wing and right stabilizer off the F-100.

The two airplanes arced down out of control. The instructor ejected from his crippled F-100. However, his automated parachute opening mechanism had been damaged in the collision, and the chute failed to open. Although he was alive during the long four mile fall, impact with the ground killed him instantly. The student pilot, seated in the rear seat under an instrument hood, failed to escape from his damaged fighter. He died when the mangled F-100 slammed into the ground and exploded.

For the 47 souls aboard the doomed DC-7 there was no chance of survival. Missing a large section of its right wing, the airliner tumbled out of control. Violent oscillations racked the airframe, and one engine after another ripped away. A series of aerial explosions began, and each of them sent showers of shredded metal flying in all directions. An historian later would explain:

Witnesses described [fire in the sky and] a rain of metal [debris] gleaming in the early morning sun [as it fell, and] the passenger plane was a spinning ball of fire that exploded intermittently as it spiraled toward the ground.

What was left of the DC-7 slammed into the desert. Potential rescuers made their way to the crash site but found that all 47 people on the airliner had perished. One awestruck witness described the horribly burned bodies:

No one was wiggling, and a lot of them were still smoking. *[and also]* What was kind of grim was to see the wristwatches of the dead still ticking.

The collision and crashes marked the worst aviation disaster in Las Vegas history. Ironically, the accident happened near the location where famed actress Carole Lombard had been killed in a fiery plane crash 16 years earlier. The next day the newspaper banner headline in the *Las Vegas Review Journal* would state:

United Airliner, Jet Plane Crash over Las Vegas, 49 Die

The Investigation: The CAB investigated this mid-air collision, and the Air Force also set up an investigative team to probe the circumstances surrounding the tragedy. In the end neither team of analysts found any aircraft or communications anomaly that could have played a role in the accident.

The weather had been clear with an estimated visibility of 35 miles. The DC-7 had been flying at about 312 knots on a heading of 23 degrees. The F-100 had been diving at an estimated 444 knots on a heading of 145 degrees. In consideration of the different headings and the descent angle of the F-100, closure speed had been on the order of 665 knots. The CAB pointed out that, technically speaking:

> The accident occurred in VFR conditions which . . . placed responsibility on the pilots of both aircraft to avoid [a] collision through visual separation.

Nonetheless, the CAB opined:

> Air Force policies did not take adequate account of human limitations to avoid [a] collision by visual means, although the limitations were recognized in other training operations and were known to the Air Force.

Both airplanes had been properly cleared into the airspace involved. Although on an IFR flight plan, the DC-7 had been flying in VFR weather. Under these conditions the airline pilots were responsible for maintaining visual separation from other traffic. The F-100 had been on a VFR flight plan, and therefore the military pilots were required to maintain visual separation.

None of the pilots had been informed of the presence of the other aircraft. Yet, the military pilots knew they were crossing Victor 8. Also, the civil pilots knew that military aircraft crossed the airway. The two aircraft had raced toward each other at a closure rate only slightly less than the speed of sound. Under these high-speed conditions, investigators agreed that the pilots had only a minimal opportunity to see the other aircraft and take evasive action.

Probable Cause: The CAB adopted an Aircraft Accident Report, File No. 1-0066, on 18 August 1958, four months after the mid-air collision. The board acknowledged:

> The Civil Aeronautics Administration did not take sufficient [civil] measures to reduce known collision exposure in visual flight conditions.

The Air Force agreed in principle that existing civil and military ATC capability had not been adequate. After an exhaustive analysis the military investigative team detailed the *Primary Cause* of the collision and crashes as follows:

> The Primary Cause of this accident was the inadequacy of the present [air traffic] control system, which allows two or more aircraft to occupy the same airspace at the same time and relies solely on the ability of pilots to see and avoid one another.

A major *Contributing Cause*, as described by insightful military investigators, pinpointed the inadequacy of the VFR see-and-be-seen concept in the modern high-speed aircraft era:

> A Contributing Cause is the high rate of closure in present-day high-speed aircraft which *exceeds the limits of human reaction time [emphasis added]*. Although required to maintain visual clearance while flying under VFR conditions, the pilots of both aircraft easily could have been preoccupied with normal crew duties which diverted their attention. A very short diversion of this nature could place the [two] aircraft too close together to allow successful evasive action.

Oh, It Was an Awful Sight!
Mid-Air Collision near Jefferson, Maryland
20 May 1958

Vickers Viscount 745, Registration N7410
Capital Airlines Flight 300
11 aboard, 11 killed

Lockheed T-33 Shooting Star, Serial No. 53-966
Maryland Air National Guard VFR familiarization flight
2 aboard, 1 killed

Synopsis: A Vickers Viscount 745 airliner descended to land in VFR conditions. A Lockheed T-33 Shooting Star jet trainer, carrying a pilot and a ground crewman, tore into the Viscount at about 8,000 feet. The collision and crashes killed the 11 people aboard the Viscount. The T-33 pilot ejected and survived, but the ground crewman died in the collision.

The Vickers Viscount 745 Flight: Capital Airlines Flight 300 was a scheduled shuttle from Chicago, Illinois, to Baltimore, Maryland, with an en route stop at Pittsburgh, Pennsylvania. The first leg of the flight went as planed on a sunny May morning. After an hour delay in Pittsburgh the Vickers Viscount 745 took off at 1050 Hours for the short trip down to Baltimore. Fortunately the airliner was almost empty. The two pilots and two hostesses carried only seven revenue passengers in the cabin:

The Vickers Viscount was the world's first commercial turboprop transport. After World War II, British European Airways (BEA) had identified the need for a medium range turboprop airliner. Vickers engineers designed a pressurized aircraft powered by four Rolls-Royce turbines mated to four-blade constant speed propellers. The new transport gained certification in 1953 and immediately began airline service with BEA.

The Viscount became popular with travelers who enjoyed a quiet and vibration-free ride. The plane could carry up to 53 passengers and cruise at almost 300 knots. Good operating economics and ease of maintenance made it an airline favorite throughout

Vickers Viscount 701, which is similar to the Vickers Viscount 745 (photo courtesy of Adrian R. Pingstone)

Europe and America. The armed forces of the United Kingdom and eight other nations purchased militarized variants.

The captain, age 38, had been flying for Capital Airlines since 1945 and had accumulated 12,719 hours of flight time. He held the requisite ATP certificate. His young first officer, age 26, was a former U.S. Navy fighter pilot who had earned a Commercial Pilot certificate. He had been employed by Capital for two years, and his logbook reflected 2,467 hours of flight experience.

The Lockheed T-33 Shooting Star Flight: A captain in the 104th Fighter-Interceptor Squadron of the Maryland Air National Guard decided to take one of his ground crewmen on a familiarization flight. The crewman met the pilot at Martin Airport near Baltimore. The pilot briefed the young man on emergency procedures, the oxygen system, and personal survival equipment. Then they climbed into a Lockheed T-33 Shooting Star. At 1107 Hours the jet trainer accelerated down Runway 14 and soared up into a crystal clear blue sky:

Lockheed had designed the tandem seat T-33 based upon the single seat F-80 jet fighter. The new plane first flew in 1948, and

Lockheed T-33 Shooting Star (photo courtesy of U.S. DOD Visual Information Center)

Lockheed built 5,691 of them before production ended 11 years later in 1959. The trainer was intended to transition proficient pilots of propeller driven planes into new jet fighters.

The popular T-33 served in the armed forces of over 30 countries. Most used the plane as a trainer, but others armed their T-33s with bombs and .50 caliber machineguns. A U.S. Navy variant, the TV2, became the first jet trainer used for carrier landings. Another variant, the CT-33, was built under license by Canadair for use in the Canadian Armed Forces. Well over fifty years after it first took to the air the T-33 remains the most popular and most easy-to-acquire jet warbird.

The pilot, age 34, had become a military aviator during World War II and had joined the Air National Guard in 1952. During his career he had logged 1,902 hours of flight time in various types of jet aircraft. On this flight he flew from the rear seat so his ground crewman could ride up front and enjoy the view.

Trouble Ahead: Washington Center radar picked up the inbound Viscount as it descended thorough 10,000 feet on airway V-44. The pilots estimated they would be on the ground in Baltimore in about 13 minutes. At 1126:48 Hours the controller cleared Flight 300 to Sugar Loaf Intersection. The captain acknowledged this clearance, and no one heard from the Viscount again.

Meanwhile, the T-33 pilot and his passenger were enjoying an aerial sightseeing tour high over northern Virginia. At 5,000 feet they had flown up the Potomac River to historic Harper's Ferry and had circled the city so the ground crewman could take photographs. Then the pilot turned east toward Baltimore and started a slow climb. The weather was clear, and the two men enjoyed the scenery as they neared the coast. The last thing the pilot remembered before impact was seeing his altimeter wind upward through 8,000 feet.

The Collision: As the two planes flew east the faster T-33 gradually overtook the Viscount from the left rear. At 1128:45 Hours the jet trainer rammed the big airliner. At 8,000 feet the T-33 right wing sawed into the left side of the Viscount fuselage and cockpit:

> This destroyed the right wing of the T-33 and shattered the nose structure of the Viscount. *[and also]* The [T-33] disintegrated in the air following the collision.

The aerial impact instantly killed the young ground crewman in the T-33 front seat. His remains later would be found in the wreckage. The pilot in the rear seat had never seen the Viscount. In fact, he thought his aircraft simply exploded. He did not recall ejecting, and he later would explain his only memory of the accident:

> The next thing I knew I was flying through the air with flames and debris all around me.

The pilot parachuted to the ground, horribly burned on his face and hands. He stumbled to a nearby farmhouse, and residents rushed him to Frederick Memorial Hospital.

After the collision the Viscount had pitched up vertically, stalled, and then entered a flat spin. The empennage broke off from the rest of the fuselage, which then ripped apart in the air. The larger pieces of wreckage all burned after they hit the ground. Debris was strewn along a path 4,500 feet long near Jefferson, Maryland.

The Brunswick Volunteer Fire Department and the Independent Hose Company rushed several fire trucks and firefighters to the main wreckage sites. Firemen soon quenched the flames, but all aboard the Viscount were beyond earthly aid. Later that day the evening edition of *The News* newspaper in nearby Frederick, Maryland, would carry a multi-line full page banner headline that announced:

----

12 Killed in Crash of Jet and Airliner, Bodies Strewn over Countryside

----

One visibly shaken firefighter told a newspaper reporter:

> Bodies were scattered everywhere! Oh, it was an awful sight! No question about anybody living through that!

This had been the second fatal accident involving Capital Airlines within six weeks. On 6 April one of Capital's Viscounts had crashed while trying to land at Midland, Michigan, killing 47 people.

The Investigation: CAB investigators reconstructed this accident. Radar data revealed the Viscount had been descending at about 235 knots. Behind and on the left side of the airliner, the T-33 had been climbing at 290 knots on a parallel course:

> The T-33 began a normal right-hand turn and continued in this turn until striking the [left] side of the airliner.

The Washington Center controller had watched the primary radar return from Flight 300. He suddenly had noticed a faint return near the Viscount, so he radioed the airliner and reported the unidentified VFR traffic. He got no response. As he stared at his scope the targets merged. Then the "blip" that represented the Viscount grew larger, then faded, then vanished.

The jet trainer had ripped into the left side of the larger plane. A conspicuity study showed that, from his seat on the right side of the cockpit, the Viscount first officer had no chance to see the onrushing T-33. In the left seat the Viscount captain, if he had turned his head far to the left and rear, *possibly* could have seen the jet approaching. However, the T-33 pilot had a clear view of the Viscount ahead:

> As for the T-33 pilot, there was no obstruction to his seeing the Viscount for well over a minute before the collision. *[and also]* The evidence is clear that the T-33 pilot had ample opportunity to see the Viscount and avoid it.

The weather had been clear with no clouds to mar visibility. After a lengthy medical recovery the T-33 pilot testified that his windshield and canopy had been clean and clear. He said he kept a sharp eye out for traffic. Investigators could not determine why he failed to see the large Viscount ahead.

The CAB found an amazing – almost unbelievable – accident history. The tragedy on 20 May marked the *third time* the T-33 pilot had been involved in a mid-air collision. The first such accident had happened during aerial gunnery training. The pilot of another plane had made a simulated firing pass, had misjudged his break, and had rammed the captain's jet. The captain had ejected and survived. His second mid-air had occurred during formation acrobatics when another jet had slammed into his plane. Once again he ejected and lived to tell about it. Yet, the board noted that these prior accidents:

> . . . in no way indicate a lack of training or patterns of behavior which are of significance to *this [emphasis added]* investigation.

In a further ironic twist of fate, the deceased Viscount first officer also had been involved in a prior mid-air collision. While flying fighters in the U.S. Navy he had misjudged his break and had side-swiped a towed aerial gunnery target.

Probable Cause: Over seven months after the accident the CAB adopted an Aircraft Accident Report, File No. 1-0074, on 6 January 1959. The Viscount had been on an IFR flight plan, but both aircraft had been flying in VFR conditions. The T-33 had been overtaking the Viscount from behind.

The Viscount pilots could not reasonably have been expected to see the jet trainer behind them. On the other hand, the T-33 pilot should have seen the Viscount ahead, and he should have remained well clear. Although a turboprop transport plane and a jet aircraft were involved, this accident did not involve a high closure rate. Board members concluded:

> Emphasis must again be made, therefore, on the fact that the obligation to see and avoid other aircraft under visual flight rules conditions constitutes a condition precedent to the use of navigable airspace.

The board spelled out the *Probable Cause* of this mid-air collision:

> Failure of the T-33 pilot to exercise a proper and adequate vigilance to see and avoid other traffic.

Turn the Other Way!
Mid-Air Collision near Fairchild AFB, Washington
8 September 1958

<u>Boeing B-52 Stratofortress, Serial No. 56-661</u> (Outcome 55)
U.S. Air Force in-flight refueling training mission
9 aboard, 8 killed

<u>Boeing B-52 Stratofortress, Serial No. 56-681</u> (Outcome 54)
U.S. Air Force instrument training mission
7 aboard, 5 killed

<u>Synopsis</u>: In VFR conditions two Boeing B-52 Stratofortress bombers approached the (1) same runway at the (2) same time. The two big aircraft rammed into each other 900 feet above the ground and crashed, killing 13 of the 16 men aboard.

The Boeing B-52 Stratofortress Flight (Outcome 55): At 1005 Hours a Boeing B-52 Stratofortress (call-sign, Outcome 55) lifted off from Fairchild AFB near Spokane, Washington. The U.S. Air Force bomber from the Strategic Air Command (SAC) carried 200,800 pounds of fuel. On an IFR clearance the crew flew over Montana and practiced in-flight refueling from a tanker. After an hour of practice the pilots turned back toward Fairchild, and at 1315 Hours they began executing jet penetrations followed by GCA approaches:

Powered by eight Pratt & Whitney turbofan engines, the gigantic B-52 served as backbone of the U.S. Air Force strategic bomber force. The aircraft had been designed to drop nuclear bombs on any location in the Soviet Union. Further, a Big Belly modification allowed it to squeeze 108 conventional 500 pound bombs into the internal bomb bay. It could carry 28 additional 750 pound bombs on underwing pylons.

Early B-52s had four .50 caliber machineguns operated by a tail gunner. These quad-fifties soon were replaced by 20mm cannons, but eventually defensive guns would be eliminated altogether and replaced by electronic countermeasures. Later, B-52s would carry precision guided munitions and cruise missiles. The bomber had an unrefueled range of over 8,000 miles. With in-

flight refueling its range was limited only by the endurance of the crew. Further modifications to the ageless bomber are projected to keep it in service until the year 2045.

On this flight the plane carried a crew of nine: four pilots, three navigators, a tail gunner, and a staff sergeant who rode along in a technical capacity. The instructor pilot, a captain, had logged 6,422 flying hours. The Air Force would document the flight experience of three of the pilots in the extremely detailed "Report of A.F. Aircraft Accident" as illustrated below. Pilot names have been deleted:

Section D—FLYING EXPERIENCE OF PILOT(S)

1. WAS OPERATOR ON INSTRUMENTS AT TIME OF ACCIDENT OR IMMEDIATELY BEFORE: Yes ... No.**X** Unknown ...

ASSIGNED DUTY ON FLIGHT ORDER NOTE: List all time to the nearest hour	(Complete items 2 through 14 for each		
	PILOT (Last Name)	CO-PILOT (Last Name)	INSTR. PILOT (Last Name)
2. Total flying hours (including AF time, student time, and other accredited time) .	3961	973	6422
3. Total rated 1st pilot and instructor pilot hours, all aircraft	2711	279	4818
4. Total weather instrument hours	304	75	700
5. Total 1st pilot and instructor pilot hours this model (F-86, B-50, C-119, etc.)	297	43	327
6. Total other (Command, a/c cmdr, co-pilot, radar control pilot) hours this model	80	129	174
7. Total 1st pilot and instructor pilot hours this model and series (F-84F, F-86D, etc.)	270	42	323
8. Total other (Command, a/c cmdr, co-pilot, radar control plt) hrs this model and series	74	121	163
9. Total pilot hours last 90 days	89	97	114
10. Total 1st pilot and instructor pilot hours last 90 days	79	28	94
11. Total pilot hours (night) last 90 days	42	24	34
12. Total pilot hours, weather and hood, last 90 days	9	17	7
13. Date and duration of last previous flight this model	3 Sep 58 10:00	26 Aug 58 6:05	21 Aug 58 10:00
14. Date and duration of last previous flight this model and series	3 Sep 58 10:00	26 Aug 58 6:05	21 Aug 58 10:00

15. INSTRUCTIONS: Attach a copy of AF Form 5 for pilot(s) involved for the previous calendar month, and for to include the flight on which the accident took place.

(excerpt from "Report of A.F. Aircraft Accident" courtesy of U.S. Air Force)

The Boeing B-52 Stratofortress Flight (Outcome 54): Only four minutes after the first bomber left the runway, another B-52 (call-sign, Outcome 54) took off at 1009 Hours, It carried 230,000 pounds of fuel, enough for over ten hours of flight. On an IFR clearance the pilots practiced instrument procedures. At 1400 Hours they began making jet penetrations and GCA approaches at Fairchild.

The B-52 aircraft commander, a major, had logged 5,785 flight hours, and the Air Force had qualified him as a command pilot. His copilot, a first lieutenant, had logged 1,299 hours. Two more pilots, two navigators, and a tail gunner completed the crew of seven men.

Trouble Ahead: Outcome 55 completed a final GCA low approach, and the pilots cancelled their IFR clearance. They switched to the Fairchild Tower frequency for a series of touch-and-go landings on Runway 23. They began a long *visual* approach.

Meanwhile, Outcome 54 had finished a series of penetrations and GCA approaches. The pilots elected to remain in the traffic pattern for one more GCA and a full-stop landing. Flying downwind, they cancelled their IFR clearance at 1812 Hours. Neither they nor their controller knew the other B-52 was already on a long final approach to the same runway. Outcome 54 rolled into the GCA approach, guided by the voice of the GCA controller:

> . . . on glide path – correction, three and three-quarters miles from touchdown – two-two-two's your heading – now dropping twenty feet below glide path – now ease your aircraft up – two-two-two's your heading . . . .

The tower controller is in visual and radio contact only with Outcome 55. The tower controller suddenly sees Outcome 54 on the same final approach course. He frantically phones the GCA controller, who radios to Outcome 54:

> [The] tower advises [you to] pull up! Break out to the right!

The pilot in Outcome 54 immediately replies:

> Roger, breaking out to the right.

The Outcome 54 pilot aborts the GCA, shoves his eight throttles forward, and banks his aircraft to the right. Meanwhile, Outcome 55 is in radio contact with only the tower. The portion of the tower ATC transcript, below, begins five minutes earlier as Outcome 55 completes a touch-and-go on Runway 23:

> Outcome 55: . . . Boeing B-52 Stratofortress
> Tower: . . . . Fairchild AFB Control Tower

Outcome 55: [Tower], fifty-five just completed [a] touch-and-go, going around for a full stop [landing], completing the mission. (1816)

Tower: Roger, fifty-five, report downwind. (1816)

Outcome 55: [Tower], fifty-five turning base, full stop. (1819)

Tower: Fifty-five, [tower], wind zero-nine-zero [at] zero-four knots, continue approach. (1819)

Outcome 55: Fifty-five, roger. (1819)

Tower: Fifty-five, tower! Another B-fifty-two [is] breaking right!

Outcome 55: Roger! Tower, tell him to turn the other way! (1821)

> *[Outcome 54 is tuned to the GCA frequency and can not hear the desperate transmissions from the tower and Outcome 55]*
>
> *[the two bombers **collide** 12 seconds later at 1821]*

The Collision: Both aircraft had slipped down the approach course for Runway 23 in the dim twilight. The Outcome 54 pilot had heeded the GCA controller's final message to "break out to the right." He added power and turned right. Two and one-half miles from the runway, 900 feet above the ground, he turned directly into the path of Outcome 55. An eyewitness explained:

> Then they hit . . . . The nose, from the wing forward, on one plane fell off, while the other plane slammed into the ground. The bomber with the missing nose became tail heavy and started to climb with its eight engines screaming like banshees. Then [it] stalled and fell off on its left wing. It hit the ground with [the] engines at full power.

Aboard Outcome 55 about half the crew had a slim chance at survival. The pilot, the copilot, and two navigators ejected. One navigator parachuted to safety. The other three had ejected too late. They died along with the five remaining crewmen who had been unable to get out of the stricken bomber.

On the other B-52, Outcome 54, two crewmembers managed to get out before the bomber plowed into the earth. One of the two copilots, a captain, parachuted to the ground. He suffered extensive trauma to all parts of his body and would be hospitalized for over a year. Without an ejection seat the tail gunner manually bailed out and parachuted to safety. The remaining five members of the crew, including a navigator who was able to eject, all perished. The next morning *The Spokesman-Review* newspaper headline in Spokane would announce the then-known casualty count:

Fairchild Counts 6 Dead, 6 Missing, as B-52s Collide

The newspaper explained the fate of the two aircraft:

> Wreckage of the two eight-jet intercontinental bombers was strewn over a two-mile area, with the main sections of each plane about one-half mile apart.

> Both planes had burned after ramming into the ground. No one on the ground was killed, but falling debris damaged many homes and businesses. Raging fires added to the destruction of property.

The Investigation: In clear VFR twilight both bombers had been making an approach to Runway 23 at the same time. Weather had not been a factor, for the lowest clouds were at 10,000 feet, and visibility had been at least 20 miles.

Investigators found that anticollision lights had not been installed on one B-52 and were not fully operational on the other. In addition, Outcome 54 had made a military-style 180 degree *continuous left turn* from the downwind leg to the final approach course. Banked to the left, the crew could not see Outcome 55 on their right.

Outcome 54 had been in radio contact with the GCA controller. Outcome 55 had been tuned to the tower frequency. Consequently, neither crew could hear radio messages to and from the other aircraft.

Boeing B-52 Stratofortress (official U.S. Air Force photo)

Clouding the issue, investigators found that procedures for breaking-off GCA practice approaches in VFR weather were not clearly defined at Fairchild AFB. In this instance:

> The tower operator failed to notify Outcome 55 of conflicting traffic *[Outcome 54]*, and the tower coordinator failed to pass [conflicting] traffic information to the [GCA] precision controller for relay to Outcome 54.

The tower officer-in-charge had been aware that both aircraft had been cleared onto final approach. For an undetermined reason he took no action to eliminate the conflict. Also, investigators noted that the tower controller had made a serious judgement error when he directed Outcome 54 to break right. The right turn had taken the bomber directly into the path of Outcome 55.

Probable Cause: The investigative team completed the "Report of A.F. Aircraft Accident" for each B-52. Both had been approaching the same runway in VFR weather. The Air Force board found many *Contributory Causes* involving air traffic control at Fairchild AFB. Controller skill and experience levels had not been adequate, and procedures for handling VFR traffic were substandard.

However, the board could not dismiss elementary and obvious facts. Both aircraft had been operating in VFR conditions. Both crews had cancelled their IFR flight plans. Accordingly, both crews had the responsibility, and the ability, to see-and-avoid all other aircraft. Therefore, the board found the **Primary Cause** of this accident to be:

Operator error, in that the pilots of both aircraft failed to maintain safe separation during visual flight in accordance with AFR *[Air Force Regulations]* 60-16.

Postscript: The Outcome 54 tail gunner, a young staff sergeant, had somehow managed to bail out. He had parachuted to safety and had walked away unscathed. Thus, this miracle-man had survived his second recent B-52 crash. Nine months earlier on 12 December 1957 he had been the *sole survivor* of a previous B-52 crash that killed the base commander and six crewmen. One literary sage mused:

Obviously, God has a special mission for that man, a reason for him to be alive.

Two Nuclear Bombs Aboard
Mid-Air Collision over Glen Dean, Kentucky
15 October 1959

Boeing B-52 Stratofortress, Serial No. 57-036
U.S. Air Force in-flight refueling practice mission
8 aboard, 4 killed

Boeing KC-135 Stratotanker, Serial No. 57-1513
U.S. Air Force in-flight refueling practice mission
4 aboard, 4 killed

Synopsis: A big Boeing B-52 Stratofortress closed on a Boeing KC-135 Stratotanker for in-flight refueling. The B-52 bomber pilots overran the tanker and rammed into it. Four men ejected from the B-52 and survived. The remaining four men in the B-52 plus all four men in the KC-135 died in the fiery crashes.

The Boeing B-52 Stratofortress Flight: A lumbering Boeing B-52 Stratofortress lifted off from Columbus AFB in Mississippi at 1430 Hours for a "Steel Trap" mission. The giant U.S. Air Force bomber carried two nuclear weapons in its bomb bay. This mishap aircraft and another B-52 practiced mock bombing runs for several hours. Then the bombers flew toward their planned refueling location high over rural Kentucky.

The B-52 carried a crew of eight: instructor pilot, command pilot, copilot, navigator, radar navigator, instructor navigator, tail gunner, and electronic warfare officer. The instructor pilot had 5,560 hours of flight experience, and the command pilot had logged 3,094 hours. The pilots, their crew, and their aircraft were assigned to the Strategic Air Command (SAC) of the U.S. Air Force.

The Boeing KC-135 Stratotanker Flight: Three hours after the two bombers took off, two big Boeing KC-135 Stratotankers climbed skyward from the same Air Force base and headed for the same rendezvous point. On IFR flight plans the Air Force tankers planned to meet the two bombers for in-flight refueling practice:

Boeing KC-135 Stratotanker (official U.S. Air Force photo)

The KC-135, commonly called a "tanker," is a military variant of the commercial Boeing 707 airliner. It is an aerial in-flight refueling platform for attack aircraft, fighters, bombers, and transports. The primary refueling method requires the receiving plane to "fly formation" under and behind the tanker. A boom operator in the tanker guides a "flying boom" (pictured above) into a receptacle on the receiving plane. The boom operator then pumps fuel through the boom and into the receiving aircraft. The tanker has a drogue which can be used to refuel aircraft fitted with refueling probes.

In 1957 the U.S. Air Force received its first KC-135. Most have since been modified by replacing the original Pratt & Whitney turbofans with four more powerful and efficient CFM-56 engines. The deck above the refueling area can accommodate up to 83,000 pounds of cargo. The KC-135 is also equipped for special high altitude aerial reconnaissance missions. It is projected to remain in service several decades into the Twenty-First Century.

Four men crewed the mishap tanker. The command pilot had logged 4,126 hours of flight time, and his young copilot had 1,112 hours of experience in the cockpit. A navigator and a refueling boom operator rounded out the crew. They flew north for 58 minutes and arrived at the rendezvous point at 1831 Hours. Their DME showed the bombers were 70 miles away and rapidly closing the range.

Trouble Ahead: At 32,000 feet the tankers flew on the reciprocal heading of the refueling track until the bombers closed within 25 miles. Then the tankers turned to the 210 degree refueling track while descending to 31,500 feet at 255 knots.

Approaching in dim twilight, the mishap B-52 pilots saw the tanker's lights when they were 12 miles away. As they closed on the tanker they failed to stabilize in the "observation" position. Closing rapidly, they angled in toward the "contact" position underneath the rear of the tanker:

> On closing to the contact position . . . it became apparent to the [command] pilot and the instructor pilot that the speed differential was too great and that the B-52 would overrun the tanker.

Realizing he was closing much too fast, the B-52 instructor pilot pulled all eight throttles back to idle. Nonetheless, inertia carried the bomber forward and swept it underneath the tanker fuselage. The bomber pilots then lost sight of the tanker above them.

The Collision: The B-52 instructor pilot started a shallow descent. Neither he nor his crew could see the tanker somewhere above them. Where was it?

> A crunching sound was heard and simultaneously all interior lighting in the B-52 went out and there was rapid decompression.

The giant B-52 rolled left and pitched up. Smoke and fire erupted on the upper deck. The instructor pilot tried to transmit a bailout order via the ICS, then realized he had no sidetone. The command pilot switched the alarm system to the "abandon" position, but the emergency alarm lights failed to illuminate. Then it got worse:

> Movement of the [flight] controls produced the sensation that they were not connected to anything.

Flames broke out in the cockpit. The bomber wallowed out of control. With no response from flight control inputs and no operative electrical system, the instructor pilot and command pilot ejected at 1846 Hours. The radar navigator and the electronic warfare officer

Boeing B-52 Stratofortress (official U.S. Air Force photo)

followed right behind them, and the four men parachuted to safety. The remaining four crewmen rode the flaming B-52 down to the ground. The impact at 1847 Hours near rural Glen Dean, Kentucky, killed them all.

After the mid-air impact the crippled tanker had spiraled earthward. Only one crewmen bailed out of the burning aircraft, and his parachute failed to open. The tanker crashed and exploded:

> The KC-135 crew, with the exception of the boom operator, were found in the forward fuselage section and had apparently made no attempt to bail out. The [body of the] boom operator was located just short of the KC-135 main impact point.

The two planes had disintegrated in the air as they fell, creating gigantic aerial fireballs visible 150 miles away. One awed witness explained to the news media:

> I looked up, and the sky was on fire!

Wreckage littered an area nine miles long and three miles wide. Fed by tons of jet fuel, huge fires raged at the main impact points. Local residents found the four surviving B-52 crewmen and took them to Critchlow's General Store in the community of Glen Dean, which had a population of about 100 people. The store owner cooked an egg and bacon supper for the erstwhile airmen, who stayed in the store for several hours. Military authorities eventually found them there and whisked them to nearby Fort Knox.

The B-52 had crashed with two nuclear bombs aboard. The next day *The Courier-Journal* newspaper in Louisville, Kentucky, would inform the public:

Two Jets Crash near Hardinsburg
Nuclear Weapon Was on Board

The accompanying newspaper article related the massive influx of military police and civil nuclear technicians:

> The tiny hamlet of Glen Dean soon became one of the busiest places in the world. . . . The [U.S.] Army set up a field mess to feed [hundreds of] military personnel.

Military disaster teams from SAC headquarters in Nebraska and the Second Air Force in Louisiana rushed to the crash site during the night. The next morning Atomic Energy Commission technicians tiptoed through the wreckage with Geiger counters in hand. Military police and air police kept curiosity seekers at bay. Helicopters ferried in hundreds of searchers from 12 military bases. Nuclear weapons specialists eventually dug the two bombs out of the rubble. One was badly mangled, but the news media would report:

> Nuclear weapons aboard the bomber did not explode. The super-secret bombs *[both wrapped in blankets for concealment]* were "displayed" to the press by the Air Force, which added that they are equipped with so many safety devices that there is "virtually no danger of their exploding accidentally."

The Investigation: Air Force investigators labored to find out why the two aircraft collided. During the mission the B-52 radar antenna tilt motor had jammed, but that had no bearing on the accident. There had been no turbulence. The surviving bomber pilots admitted they had spotted the tanker lights while still 12 miles away. The night VFR weather had not been a factor. Investigators spelled out four safety problems, but they had surfaced *after* the mid-air collision:

1. [The B-52] aerodynamic hatch lifters, designed to ensure emergency separation of [hatches for] the electronic warfare officer, radar observer, and navigator, were inadequate.

2. [The B-52] gunner was not wearing his parachute [as required] during refueling operations.

3. [B-52] emergency interphone and alarm system power failed immediately upon aerial impact.

4. [With regard to the tanker], means of emergency egress from the KC-135 are inadequate.

Investigators agreed the cause of the collision lay in the B-52 crew's failure to stabilize in the contact position. The surviving B-52 instructor pilot admitted he had been going too fast as he angled toward the tanker. He had not been able to slow down in time. He had pulled all eight throttles back to idle, but the bomber's momentum had carried it forward and under the tanker. The bomber pilots then lost sight of the tanker above. Instead of *jamming* the nose down, they had tried a gentle descent. The descent was too little and too late, and the two huge planes collided with tragic results.

Probable Cause: The Second Air Force completed a "Report of A.F. Aircraft Accident" for this mid-air collision. Both flight crews had been qualified for the mission. Nuclear safety devices were adequate, and there had been no relevant material failures. The Air Force realized that "human factors" prompted the accident.

While in-flight refueling, a B-52 pilot is "flying formation" under and behind the rear of a tanker. The pilot has to get close enough to allow the tanker boom operator to "fly" the boom into the refueling receptacle above and behind the B-52 cockpit. When sliding in toward the contact position, and while maintaining this "formation" position during refueling, a pilot must rely on visual cues alone:

SAC [procedures] are inadequate in providing a definite procedure to be followed when overrunning occurs.

In the instant case the B-52 pilots had overrun the tanker and had lost sight of it. Obviously the tanker pilots could not see the B-52 under them. The B-52 pilots could not see the tanker somewhere above them. Using vague wording the Air Force found the ***Primary Cause*** of the accident to be "supervisory error" in that:

The instructor pilot of the B-52 failed to take necessary action to avoid collision with the KC-135 tanker.

In more specific language the Air Force pointed out that "operator error" was a ***Contributing Cause***:

The [B-52 instructor] pilot did not maintain visual reference when [he was] in dangerously close proximity to the KC-135.

A Routine Check-Flight
Mid-Air Collision near Cheyenne, Wyoming
15 December 1959

<u>North American F-86 Sabre, Serial No. 52-3662</u>
Wyoming Air National Guard check-flight
1 aboard, 0 killed

<u>Beechcraft C-35 Bonanza, Registration N1839D</u>
Commercial business and public relations flight
1 aboard, 1 killed

<u>Synopsis</u>: Two North American F-86 Sabre fighters overran a civil Beechcraft C-35 Bonanza. The Sabre wingman rammed the Bonanza, destroying both aircraft. The Sabre pilot ejected and survived. The aerial impact threw the Bonanza pilot out of his mangled aircraft, and he fell to his death.

The North American F-86 Sabre Flight: Ten days before Christmas two North American F-86 Sabre fighters roared down the runway at Cheyenne Municipal Airport, Wyoming, at 1420 Hours. Both planes and both pilots were assigned to the 187th Fighter Interceptor Squadron of the Wyoming Air National Guard, which was based at the joint-use civil and military airport:

Because flying formation requires the wingman's undivided attention to the leader, the responsibility to see and avoid other aircraft was entirely that of the formation leader.

This VFR flight had been scheduled as a tactical evaluation for one of the squadron pilots. He would lead the flight-of-two and conduct (1) ground controlled intercepts, (2) a simulated instrument penetration, and (3) an ILS low approach. A check-pilot would fly as wingman and evaluate the performance:

The F-86 Sabre, often called the SabreJet, had been developed as a high-altitude day fighter. It had become the premier United States fighter during the Korean War in the early 1950s. The subsonic Sabre could carry bombs, rockets, and napalm in a ground attack role. It sported six .50 caliber machineguns for

North American F-86 Sabre (official U.S. Air Force photo)

strafing or air-to-air combat. Later variants included an all-weather night fighter and the FJ4 Fury model for the Navy and Marine Corps. A single General Electric J-47 turbojet powered the aircraft. North American turned out over 9,500 Sabres during a long production run.

The pilot being evaluated was a first lieutenant, age 30, a reserve officer. When not flying with the Air National Guard he was employed as a relatively new commercial pilot, and he held the requisite Commercial Pilot certificate. He had logged 1,400 hours, over 800 of which were in civil airplanes. The check-pilot, a captain, was assigned full-time to the Air National Guard as an air training instructor. His logbook reflected 2,450 hours of flight experience.

The Beechcraft C-35 Bonanza Flight: With only the pilot aboard, a Beechcraft C-35 Bonanza took off from St. Cloud, Minnesota, at daybreak on a business trip. After a stop at Dickinson, North Dakota, the pilot took off again at 1235 Hours and flew toward his final destination at Denver, Colorado. He air-filed a VFR flight plan that specified a cruising altitude of 8,500 feet, 3 hours and 15 minutes en route, and 5 hours of fuel on board:

The famous Bonanza had become a civil aviation institution. Designed during the waning years of World War II, the plane was considered the Cadillac of the general aviation light aircraft fleet. It carved out a niche at the very top of the single engine market, as extolled by an industry publication:

> *"The Model 35 Beechcraft Bonanza became a legend and set the standard for what a four-place airplane should be."*

The pilot, age 37, was president of Scenic Outdoor Advertising Inc., and he was one of the Bonanza owners. He had earned a Private Pilot certificate and had logged 325 hours of flight time. On this trip he planned to visit business clients, and he had packed the Bonanza full of gift-wrapped Christmas presents.

Trouble Ahead: Around 1500 Hours the Sabre pilots finished the intercept portion of their check-flight. After a VOR penetration and a low pass down Runway 26 the two jets climbed and began to circle around for a final landing.

Meanwhile, the Bonanza droned straight and level toward Denver at 9,000 feet msl, roughly 2,850 feet above the terrain. The pilot radioed Cheyenne to check the winds aloft forecast, and he confirmed he was VFR and en route to Denver. He was told the most favorable winds would be between 8,000 and 11,000 feet. The pilot acknowledged this information, and no one heard from him again.

The portion of the military UHF radio transcript, below, begins as the Sabre flight leader completes his intercepts:

Lead: . . . . F-86 Sabre flight leader (the lieutenant)
CkPilot: . . . . F-86 Sabre wingman (the check-pilot)
Tower: . . . . . Cheyenne Airport Control Tower

[the Sabre flight leader requests a VOR penetration followed by an ILS and low approach on Runway 26]

Tower: . . . maintain VFR, report the VOR outbound at two-zero-thousand feet, report the outer marker inbound. (1500)

[17 minutes pass; the Sabre flight leader completes the instrument penetration]

Lead: . . . [we are] at the middle marker inbound, on the go. (1517)

<u>Tower</u>: Roger, cleared for low approach, Runway two-six. (1517)

[the two Sabres complete a low pass over the runway]

<u>Lead</u>: Cheyenne Tower, request simulated flameout pattern. (1519)

<u>Tower</u>: . . . SFO approved, maintain VFR. (1519)

<u>CkPilot</u>: . . . and I'm low on fuel, so we need to forget the SFO and go ahead and land. (1519)

<u>Lead</u>: Let's enter on initial and join on the turn. (1519)

<u>CkPilot</u>: Roger. (1519)

[the tower controller turns his attention to a Lockheed T-33 military trainer that is preparing to land]

[the Sabres climb and turn left; watching his leader, the check-pilot closes to within 4 to 5 feet of wingtip separation]

[the Sabre check-pilot __collides__ with the Bonanza at 1520]

The Collision: At 9,000 feet and 312 knots the two Sabres completed their climb and turn, and then they flew east. The Bonanza droned southeast on a heading of 154 degrees at 139 knots, four miles south of the airport. The Sabre formation quickly overran the slower Bonanza. The Sabre check-pilot, properly concentrating on maintaining close formation, never saw the Bonanza ahead:

> Initial contact occurred when the F-86 nose structure contacted the fuselage of the [Bonanza] just behind the rear cabin window. . . . The F-86 penetrated and cut through the Bonanza fuselage [and] sheared off the Bonanza fuselage.

The collision destroyed the Sabre, which exploded into flame. The check-pilot ejected, cleared the fireball, and parachuted down to the ground with only minor injuries. In the subsequent words of the local news media:

> The jet disintegrated, scattering pieces of fuselage like confetti over a quarter-mile area.

The mid-air collision had torn off the rear half of the Bonanza fuselage. The aerial imbalance ripped off the engine and threw the

Beechcraft B-33 Bonanza, similar to the Beechcraft C-35 Bonanza (photo courtesy of Adrian R. Pingstone)

forward half of the airplane into a violent swirling tumble. Wild oscillations tossed the hapless pilot and his Christmas presents through the ragged hole where the rear of the plane should have been. Still alive, he free-fell to the ground. The next day the *Wyoming State Tribune* newspaper headline would report:

Planes Collide in Mid-Air, One Dead

An accompanying newspaper article would graphically describe the Bonanza pilot's fate:

> [The pilot's] mutilated body was found almost a quarter of a mile south of the . . . front half of the fuselage. [His] body had gouged out a hole about six inches deep and a yard in diameter in the hard prairie turf.

The main wreckage of the Sabre fell into a stock car racetrack infield. The front half of the Bonanza tore through electrical power lines as it fell, and pieces of shredded Christmas presents were found scattered along a mile-long path.

Four pumpers from the Cheyenne Fire Department sped to the main wreckage sites, but there was no fire. Only smouldering rubble remained. The Sabre had carried 24 air-to-ground rockets with high-explosive warheads. None had detonated, and Air National Guard searchers eventually found and recovered all of them.

The Investigation: The CAB investigated this tragic accident. The Bonanza pilot was dead and could no longer offer input, but the Sabre pilot had survived. He told investigators he had visually concentrated on maintaining his position as wingman. He said he never saw the Bonanza. In fact, he *assumed* he had hit the T-33. Before the collision he had heard its UHF radio transmissions and knew it was somewhere near the airport:

> During the final 30 second period [before the collision] the F-86s were positioned 110 degrees to the right rear of the Bonanza, while the Bonanza was 26 degrees to the left of the nose of the jet formation leader.

The accident had occurred in excellent flying weather, with thin cirrus clouds high overhead and an estimated visibility of 90 miles. The flight leader, the lieutenant, had been responsible for maintaining separation from all other aircraft. He acknowledged this to investigators and told them he understood his responsibility. He said that during the 30 seconds before the collision he had scanned the left quadrant, then straight ahead, and then the right quadrant:

> [He said that] when he returned his vision forward he saw an aircraft *[the Bonanza]* immediately in front of him and made a violent pull-up to avoid it.

Cheyenne Municipal Airport was the home of the Air National Guard squadron. It also served three scheduled carriers and heavy general aviation traffic. The airport had a conventional five-mile-radius control zone, and the collision had occurred inside this area. The Sabre radios transmitted and received only on military UHF frequencies. The Bonanza pilot communicated only on civil VHF frequencies. Cheyenne Tower did not simulcast on UHF and VHF, so the two Sabre pilots and the Bonanza pilot could not hear radio messages to or from each other.

Probable Cause: The CAB adopted the Aircraft Accident Report, File No. 2-1774, in August 1960 eight months after the accident. In crystal clear weather the Sabre flight had overtaken the Bonanza from its right rear. For 60 seconds before the collision the Bonanza had been "well within the forward visual quadrant" of the Sabre leader:

> The board concludes that there was an adequate opportunity for the jet formation leader to have seen the [Bonanza].

Because the two Sabres overtook the Bonanza from behind, the board stressed that the civil pilot had no opportunity to see the two military fighters bearing down on him. The board spelled out the *Probable Cause* of this mid-air collision:

> The board determines that the Probable Cause of this accident was that, during an overtaking situation, the jet formation leader failed to see the [Bonanza] in time to lead his wingman off [of the] collision course.

Two Planes IFR at the Same Altitude
Mid-Air Collision over Rio de Janeiro, Brazil
25 February 1960

Douglas R6D Liftmaster, Bureau No. 131582
U.S. Navy logistics flight
38 aboard, 35 killed

Douglas DC-3, Registration PP-AXD
Consorcio Real Aerovias Nacional passenger flight
26 aboard, 26 killed

Synopsis: *A military Douglas R6D Liftmaster and a Douglas DC-3 airliner made IFR approaches toward separate airports in the same city. The two aircraft collided head-on in the clouds at 5,000 feet. The accident killed 35 of the 38 people on the R6D and all 26 people on the DC-3.*

The Douglas R6D Liftmaster Flight: Dwight D. Eisenhower, the President of the United States, wanted an American military band to perform during his visit to Brazil. A U.S. Navy plane was scheduled to ferry 19 members of the Navy band to Rio de Janeiro. In addition to band members and musical equipment the aircraft would carry 12 antisubmarine warfare specialists and a crew of seven.

The Douglas R6D Liftmaster left the naval station at Buenos Aires, Argentina, at 0827 Hours for the five hour flight to Galeao International Airport. This new airport was located on Governador Island in Guanabara Bay at Rio. The U.S. Navy later would report:

The flight was uneventful until reporting to Rio [Approach Control] at Bagre Intersection at 1556Z *[1256 local time]* at 3,900 meters *[about 13,000 feet].*

The R6D (known as the C-118 in the U.S. Air Force) was a military variant of the civil Douglas DC-6 airliner. The aircraft and crew were assigned to Navy Fleet Tactical Support Squadron VR-1. Both experienced pilots had logged over 5,000 hours of flight time. Their aircraft had been accepted from the factory in 1953 and had flown 10,078 hours since that time. Each of its four huge Pratt & Whitney R-2800 radial engines pumped out 2,465 horsepower.

Douglas DC-6, the civil equivalent of the military Douglas R6D Liftmaster (photo courtesy of U.S. National Weather Service)

The Douglas DC-3 Flight: At 1200 Hours at Campos, Brazil, 22 passengers climbed into the cabin of a Douglas DC-3. Consorcio Real Aerovias Nacional, a Brazilian airline, planned to whisk them to Rio on a VFR flight plan. The DC-3 pilots intended to land at Santos Dumont Airport in the old part of the city, eight miles south of the newer Galeao Airport.

The DC-3 captain, age 33, had logged 7,622 hours of flight time. He and his first officer, age 26, were fully instrument rated. They were backed up by two additional crewmen, a radio operator and his assistant. The twin engine DC-3 had flown 26,184 hours since acceptance and had been meticulously maintained. The pilots taxied to the runway, took off from Campos, and flew south into increasingly cloudy skies.

Trouble Ahead: The weather turned progressively rotten as the two planes neared their destination. Rio Approach Control estimated one mile visibility in light rain with a solid overcast somewhere between 300 feet and 1,000 feet.

Nearing Rio at 1301 Hours, the DC-3 pilots radioed Approach Control and reported they could no longer maintain VFR flight. The controller issued them an IFR clearance, and they forged ahead. Five minutes later they reported overflying Santos Dumont radio beacon, located at the airport. At 1,500 meters (about 5,000 feet) the pilots turned left for an ADF approach to Santos Dumont Airport:

Both aircraft were operating in instrument weather conditions, [and] the weather was solid overcast.

Already on an IFR flight plan, the Navy R6D approached the city from the south. The Navy pilots radioed Rio Approach Control and reported their position at Bagre Intersection at 3,900 meters (about 13,000 feet). Per their IFR clearance, the pilots continued inbound from the Ilha Rasa radio beacon toward the Santos Dumont beacon and descended toward 1,500 meters (about 5,000 feet).

Unfortunately the R6D was headed for the (1) same location and (2) same altitude as the DC-3. The portion of the radio transcript, below, contains messages between ATC and the R6D pilots. The transcript begins as the Navy pilots make their initial radio contact with Approach Control:

Navy 582: . . . . . . Douglas R6D Liftmaster
AppControl: . . . . Rio de Janeiro Approach Control

Navy 582: Rio Approach, this is Navy five-eight-two, over. (1256)

AppControl: Navy five-eight-two, this is Rio, go ahead. (1256)

Navy 582: Rio Approach, Navy five-eight-two is inbound at Bagre Intersection [at] three-thousand-nine-hundred meters *[about 13,000 feet]* for landing at Galeao. (1256)

AppControl: Navy five-eight-two, begin descent to eighteen-hundred meters *[about 6,000 feet]* at Ilha Rasa [radio beacon]. (1256)

Navy 582: Roger, eighteen-hundred at Ilha Rasa. (1256)

AppControl: Navy five-eight-two, what is your altitude now? (1300)

Navy 582: Rio Approach, Navy five-eight-two is passing through two-thousand-five-hundred meters *[about 8,200 feet]*. (1300)

AppControl: Navy five-eight-two, Rio, roger. (1300)

[meanwhile, the Brazilian DC-3 overflies Porte Das Caxias at 1301 and reports that it can no longer maintain VFR flight; Approach Control issues the DC-3 pilots an IFR clearance]

Navy 582: Rio Approach, this is Navy five-eighty-two, beyond Ilha Rasa – aaahhh – say again our clearance? (1304)

AppControl: Navy five-eight-two, report Ilha Rasa and descend to fifteen-hundred meters *[about 5,000 feet]* at Ilha Rasa. (1304)

Navy 582: Roger. (1304)

[at 1306 the Brazilian DC-3 reports over Santos Dumont radio beacon at 1,500 meters, and the pilots enter a left procedure turn for an ADF approach to Santos Dumont Airport]

Navy 582: Approach Control, Navy five-eight-two is over Santos Dumont [radio beacon], we're inbound for Galeao at fifteen-hundred meters *[about 5,000 feet]*. (1307)

AppControl: Navy five-eight-two, Rio Approach, roger. (1307)

[both aircraft are now flying IFR at the same altitude]

[the R6D and the DC-3 collide over Sugar Loaf Mountain at 1307 at about 5,000 feet, about 2.5 miles south of Santos Dumont radio beacon]

The Collision: Slowing to about 145 knots, the R6D had flown north toward Santos Dumont radio beacon. At about 120 to 140 knots the DC-3 had been negotiating the procedure turn for an ADF approach to Santos Dumont Airport. Both aircraft were flying at about 5,000 feet. During the turn in IMC weather the DC-3 blindly rammed almost head-on into the R6D:

> It is extremely remote that either pilot saw the other aircraft prior to the collision . . . since both aircraft were in actual instrument conditions.

The wings and engines of the two planes tore into each other. The entire R6D empennage plus the rear of the cabin broke away. The forward fuselage plunged vertically down into Guanabara Bay. All 35 people in the front part of the Navy aircraft were killed by the mid-air collision or by impact with the sea below:

> Thirty-eight feet of the aft fuselage and empennage separated from the aircraft.

Three lucky men of the Navy antisubmarine team were sitting in the rear of the R6D cabin with their seat belts fastened. When the airframe tore apart they ended up in the rear section of the fuselage. They lived through the collision and impact with the water below, as described in an independent report:

> The three men who miraculously survived without serious injury were in rear-facing seats in that part of the R6D which dropped into the water in an oscillating motion similar to a falling leaf.

All aboard the Brazilian DC-3 would die, for the mid-air collision had ripped off the entire right wing. Minus the wing, the DC-3 arced over Sugar Loaf Mountain and tumbled into the bay a mile below:

> The severed [DC-3 wing was] cut off at an angle of approximately 23 degrees. Red and blue paint marks exactly corresponding to the paint scheme of the R6D propeller tips were found on the severed section [of the wing].

The Investigation: The U.S. Navy dispatched an investigative team to Rio de Janeiro. While divers labored to retrieve wreckage from the bottom of Guanabara Bay the investigators began checking electronic navaids, ATC facilities, and human factors:

> Brazilian tower operators and approach controllers communicate in Portugese with Brazilian aircraft and in English with all others.

Investigators learned that controllers at Rio used a single VHF frequency for all IFR traffic. This included planes en route, planes in contact with Approach Control, and planes in contact with the two control towers. The approach controller worked in a small room in the Santos Dumont Airport tower. He had been talking with the R6D and DC-3 pilots on the single congested frequency.

Although ICAO regulations mandated English, the controller used both English and Portugese. The R6D pilots had been unable to understand the Portugese messages to and from the DC-3 and other aircraft. With respect to the controller the Navy investigation noted:

> His English speaking ability is considered to be very poor.

Alarmed investigators learned that neither aircraft had accurately reported ADF station passage over Santos Dumont radio beacon. The DC-3 pilots had reported over Porte Das Caxias at 1301 Hours, and five minutes later they reported over Santos Dumont. To cover that

Douglas DC-3 on the dirt airstrip at Dong Ha, Republic of Vietnam, in 1966, with "Air America" – a clandestine airline owned by the U.S. CIA – painted on the aft fuselage (photo by the author, Marion F. Sturkey)

20 nautical mile distance in only five minutes the DC-3 would have had to fly at 240 knots.

The R6D pilots had reported passing over Santos Dumont radio beacon at 1307 Hours. In reality, however, they had never reached it. Headed straight *north*, they collided with the DC-3 over Sugar Loaf Mountain while still over two miles *south* of the navaid.

What went wrong? How could that be? Why had pilots in both planes reported station passage before reaching the navaid? Investigators found a clue printed in Portugese on the ADF approach chart for Santos Dumont Airport:

> The Brazilian let-down plate *[the approach chart]* for Santos Dumont Airport notes the possibility of receiving a "false-cone" [from the Santos Dumont radio beacon transmitter]. Interviews with Brazilian pilots revealed that some had experienced the false station passage on numerous occasions.

The culprit turned out to be the massive steel cable-car system that carried tourists to the restaurant and scenic overlook high atop the granite massif of Sugar Loaf Mountain. Tests proved the heavy steel cable network electro-magnetically interfered with the radio beacon transmitter over two miles to the north. This interference often caused cockpit ADF needles to rotate. On such occasions pilots would erroneously report "station passage" long before actually overflying the radio beacon.

Probable Cause: In October 1960, eight months after the accident, the U.S. Naval Aviation Safety Center commander concurred with findings of the investigative team. He stressed that ATC "personnel, navigation facilities, weather, and [the] language barrier" all contributed to the accident.

ADF radio beacons were the only navaids at Rio. The crucial Santos Dumont beacon was prone to "false-cones." ATC transmitted most messages in Portugese, and proficiency in English was poor. There was no surveillance radar or precision approach capability.

The Navy report stated that "the exact cause of the accident could not be determined." In lieu of a *Probable Cause* the report detailed five *Contributing Causes*:

[First], the fact that ATC did not mandate 300 meter *[apx. 1,000 feet]* altitude separation.

[Second], the number of changes in instrument clearances for both the R6D and DC-3 in a short period of time.

[Third], the poor English used by the controller.

[Fourth], the lack of radar . . . and the [confusing] conditions existing in the Approach Control Center.

[Fifth], the reported arrival of the DC-3 at [Santos Dumont radio beacon] within 5 minutes after reporting at Porte Das Caxias, a distance of 20 nautical miles, *[which would have required the DC-3 to fly at roughly twice its approach speed]* created a state of confusion and anxiety in the mind of the controller, resulting in his *failure to provide altitude separation [emphasis added]* for both aircraft.

Radar Service Terminated
Mid-Air Collision over New York City
16 December 1960

<u>Douglas DC-8, Registration N8013U</u>
United Airlines Flight 826
84 aboard, 84 killed (plus 8 killed on the ground)

<u>Lockheed L-1049 Super Constellation, Registration N6907C</u>
Trans World Airlines Flight 266
44 aboard, 44 killed

<u>Synopsis</u>: A new Douglas DC-8 and a Lockheed L-1049 Super Constellation descended for landing at separate airports in the same city. ATC had radar contact with both aircraft, and ATC controlled both flights. Nonetheless, the two airliners collided. The 128 people on board, plus eight more on the ground, died in the collision and fiery crashes.

The Douglas DC-8 Flight: United Airlines Flight 826 took off from Chicago, Illinois, at 0911 Hours and sped toward Idlewild Airport (also called New York International Airport, and later renamed JFK International Airport) in New York. The Douglas DC-8 carried a light load of only 77 passengers and a crew of seven. The four-jet airliner had been delivered to United in December of the previous year and had logged only 2,434 hours in the air:

The DC-8, the largest narrowbody subsonic airliner ever built, first flew in 1958. It entered service with Delta and United the next year. Its four Pratt & Whitney turbojets soon were replaced with new turbofans, making the DC-8 the world's first airliner powered by the new-technology engines. Fuselage stretches over the years allowed the transport to seat up to 269 passengers at takeoff weights up to 355,000 pounds. It would remain the world's largest airliner until the Boeing 747 entered service. As late as August 2006 an impressive 143 DC-8s still remained in service, most of them configured as freighters.

The DC-8 captain, age 46, had been flying with United for 19 years. He had logged 19,100 hours and had checked-out in the new

Douglas DC-8 (photo courtesy of U.S. NASA)

DC-8 six months before this final flight. He, his first officer, and his flight engineer each held ATP certificates. Four stewardesses catered to the needs of passengers in the cabin.

The flight plan specified flight level 270, a speed of 478 knots, and 1 hour and 29 minutes en route. At 1022 Hours the controller at New York Center confirmed: "radar contact."

The Lockheed L-1049 Super Constellation Flight: Shortly before the United flight left Chicago, Trans World Airlines (TWA) Flight 266 took off from Columbus, Ohio. The airliner was bound for La Guardia Airport, a different airport in New York City. The Lockheed L-1049 Super Constellation carried a slim passenger load of 39 plus a crew of five. For this flight the Constellation weighed 101,144 pounds with its passengers and 2,600 gallons of fuel.

The pilots planned on a flight time of 1 hour and 32 minutes. Toward the end of the trip they approached the eastern seaboard at 19,000 feet and followed mandates of New York Center. At 1019 Hours as the "Connie" descended through 11,000 feet the controller voiced the comforting words: "radar contact."

Trouble Ahead: The DC-8 was equipped with a FDR, but neither airliner had a new-technology CVR. Nonetheless, ATC recorded all radio transmissions and tracked both aircraft on radar. The state of transponder technology in the year 1960 did not provide an altitude display for the controller. He relied on pilot radio reports in order to afford vertical separation to aircraft under his control.

The portion of the ATC transcript, below, contains radio traffic between the controllers and pilots. At no time were the DC-8 pilots and Constellation pilots in contact with the same controller. They could not hear radio traffic to or from each other:

> United 826: . . . . United Airlines Flight 826 (Douglas DC-8)
> TWA 266: . . . . TWA Flight 266 (Lockheed L-1049)
> Center: . . . . New York Center (ARTCC)
> AppControl: . . . . . . La Guardia Approach Control

Center: United eight-two-six, [your] clearance limit is Preston Intersection . . . maintain flight level two-five-zero. (1015)

> *[Center reports <u>radar contact</u> with the Constellation at 1019]*

> *[Center reports <u>radar contact</u> with the DC-8 at 1022]*

> *[Center reports the Idlewild Airport weather to the DC-8; the pilots begin a descent at 1024:37]*

Center: United eight-twenty-six, cleared to proceed on Victor thirty until intercepting Victor one-twenty-three and that way to Preston, it'll be a little bit quicker. (1025:09)

> *[Center clears the Constellation to descend to 9,000 feet and to contact La Guardia Approach Control]*

AppControl: TWA two-sixty-six, maintain nine-thousand, report the zero-one-zero [radial of] Robbinsville [VOR], ILS Runway four, landing Runway four . . . . (1026:22)

> *[Center clears the DC-8 to descend to 11,000 feet at 1026:49]*

> *[the Constellation reports passing the 010 radial of the Robbinsville VOR; La Guardia Approach Control clears the pilots to descend to 8,000 feet at 1028]*

Center: United eight-twenty-six, I show you crossing the centerline of Victor thirty at this time. (1028:41)

> *[the DC-8 pilots acknowledge they are on Victor 30; they ask Center how far they are from Victor 123]*

Center: United eight-twenty-six . . . I show you fifteen, make it sixteen, miles [from] Victor one-twenty-three. (1028:56)

[at 1029:49 La Guardia Approach Control clears the Constellation to descend to 6,000 feet for radar vectors to the ILS course for Runway 4]

Center: United eight-two-six, descend to and maintain five-thousand feet . . . looks like you'll be able to make Preston at five-thousand [feet]. (1030:07)

[at 1032:09 La Guardia Approach Control tells the Constellation to turn right to a heading of 130 degrees to intercept the ILS]

Center: United eight-two-six, if holding is necessary at Preston, [use a] southwest one minute pattern, right turns, the only delay will be in descent. (1032:16)

United 826: Roger, no delay, we're out of seven-[thousand feet].

[La Guardia Approach Control clears the Constellation to descend to 5,000 feet at 1032:16]

[both aircraft are now cleared to the same altitude, 5,000 feet]

AppControl: TWA two-sixty-six, traffic at two-thirty, six miles, northeast-bound. (1032:47)

[the Constellation pilots acknowledge the "traffic advisory"]

[in response to a query from La Guardia Approach Control, the Constellation reports descending through 5,500 feet at 1033:08]

Center: United eight-twenty-six, roger, and you [have] received the holding instructions at Preston, radar service is terminated, contact Idlewild Approach Control . . . . (1033:20)

AppControl: TWA two-sixty-six, turn left now, heading one-three-zero. (1033:21)

TWA 266: Roger, heading one-three-zero. (1033:23)

AppControl: TWA two-sixty-six, roger, that appears to be jet traffic *[meaning, it is moving fast]* off to your right now, three o'clock at one mile, northeast-bound. (1033:26)

<u>United 826</u>: Idlewild Approach control, United eight-twenty-six, approaching Preston at five-thousand. (1033:28)

*[4 seconds later the two airliners **collide** at 1033:32]*

The Collision: The two aircraft knifed into each other at 5,175 feet in the clouds over Staten Island. The No. 4 engine on the DC-8 slashed through the Constellation cabin like a giant meat cleaver. The mid-air impact tore the Constellation apart, and no one aboard had a chance to survive:

> The Constellation broke into three main sections following the collision.... Numerous pieces of aircraft structure were strewn over a wide area in the vicinity of Miller [Army Airfield].

The rear of the Constellation cabin and the empennage tumbled earthward. The forward part of the cabin and the left wing followed. Two sections of the severed fuselage fell down onto Miller Field, and smaller pieces of debris rained down into the surrounding neighborhood. An intense fire consumed all remnants of the Constellation fuselage. There had been 44 people on board, and the collision and fiery crash killed them all.

Meanwhile, on the DC-8 the impact had torn off the outer right wing and the No. 4 engine. The high-speed collision also ripped many enhanced lift devices off the left wing. Pieces from the DC-8 showered down onto Miller Army Airfield a mile below:

> Both main landing gear doors, the outboard section of the right wing, and the No. 4 engine impacted in the Miller Field area.

Despite the damage the pilots still had partial control over their DC-8. They made a valiant attempt to save their airliner, but its wounds proved to be too great. The craft staggered eight miles to the northeast, sinking lower and lower:

> The DC-8 continued on a northeasterly heading and crashed into the heavily populated area of Brooklyn.

The airliner rammed through tall buildings and exploded at the intersection of Sterling Place and Seventh Avenue. Flaming wreckage cartwheeled in all directions. The left wing burned in the street.

Part of the fuselage burned atop a nearby building. The right wing and the No. 3 engine tore through the Pillar of Fire Church, sending a literal *pillar of fire* hundreds of feet into the snowy sky. Ironically, the revisionist Methodist sect had drawn its name from a biblical miracle related in *Exodus 13:21*. The ancient Israelites had relied upon a "pillar of fire" to lead them out of enslavement in Egypt:

> [A] 15-foot section [of the left wing] cut through and came to rest in a building at 126 Sterling Place, leaving two feet of the wing protruding through the roof.

The crushed rear of the DC-8 cabin tumbled down the street and threw an 11 year old boy out into a snowbank. Two policemen found him there, horribly burned and in shock. Yet, he sat up and spoke through blackened lips to the officers:

My name is Stephen. Mommy? Daddy?

The young boy hailed from Wilmette, Illinois. He was on his Christmas vacation and had been going to meet his grandparents, who were waiting for him at Idlewild. Instead, an ambulance rushed him to nearby New York Methodist Hospital.

Against the backdrop of the two fiery crashes in the city and the tremendous loss of life, the desperate efforts to save Stephen became a non-stop media event. Alerted by television and radio, compassionate New Yorkers flocked to the hospital to donate blood. Others donated skin for grafting to cover the boy's terrible burns. Noted aviation authority Robert J. Serling would write in *Loud & Clear*:

> He had burns over 80 percent of his body. He fought hard for his life, and so did a platoon of doctors. He managed to talk to CAB investigators, who had tears in their eyes as they listened to his story. He said he remembered looking out of the plane's window at the snow-covered city below. "It looked like a picture from a fairy book," he told them. "It was a beautiful sight." . . . A nurse burst into tears. So did a [newspaper] reporter.

Although the entire nation rooted for Stephen, his life slipped away the following day. His death completed the casualty toll. He was the 136th fatality. Everyone aboard the two airliners, plus eight more on the ground, had perished in the collision and crashes.

Yet, it could have been much worse. The main fuselage of the DC-8 had fallen only 250 feet from Public School No. 41. Hundreds of awestruck schoolchildren had rushed to the windows to watch orange flames and back smoke boiling skyward.

The Investigation: The FDR on the DC-8 fixed the time of the collision at 1033:32 Hours. Two controllers, each responsible for separate areas of airspace, had cleared their flights to descend to the same altitude, 5,000 feet. La Guardia Approach Control had been providing radar vectors to guide the Constellation toward the ILS glidepath for Runway 4:

> At no time prior to the one-mile advisory was any information furnished to Flight 266 *[the Constellation]* which would have alerted it to a conflict.

ATC handled the DC-8 differently. The pilots made position reports to ATC, but they were responsible for their own navigation. They did not receive ATC radar vectors. Twelve seconds before the collision, New York Center had handed-off the jetliner to Idlewild Approach Control. Still, long before the handoff Center had issued a clearance limit. That aerial "fix" in the sky, Preston Intersection, was a point beyond which the DC-8 pilots were not authorized to fly:

> Preston Intersection is defined as the intersection of the 346 degree radial of the Colts Neck [VOR] and the 050 degree radial of the Robbinsville [VOR].

Normally the DC-8 pilots would have tuned one of their VOR receivers to Colts Neck VOR and the other receiver to Robbinsville VOR. The cockpit instruments simultaneously would have displayed the radials from both VOR stations, easily allowing the pilots to identify the intersection. However, investigators found the DC-8 pilots had radioed their maintenance base to report a problem. One of their two VOR systems was inoperative, and ATC had not been aware of this equipment failure:

> *With only one operative VOR it is still possible to identify a VOR intersection, but it is more difficult. Pilots have to switch their*

*single operative receiver back and forth from one VOR frequency
to the other, constantly checking and cross-checking the radials.
This procedure consumes a lot of time, and it is not as accurate
as a simultaneous cockpit display.*

Investigators learned that difficulties had continued to mount for
the DC-8 pilots. At 1025:09 Hours, ATC had issued them a revised
clearance for a new route. This had shortened the distance to Preston
Intersection. It reduced the time available for the pilots to repeatedly
re-tune their only operative VOR receiver to establish the location of
the intersection. Investigators opined:

> The crew committed a primary error by apparently failing to
> record and note the time and distance required to comply with
> their new clearance.

Both flights had been cleared to 5,000 feet, so identification of
Preston Intersection had been crucial. Tragically the DC-8 pilots did
not stop (that is, enter the mandated *holding pattern*) at the intersection. They (1) sped through it, (2) left the airspace reserved for them,
and (3) flew into airspace reserved for the Constellation.

Probable Cause: The CAB adopted its Aircraft Accident Report,
File No. 1-0083, on 12 June 1962 roughly 18 months after the
accident. The board stressed that the DC-8 did not enter a holding
pattern at Preston Intersection, as ATC had instructed:

> When radar service was being terminated at 1033:20 [Hours],
> Flight 826 *[the DC-8]* had already proceeded eight or nine miles
> beyond Preston [Intersection].

The board found that ATC clearances and ATC handling of the
two flights had been proper. The La Guardia controller who vectored
the Constellation had seen the radar return from the DC-8. He
watched the two "blips" converge on his scope. Yet, the controller
had no way to know the altitude of the DC-8, so he radioed routine
advisories to the Constellation. At the time of the last advisory at
1033:26 Hours the collision had been a mere six seconds away. The
board found the following *Probable Cause* of the accident:

[United] Flight 826 *[the DC-8]* proceeded beyond its clearance limit and the confines of the airspace allocated to the flight by Air Traffic Control.

The board found two **Contributing Factors**:

[First was] the high rate of speed of the United DC-8 as it approached Preston Intersection.

[Second was] the change of clearance which reduced the en route distance along Victor 123 by approximately 11 miles.

Lockheed C-121 used by Dwight D. Eisenhower, President of the United States – a military variant of the civil Lockheed L-1049 Super Constellation airliner (photo courtesy of U.S. DOD Visual Information Center)

VFR-On-Top?
Mid-Air Collision near Tallahassee, Florida
8 June 1962

Lockheed L-18 Lodestar, Registration N45W
Winn-Dixie Stores Inc. business flight
4 aboard, 4 killed

Lockheed T-33 Shooting Star, Serial No. 51-4532A
U.S. Air Force instrument training flight
2 aboard, 0 killed

Synopsis: A military Lockheed T-33 Shooting Star on an IFR flight plan climbed in VFR conditions. The T-33 overtook a civil Lockheed L-18 Lodestar, also on an IFR flight plan, and rammed into it. The mid-air collision and crash killed all four people on the Lodestar. The two T-33 pilots ejected and survived.

The Lockheed L-18 Lodestar Flight: Early on a Friday morning four people climbed aboard a Lockheed L-18 Lodestar at Montgomery, Alabama. Winn-Dixie Stores Inc. owned the plane, and the two pilots were Winn-Dixie employees. Two female passengers and their small dog settled into the Lodestar cabin for a business trip to Jacksonville, Florida:

The Lodestar had evolved from the smaller Lockheed L-14 Super Electra. The factory lengthened the fuselage by five feet to create a compact 17 passenger airliner, and the new airplane entered service in 1940. By then most airlines had decided to purchase the larger Douglas DC-3, and initial Lodestar sales were slow.

Lockheed turned out 625 Lodestars, most of them during World War II. The twin engine and twin tail transport, fitted with a variety of Wright and Pratt & Whitney radial powerplants, served in the armed forces of the United States and seven allied nations. Known as the C-56 through C-60 in the U.S. Army Air Force and as the R-50 in the U.S. Navy, Lodestars served as troop transports, cargo haulers, SAR craft, and antisubmarine hunter-killers. After the war most became surplus and got converted into civil executive transports.

Lockheed C-60, a military variant similar to the civil Lockheed L-18 Lodestar (official U.S. Air Force photo)

The Lodestar took off from Dannelly Field near Montgomery at 0706 Hours on an IFR flight plan. ATC cleared the pilots via V-7 to Tallahassee VOR, then via V-22 to Jacksonville at 8,000 feet. The pilots estimated 1 hour and 30 minutes en route.

The Lockheed T-33 Shooting Star Flight: Seven minutes after the Lodestar left Montgomery, a Lockheed T-33 Shooting Star climbed skyward from Moody AFB near Valdosta, Georgia. The U.S. Air Force jet trainer carried an instructor pilot, an Air Force captain, in the rear seat. A relatively new pilot, a second lieutenant, manned the front cockpit. On an IFR flight plan they flew toward Tallahassee Municipal Airport to practice instrument approaches.

The jet trainer arrived over Tallahassee VOR at 20,000 feet. The student executed a jet penetration followed by two ILS approaches to Runway 36. Then the instructor told him to return to Moody.

Trouble Ahead: The IFR Lodestar motored southeast at 8,000 feet on V-7. The pilots saw a few scattered clouds, but they were flying in VMC weather. They radioed Jacksonville Center and estimated they would reach Tallahassee VOR at 0757 Hours.

Meanwhile, with the student pilot flying the plane and handling the radios, the T-33 climbed toward the same navaid. Tallahassee Approach Control cleared the jet pilots to the VOR at 3,000 feet and told them to expect further routing back to Moody. The instructor then took over the radio. He told Approach Control he was in VFR conditions and requested a VFR climb to 20,000 feet.

The portion of the ATC radio transcript, below, contains UHF radio traffic between Approach Control and the T-33. The transcript begins with the initial clearance to 3,000 feet:

> Air Force 532: . . . . Lockheed T-33 Shooting Star
> AppControl: . . . . . . Tallahassee Approach Control

AppControl: [Air Force jet five-three-two], cleared from present position direct to Tallahassee Omni, maintain three-thousand feet, anticipate routing, Victor 22 [to] Greenville [Intersection], direct to Valdosta Omni. (0750)

[the student pilot incorrectly reads-back the clearance]

[the instructor pilot takes over the radio, reports he is VFR, and requests a VFR climb]

AppControl: You wish a VFR climbout to altitude and [you wish to] cancel your IFR [flight plan]? (0750)

Air Force 532: Aaahhh – negative – [Air Force jet] five-three-two [is] requesting a VFR-on-top from present position to – aaahhh – back to – aaahhh – Moody, we're VFR at this time, we'll climb VFR and proceed [to] Moody VFR-on-top if it's all right.

AppControl: Roger, what – what altitude do you anticipate being at – aaahhh – VFR-on-top?

Air Force 532: We're VFR-on-top at this time at four-thousand feet, and we're gonna – we'll climb to twenty-thousand [feet]. (0754)

[ATC asks the pilot if he plans to fly direct to Valdosta VOR]

Air Force 532: Affirmative. (0754)

[the controller erroneously believes the pilot has cancelled his IFR flight plan; at 0754 the controller notifies Moody Approach Control that the IFR flight plan has been cancelled]

[the climbing T-33 collides with the cruising Lodestar at 0756]

The Collision: At about 182 knots the Lodestar cruised toward the VOR at 8,000 feet on a heading of 121 degrees. The T-33 climbed toward the VOR at 270 knots on a heading of 114 degrees. With a closure rate of about 88 knots the T-33 overtook the slower Lodestar from behind:

> Neither the [Lodestar] captain nor the copilot could have seen the T-33 coming up behind them. *[and also]* The two aircraft collided in mid-air at approximately 8,000 feet at a point approximately 4.3 miles northeast of the Tallahassee VOR.

The T-33 rammed into the civil transport with tragic results. The Lodestar No. 1 propeller chewed into the T-33 right wing. The Lodestar right wing ripped away, sending the transport tumbling toward the ground. The T-33 empennage snapped off, throwing the jet trainer into a violent inverted flat spin.

The T-33 pilots ejected and parachuted down to safety, but the four Lodestar occupants were doomed. Minus a wing, their plane tumbled out of control and exploded when it slammed into the ground. The two pilots, two passengers, and the dog perished in the pyre on Welaunee Plantation. That evening the *Tallahassee Democrat* newspaper in Tallahassee, Florida, carried a banner headline:

----

4 Die In Mid-Air Plane Crash Here

The newspaper reporter described the four victims:

> The passengers were burned and mangled. . . . Two bodies were found near the wreckage, and the two others were about 200 yards away. The body of a small dog was [found in] the wreckage.

The Investigation: Investigators examined radio messages between ATC and the two aircraft. The Lodestar used a civil VHF frequency, and the T-33 operated only on UHF. Pilots in each plane could not

Lockheed T-33 Shooting Star (photo courtesy of U.S. DOD Visual Information Center)

hear radio transmissions to and from the other aircraft. The Lodestar had been on course and on time. However, the T-33 pilot's route, altitude, and intentions fell into dispute:

> The controller was obviously under the impression that the T-33 was returning to Moody VFR, since he informed Approach Control that [the T-33 pilot] had cancelled his IFR clearance. The T-33 pilot was under the belief that his flight was operating under an instrument flight plan at all times.

In any event the T-33 pilot had said he was climbing in VFR weather. Under such conditions, regardless of the type of flight plan, the pilot is responsible for visual separation from all other aircraft. Unfortunately radar at Jacksonville Center and at Moody Approach Control could not paint aircraft flying below 10,000 feet in the Tallahassee area. Consequently, ATC was unable to offer radar traffic advisories to either plane.

The CAB conducted a conspicuity study while flying in another T-33. Investigators duplicated the airspeed, climb rate, climb angle, closure speed, and closure angle at the time of the accident:

> The student pilot in the front seat of the T-33 should have seen the Lodestar . . . . *[and also]* The Lodestar was in a position 17 degrees above the student pilot's eye [level] and 13 degrees to his

left, [and it] could have been viewed [through] the front canopy and windscreen.

There had been obstructions to the instructor's visibility from the T-33 rear seat. Yet, the pilot in the front seat should have had a clear view of the Lodestar. He had not used the instrument hood at any time during the flight. Investigators could not determine why he did not see the large transport as he slowly overtook it from behind.

Probable Cause: The CAB adopted an Aircraft Accident Report, File No. 2-0760, the year following the accident. The board concluded the powerplants and systems of the planes had not played a role in the collision. Weather had not been a factor, for the visibility had been estimated at 12 miles.

ATC thought the T-33 pilot had cancelled his IFR flight plan, but the pilot insisted he had not done so. Yet, that issue was moot because of the VFR conditions. Accordingly, the board detailed an elementary *Probable Cause* of this mid-air collision:

The board determines the Probable Cause of this accident was failure of the T-33 pilots to observe the Lodestar while climbing through its flight altitude.

Beware of the Whistling Swans
Bird Strike near Ellicott City, Maryland
23 November 1962

Vickers Viscount 745, Registration N7430
United Airlines Flight 297
17 aboard, 17 killed

Flock of Whistling Swans *(Olor columbianus)*
Migratory flight near feeding area
N.A. / N.A.

Synopsis: At 240 knots a Vickers Viscount 745 flew into a flock of Whistling Swans. One large swan penetrated the horizontal stabilizer, causing both sides of the stabilizer and elevator to tear off of the aircraft. The Viscount rolled inverted and crashed, killing all 17 people on board.

The Vickers Viscount 745 Flight: Friday morning dawned cold and clear in the northeast United States. Shortly before noon in Newark, New Jersey, a slim load of 13 passengers boarded United Airlines Flight 297, a Vickers Viscount 745 turboprop airliner. At 1139 Hours the aircraft accelerated down the runway and headed south toward Washington National Airport in the nation's capital. The pilots estimated an hour en route at 10,000 feet and 260 knots.

A crew of four staffed the Viscount. The captain, age 39, and his first officer, age 32, both held ATP certificates. The two stewardesses had an unusually light workload. Both were new employees who had been flying with United for less than five months.

The Whistling Swan Flight: From 0815 Hours to 1705 Hours, U.S. Weather Bureau radar at Washington National Airport displayed targets described in the radar log as "angels." The CAB explained:

[*Angels* means radar] contacts of unknown origin, not associated with precipitation, [and assumed] to be birds or insects.

Angels were scattered over a 30 mile area and moving northeast at 30 to 40 knots. ATC took note of these radar returns and issued "traffic advisories" to all airline flights in the area. Airline pilots told

Vickers Viscount 802, similar to the Vickers Viscount 745 flown by United Airlines (photo courtesy of Adrian R. Pingstone)

their controllers the traffic was actually "large flocks of birds." In addition a private pilot flying near Beltsville, Maryland, radioed ATC and reported hundreds of "very large white birds" at 5,500 feet. The birds later would be identified as Whistling Swans:

> *Whistling Swans (Olor columbianus), also called Tundra Swans, breed in the tundra areas near Hudson's Bay in Canada. They are the largest and most visually inspiring birds of the Arctic, and healthy adults have few natural predators. They winter from the Chesapeake Bay marshes south to the Carolina wetlands along the eastern coast of the United States.*

> *Exclusive of their black bills, legs, and feet, Whistling Swans are snowy white. Wingspans of seven feet are common, and adult males often weigh over 20 pounds. Migrating "whistlers" draw their name not from their melodious call, but instead from the high pitched whistling noise created by the powerful beating of their wings in flight. According to <u>National Geographic</u>, when migrating the swans often fly at altitudes of over 20,000 feet.*

Trouble Ahead: New York Center soon handed-off Flight 297 to Washington Center. At 1214 Hours as the Viscount neared the airport the new controller cleared the pilots to descend to 6,000 feet. Six minutes later Center told them to contact Washington Approach Control. ATC assumed the Viscount would be landing in about 15 minutes. Meanwhile, more and more pilots were reporting tremendous flocks of birds north of Washington:

As a result of the numerous radar contacts [and] pilot reports and ground observer sightings of flocks in the general area . . . there is no doubt that a definite hazard of in-flight collision with birds existed at the time of the accident.

The part of the ATC radio transcript, below, begins at 1219 Hours as the Viscount descends toward Washington National Airport:

United 297: . . . . United Airlines Flight 297
Center: . . . . Washington Center (ARTCC)
AppControl: . . . Washington Approach Control

Center: . . . be advised there's been numerous reports of considerable amounts of ducks and geese around this area. (1219)

United 297: Two-niner-seven, roger on the birds. (1219)

[Center hands-off the flight to Washington Approach Control]

AppControl: United two-nine-seven, radar contact, turn left heading two-zero-zero [for] radar vector [to] omni final approach course via Alexandria Intersection, landing [on] Runway three-six, wind [is] northwest [at] ten, altimeter three-zero-three-seven. (1222)

United 297: Roger, left two-zero-zero for Runway thirty-six. (1222)

AppControl: Two-ninety-seven, now turn farther left to one-eight-zero degrees. (1223)

*[the Viscount **collides** with Whistling Swans]*

[the Viscount drops off Approach Control radar at 1224]

The Collision: The pilots had followed the controller's instructions as they approached the airport at 240 knots. Suddenly at least two big swans had struck the horizontal stabilizer. One struck the airfoil a glancing blow and did not penetrate, but the other swan dealt a fatal injury to the airliner:

On the left horizontal stabilizer . . . the bird penetrated the leading edge . . . . The bird fractured the spar web, partially separating it from the top and bottom caps, and then made final contact with the lower leading edge of the elevator . . . .

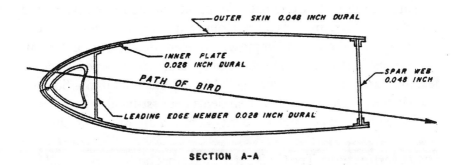

CROSS SECTIONAL VIEW OF STABILIZER IN BIRD STRIKE AREA

Cross-section diagram showing the path of the Whistling Swan, left to right, through the left horizontal stabilizer (diagram courtesy of U.S. CAB)

The entire left horizontal stabilizer and elevator broke off of the aircraft. The weakened right horizontal stabilizer and elevator then failed and peeled away in flight. Loss of pitch control caused a "violent nose-down pitching moment." The aircraft rolled inverted, and airspeed increased to 365 knots. Vertical acceleration soared to negative 3 G's as the plane dove toward the ground. The crash six miles south of Ellicott City, Maryland, instantly killed all 17 people on board the Viscount.

One nearby resident said that when the plane hit the ground, "I thought at first we had been bombed." An intense post-impact fire destroyed most of the crumpled fuselage and wings. The next day the *Baltimore Sun* newspaper headline would report:

---

17 Die in Airliner Crash

The crash had thrown debris and body parts high into the air. The newspaper article took note:

[Above] where the plane crashed, clothing hung in rags on tree limbs as high as 30 feet above the ground.

An extraordinary irony had taken place at the nearby Maryland State Police Academy. At 1300 Hours a police instructor had been scheduled to make a two hour presentation entitled "Downed Aircraft Procedures." At that moment word of the crash had reached academy headquarters. The instructor loaded his 41 students into a bus and raced to the accident scene. The students got first-hand experience in traffic control and perimeter security at a *real* aircraft crash site.

The Investigation: The CAB determined that both the right and left sides of the horizontal stabilizer and elevator had separated from the aircraft prior to ground impact. Parts of the stabilizer and elevator were strewn along a path 2,908 feet long:

> Bird remains were found on both horizontal stabilizers. *[and also]* A partial bird carcass, consisting of a large piece of skin covered with white feathers, was found . . . .

The CAB took specimens from the horizontal stabilizer, plus the bird carcass, to the Fish & Wildlife Service of the U.S. Department of the Interior. Analysts identified the bird as *Olor columbianus*, commonly called a Whistling Swan. Metallurgical analysis revealed that swan impact had seriously weakened the horizontal stabilizer, to which the elevator is attached:

> [The aircraft] struck two birds, one on the left horizontal stabilizer and one on the right. The bird strike on the left horizontal stabilizer . . . resulted in penetration [and] weakened the structure so that the normal download initiated immediate failure of the horizontal stabilizer and elevator. . . . [The resulting] breakup rendered the aircraft uncontrollable.

Collisions with birds had been a nuisance for many years. Bird strikes had damaged fuselages, windshields, antennae, cowlings, and wings. However, exclusive of stronger windshields, no other "bird-proofing" measures had been deemed necessary. Until 1960, a bird strike had never caused loss of a transport category aircraft in the United States. CAA Report No. 62, issued in 1949, had stated:

> No record exists of any fatality in air carrier operation in the United States caused by collision of aircraft with birds.

However, in 1960, the situation had changed. A new Lockheed L-188 Electra four-engine turboprop had flown into a flock of starlings on takeoff from Logan International Airport in Boston, Massachusetts. Three of the four engines had ingested multiple birds. The engines lost power only 150 feet above the ground. The Electra stalled and crashed, killing 62 people:

> This [1960 Electra] accident clearly demonstrated that even small birds, if in sufficient numbers, could . . . render a modern aircraft uncontrollable.

Probable Cause: Four months after the crash of Flight 297 the CAB adopted the Aircraft Accident Report, File No. 1-0034, on 19 March 1963. This accident differed from the Electra crash in 1960. In the instant case, a high speed collision with large swans had destroyed the structural integrity of the airframe. Investigators learned that each November the largest concentration of Whistling Swans on the North American continent is found in the Chesapeake Bay area, which is adjacent to the Viscount crash site.

The board noted that modern jet and turboprop transport category aircraft have higher cruise speeds than older designs. Higher speed equates to more vulnerability to structural damage from bird strikes. The board reasoned:

> Any project established to evaluate and attempt [to solve the problem of] catastrophic bird strike damage will be a colossal undertaking, but the problem is no less colossal.

The board outlined the *Probable Cause* of the loss of United Airlines Flight 297 as follows:

> The board determines that the Probable Cause of this accident was a loss of control following separation of the . . . horizontal stabilizers, which had been weakened by a collision with a Whistling Swan.

Fatal Optical Illusion
Mid-Air Collision over Carmel, New York
4 December 1965

Boeing 707, Registration N748TW
Trans World Airlines Flight 42
58 aboard, 0 killed

Lockheed L-1049 Super Constellation, Registration N6218C
Eastern Airlines Flight 853
54 aboard, 5 killed

Synopsis: A Boeing 707 and a Lockheed L-1049 Super Constellation collided in flight due to a cruel optical illusion. The Boeing plane, with 25 feet of one wing torn off, landed safely. Without elevator or rudder control the Constellation made a controlled crash, killing the captain and four passengers.

The Boeing 707 Flight: After a flight from San Francisco, California, Trans World Airlines (TWA) Flight 42 approached JFK International Airport in New York. Under the control of New York Center the Boeing 707 descended to 11,000 feet and cruised above a solid cloud layer:

The Boeing 707 is a four engine jet airliner developed in the 1950s. It became the first commercially successful jet transport and is credited with starting the so-called Jet Age. Pan Am had inaugurated airline passenger service with the new aircraft in 1958 on a New York to Paris flight.

The factory turned out over 1,000 Boeing 707s for the world's airlines. Over the years the planes came in many variants with maximum takeoff weights ranging up to 336,000 pounds. The Western World's armed forces use military variants as a tankers, freighters, surveillance planes, command-and-control platforms, and VIP transports. The United States selected the Boeing 707 as the first jet presidential transport, Air Force One.

The TWA captain, age 45, had logged 18,842 flight hours and held the required ATP certificate. His first officer, age 42, and his

Boeing E-6A TACAMO, a military variant of the civil Boeing 707 airliner (photo courtesy of U.S. DOD Visual Information Center)

flight engineer, age 41, had the proper qualifications and experience. Four "hostesses" in the cabin had received their most recent emergency procedures refresher training the previous month:

> TWA [Flight] 42 . . . approached the Carmel VORTAC from the northwest [at 11,000 feet]; they were flying in clear air above an overcast with no restrictions to [horizontal] visibility.

The Lockheed L-1049 Super Constellation Flight: A triple-tail Lockheed L-1049 Super Constellation, Eastern Airlines Flight 853, clawed its way skyward from Logan International Airport at Boston, Massachusetts. The famous Boston to Newark "shuttle" climbed to 10,000 feet for the short flight to the airport at Newark, New Jersey. Thirty-two minutes after takeoff the Boston Center controller handed-off the flight to New York Center.

The captain, age 42, had been a military bomber pilot before joining Eastern in 1953. He had earned an ATP certificate and had logged 11,508 flight hours. His young first officer, age 34, had 8,090 hours of flying experience, and he and the flight engineer each held a Commercial Pilot certificate. Powered by four Curtis-Wright radial engines mated to Hamilton Standard propellers, their Constellation had accumulated 32,883 flying hours.

Trouble Ahead: ATC maintained radio and radar contact with both IFR airliners. The Boeing 707 flew at 11,000 feet and 355 knots, skimming the tops of the clouds. At 10,000 feet the Constellation flew through occasional cloud tops at 215 knots. Their courses crossed at Carmel VORTAC, and controllers watched the radar targets slowly converge on their scopes:

> No traffic [advisory] was given to either crew and none was required, since a standard vertical separation minimum of 1,000 feet was being provided.

The top of the cloud layer *sloped upward* north of Carmel. Suddenly the Constellation first officer spotted the converging Boeing jet at his two o'clock position, just above the clouds. It *appeared* to be flying at his altitude and headed right at him. He shouted: "Look out!" The two pilots pulled back – hard – on their control columns to climb and avoid a collision.

Seconds later the Boeing 707 captain saw the onrushing Constellation. He instinctively took evasive action:

> He also believed he was on a collision course, made an immediate right bank, and pulled back on his [control column].

The Constellation intra-cockpit and ATC radio transcript, below, has been *reconstructed* from technical sources. The transcript begins as Center picks up the Constellation on radar. The captain is flying the airliner on autopilot:

Capt: . . . . Captain, Eastern Flight 853
FO: . . . . . First Officer, Eastern Flight 853
FltEngr: . Flight Engineer, Eastern Flight 853
Radio-: . . (prefix) Radio transmission, Eastern Flight 853
PA-: . . . . (prefix) Public Address system, Eastern Flight 853

[New York Center identifies the Constellation on radar at 10,000 feet at 1610:20]

[the New York Center controller watches the Constellation radar target pass over Carmel VORTAC at 1618]

[the First Officer suddenly sees the Boeing 707; he shouts]

FO: Look out! (1618:35)

*[the Captain and First Officer pull back on their control columns and climb, but the airliners **collide** at 1618:43]*

[the Flight Engineer watches hydraulic pressure drop to zero; the Constellation wallows out of control]

FltEngr: Pressure and quantity!

[the elevator and rudder controls are inoperative; the airliner begins to spiral earthward]

FO: How about power! *[shouted in desperation]*

[the Captain jams the throttles forward; the aircraft climbs, then stalls, then dives toward the ground again]

[the Captain makes an announcement to his passengers]

PA-Capt: This is the captain – please fasten your seatbelts – we're out of control.

Radio-FO: Mayday! Mayday! Mayday! Eastern eight-fifty-three! . . . mid-air collision! . . . out of control! (1621)

[after a series of climbs and dives the Captain adjusts engine power to keep the nose level, but the plane still descends]

[the Captain radios ATC and explains that (1) his flight controls are inoperative, (2) he can not reach any airport, and (3) he can not stop his descent; he concludes by remarking]

Radio-Capt: . . . keep an eye on us, please, to see where we wind up.

[nearing the ground, the pilots see a small field ahead]

Capt: How about that field?

FO: Let's do it!

PA-Capt: Brace yourselves!

*[the pilots make a **controlled-crash** in the field at 1628:15]*

The Collision: Inexplicably, it seemed, the evasive maneuvers had drawn the two airliners closer together. They collided at about 11,000 feet almost directly over Carmel VORTAC:

As [the Boeing 707] passed under and toward the rear of [the Constellation] the No. 1 engine cowling and a portion of the left wing of the [Boeing 707] struck the underside of the main fuselage of the [Constellation].

The impact had torn the outer 25 feet off the left wing of the Boeing jetliner. Flying debris had gouged holes all along the fuselage and what remained of the left wing.

The Boeing captain and first officer pulled their wounded aircraft out of a steep dive. Despite loss of the outer left wing the sturdy Boeing was still flyable – more or less. The pilots managed to make an emergency landing on Runway 31 Left at JFK Airport at 1640 Hours. No one on board was seriously injured. The next day an *Associated Press* newspaper headline would explain:

Pilot of Crippled Plane Did "Miraculous Job" Landing with 30-Foot Section of Wing Gone

----

Meanwhile, the critically wounded Constellation had fared much worse. After the collision its upward inertia had continued the climb for several seconds. Then its airspeed decayed, and the airliner stalled and dove toward the ground. The pilots pulled back on their control columns, but their efforts had no effect:

This [mid-air collision] tore out the hydraulic boost package and the control cables to the rear empennage of [the Constellation], effectively rendering the elevator and rudder flight controls inoperative. . . . The aircraft entered a tight left spiral.

The Constellation pilots had no control over the elevator or the rudder. Even the trim tabs failed to work. In desperation they found

Lockheed C-69, the original military Constellation variant; a later "stretched" model would become the civil Lockheed L-1049 Super Constellation airliner (official U.S. Air Force photo)

that engine power had some effect. More power would bring the nose up. Less power would allow the nose to fall. Yet, with the nose level they still were descending at about 500 feet per minute. There was nothing they could do about it. Nothing!

As they neared the ground the pilots spotted a partially wooded field straight ahead. They knew they would only get this one slim chance. Using his four throttles and ailerons, gear and flaps up, the captain managed to pancake his battered flying machine into the field. The *Associated Press* later would comment:

> That he was able to bring his [damaged] aircraft into the small field under twilight conditions is little short of miraculous.

The CAB later would explain the crash-landing:

> The first impact was with a tree which was broken 46 feet above the ground. Nearly 250 feet farther the left wing contacted a large tree and [ripped off] of the aircraft. Contact with the ground was made [and] the fuselage [broke] into three main pieces . . . .

Fuel from ruptured wing tanks ignited and sent flames billowing up into the sky. Passengers in the cabin, badly injured, leapt out of the wreckage through holes in the torn-apart fuselage.

The first officer, critically hurt, crawled or fell out of a cockpit emergency exit and collapsed onto the ground. The flight engineer would have no recollection of how he managed to crawl to safety and escape the advancing flames.

Instead of trying to escape, the captain crawled into the burning cabin to assist his passengers. He helped as many as he could before succumbing to toxic gasses from the fire. A newspaper article would explain his devotion to duty:

> [He] sacrificed his life while helping to evacuate passengers after the crash . . . His body was found midway down the smashed and charred cabin of the Constellation.

The captain and four of his passengers died in the fire. Forty-eight others suffered terrible burns and serious body trauma, but they survived. The remaining passenger somehow escaped with only singed eyebrows.

The Investigation: CAB investigators found that no aircraft systems contributed to this accident. The altimeters had functioned properly. ATC handling had been correct. The two aircraft should have passed each other with 1,000 feet of vertical separation. Yet, the three surviving pilots (two from the Boeing, one from the Constellation) each insisted the two aircraft had been at the same altitude.

The pilots had fallen prey to a sinister and rare optical illusion. The two aircraft had – in fact – 1,000 feet of vertical separation, but the cloud tops *sloped upward* north of Carmel. The CAB noted:

> Aircraft pilots are susceptible to many types of flight conditions which may result in spatial disorientation and optical illusion. These illusions or disorientations result from reliance on the physiological sensing elements of the body, which can give false or conflicting information to the senses.

The pilots in both airliners had not been able to see the ground, for it had been hidden below the cloud layer. The Constellation pilots had spotted the Boeing jet skimming along just over the tops of *higher* clouds. Without a visible horizon they had no way to know the cloud tops were not level. The CAB would explain the cruel visual phenomenon:

> With higher clouds behind [the Boeing], the [Constellation] first officer would receive an impression of an aircraft on or very near the apparent horizon. . . . He had no visual aid to assist him in determining the true horizon, and the build-up of clouds to the north would present a *false horizon [emphasis added]* on which to base his analysis of [vertical] separation.

Visually sensing that both planes were flying at the same altitude, the Constellation pilots had climbed to avoid a collision. Tragically they zoomed up 1,000 feet and collided with the other plane.

Probable Cause: The CAB adopted the Aircraft Accident Report, File No. 1-0033, on 13 December 1966 slightly over a year after the accident. The board found that both IFR airliners, in radio and radar contact with New York Center, had been vertically separated by 1,000 feet. Yet, sloping cloud tops visually tricked the pilots into believing they were at the same altitude.

The supreme irony was that the vigilance of both crews prompted the collision. If the pilots had not seen each other, all would have been well. They had proper vertical separation. If no evasive action had been taken the accident never would have happened, and the two aircraft would have passed each other with 1,000 feet of vertical separation. Consequently, the CAB found the *Probable Cause* of this unique accident to be:

> Misjudgement of altitude separation by the crew of [the Constellation] because of an optical illusion created by the up-slope effect of cloud tops, resulting in an evasive maneuver by the [Constellation] crew and a reactionary evasive maneuver by the [Boeing] crew.

Broken Arrow
Mid-Air Collision near Palomares, Spain
17 January 1966

Boeing B-52 Stratofortress, Serial No. 58-256
U.S. Air Force "Operation Chrome Dome" mission
7 aboard, 3 killed

Boeing KC-135 Stratotanker, Serial No. 61-273
U.S. Air Force "Operation Chrome Dome" mission
4 aboard, 4 killed

Synopsis: While attempting in-flight refueling a Boeing B-52 Stratofortress – carrying four nuclear bombs – rammed into a Boeing KC-135 Stratotanker. Both aircraft disintegrated in the air. Three nuclear bombs fell onto land, and another fell into the sea. Four crewmen parachuted to safety; the other seven died in the mid-air collision.

The Boeing B-52 Stratofortress Flight: A U.S. Air Force crew climbed into a Boeing B-52 Stratofortress at Seymour Johnson AFB in North Carolina. They took off, flew east over the Atlantic Ocean, and followed the Operation Chrome Dome route over the Mediterranean Sea. After skirting the northern coast of Africa and reaching the Middle East the bomber turned to the west and headed for home. In the cavernous bomb-bay the B-52 carried four thermonuclear B-28 hydrogen bombs:

Operation Chrome Dome dates back to 1957 after the Soviet Union launched its Sputnik satellite. Shortly thereafter the U.S. Strategic Air Command (SAC) began keeping fully armed B-52 bombers circling the Earth 24 hours per day. The nuclear bombers were poised to strike targets throughout the Soviet empire in response to any Soviet attack upon the United States. For almost 40 years this B-52 airborne deterrent, in conjunction with submarine-launched and land-based ICBMs, provided an effective shield against Soviet aggression.

Seven men crewed the giant B-52: instructor pilot, copilot, radar navigator, navigator, electronic warfare officer, tail gunner, and staff

Boeing B-52 Stratofortress (official U.S. Air Force photo)

pilot. The instructor pilot, a captain, and his crew were assigned to the 68th Bombardment Wing at Seymour Johnson AFB. The B-52 (call-sign, Tea 16) had already been refueled on the eastbound leg. Another in-flight refueling would be required during the long trip back to the United States.

The staff pilot, a major, could relieve the other two pilots during routine point-to-point portions of the 23 hour flight. However, he was not qualified to conduct in-flight refueling operations.

The Boeing KC-135 Stratotanker Flight: After a briefing at 0725 Hours a Boeing KC-135 Stratotanker took off from Moron air base in Spain for an anticipated 1 hour and 50 minute refueling mission. The tanker crew planned to rendezvous with the B-52 at 30,000 feet in the Saddle Rock Air Refueling Area over the Mediterranean Sea, south of the Spanish coast.

The tanker (call-sign, Troubadour 14) carried a crew of four: pilot, copilot, navigator, and refueling boom operator. All were assigned to the 340th Bombardment Wing at Bergstrom AFB in Texas. They were TDY to the Spanish Tanker Task Force at Moron.

Boeing KC-135 Stratotanker refueling a General Dynamics F-16 Fighting Falcon (photo courtesy of U.S. DOD Visual Information Center)

Trouble Ahead: The tanker flew west on a 256 degree track as the B-52 bomber closed in. There was no turbulence, and the early morning weather was clear; however:

> The pilot of the [B-52] apparently thought adequate separation . . . was being maintained when he observed the altimeter reading 225 [feet] below the refueling altitude and the vertical [speed] indicator [showing] a 300 [foot] per minute rate of descent.

As the KC-135 flew straight and level, the B-52 angled in toward the refueling position under the rear of the tanker. The tanker boom operator waited for the B-52 to stabilize in the "contact" position. Then he would "fly" the boom into the refueling receptacle on top of the B-52 fuselage, above and behind the cockpit. Unfortunately the B-52 was angling in (1) much too fast and (2) much too high.

The Collision: The three pilots in the B-52 cockpit and the boom operator in the tanker watched the out-of-control rendezvous. None of them called for a break-away:

> The B-52 aircraft continued to close, and a collision resulted. The B-52 pitched down and left, followed by a large explosion.

The B-52 had struck the refueling boom and had rammed it back into the tanker. The KC-135 exploded into a gigantic aerial fireball as over 300,000 pounds of jet fuel erupted into flame. Both aircraft broke apart in the air:

> The accident developed with such speed that the KC-135 crew was not able to affect egress from their aircraft.

The four tanker crewmen fell earthward in the flaming wreckage. Remains of the burning aircraft tumbled to the ground near the rural farming and fishing community of Palomares, Spain. Bodies of the crewmen would be found in the wreckage.

Some airmen in the crippled B-52 had better luck. Two of the pilots and the radar navigator ejected. The third pilot manually bailed out through the radar navigator's open hatch. The other navigator ejected only moments before striking the earth, and:

> [He] was killed by impacting the ground about 25 yards from the cockpit section [of the B-52]. The gunner and [electronic warfare officer] did not eject and were fatally injured in the crash.

The three unscathed pilots parachuted down into the sea, and Spanish fishermen soon plucked them out of the water. The radar navigator, horribly burned, parachuted down onto dry land. The fate of the four nuclear bombs is explained in a technical report:

> [The bomber had] disintegrated, and down went the B-52's four hydrogen bombs, creating the first U.S. nuclear weapons crisis near a populated area. Each 1.5 megaton bomb packed 100 times more explosive power than the bomb dropped on Hiroshima in World War II.

Each bomb had been equipped with a parachute. One chute deployed and allowed its bomb to float down into a dry riverbed. A second chute also deployed, allowing another bomb to drift down into the Mediterranean Sea.

The other two chutes failed to deploy. The remaining two bombs fell free from the mangled bomber and screamed down near the village of Palomares. The non-nuclear high explosives in both bombs detonated upon impact. The blasts scattered radioactive plutonium over 650 acres of Spanish farmland. Independent technical analysts later would explain the danger:

> Plutonium is extremely toxic and lingers in the environment for ages. At takes 24,000 years for half of a given amount to decay, [and] it can cause cancer and other serious health problems.

Nuclear specialists from the U.S. Armed Forces and the Spanish Civil Guards rushed to Palomares. They recovered the bomb that landed in the riverbed and found it to be only slightly damaged. They cordoned-off the area where high explosives in two additional bombs had detonated. A massive decontamination effort began. Meanwhile, a flotilla of 33 U.S. Navy ships gathered to search for the fourth nuclear bomb that had fallen into the sea.

The Investigation: The U.S. Air Force dug into the aeronautical aspects of this accident. Neither mechanical anomalies nor weather were deemed to be factors. The board noted that the aerial explosion prevented anyone in the tanker from manually bailing out. Board members learned the two crewmen who failed to get out of the B-52 had not received training in the egress procedure.

It became clear the B-52 had closed on the tanker far too rapidly. The bomber pilot had been unable to stabilize in the contact position. Investigators found that existing SAC in-flight refueling procedures were adequate, but they had not been followed:

> If [SAC] directives had been followed by all personnel involved, this accident would not have occurred.

Investigators interviewed the three surviving pilots B-52 pilots. The instructor pilot (the aircraft commander) should have stayed in his command position, the left seat, for the delicate and demanding in-flight refueling operation. However, the staff pilot was a major, and he outranked the instructor pilot, a captain. The staff pilot had *persuaded* the instructor to vacate the left seat. The staff pilot then occupied the seat and attempted to rendezvous with the tanker. The danger had become obvious moments before the collision. Yet, neither the instructor nor the copilot had spoken up or taken any action to prevent the planes from colliding.

Probable Cause: Only two weeks after the accident the Air Force completed a comprehensive Accident/Incident Report on 31 January 1966. The board found the *Primary Cause* of the tragedy to be:

> Supervisory error, in that the instructor pilot . . . failed to take necessary action to prevent under-run and subsequent collision with the tanker.

The Air Force outlined four *Contributing Cause* factors:

> [First], pilot error, in that the [staff] pilot of the B-52, while attempting to move into the refueling position, failed to maintain sufficient separation between the two aircraft to avoid [the] collision and subsequent break-up and crash of both aircraft.

> [Second], crew factor, in that a break-away was not called [for] by any crew member, bomber or tanker, when it became apparent that a hazardous under-running condition existed.

> [Third], violation of [SAC directives] in that the instructor pilot of the B-52 permitted a staff-qualified pilot to occupy a primary crew position and allowed him to attempt air refueling on a Chrome Dome sortie.

> [Fourth], supervisory factor, in that sufficient emphasis was not placed on the importance of compliance with the restrictions of [SAC directives] as [they] pertain to primary crew members occupying the [front] seats during critical phases of flight.

Postscript: Eighty days after the accident the submersible *Alvin* finally located and recovered the bomb that had fallen into the sea. The high explosives had not detonated, and no lethal radiation had been released. Still, this had been a close call. In the technical white-paper, *The Effects of Nuclear Weapons*, the U.S. Atomic Energy Commission had warned the U.S. Armed Forces in 1962:

> Nuclear weapons are designed with great care to explode only when deliberately armed and fired. Nevertheless, there is always a possibility that, as a result of accidental circumstances, a nuclear explosion will take place *inadvertently [emphasis added]*.

The recovered B-28 hydrogen bomb that descended by parachute to the ground near Palomares, Spain; note the bent and dented exterior casing (official U.S. Air Force photo)

Around the village of Palomares a virtual army of almost 2,000 American technicians and Spanish laborers toiled to decontaminate the farmland. They stripped over 1,750 tons of radioactive topsoil from the ground and packed the contaminated dirt into thousands of steel drums. They shipped these containers to the Savannah River Plant, a nuclear fuel facility in South Carolina. The drums were entombed in reinforced concrete and buried in deep clay pits.

Years later the United States government would settle about 500 claims from Spanish citizens whose health allegedly had been damaged by radiation. The cost of decontamination and compensation would total over 120 million dollars.

"Broken Arrow" is the U.S. Armed Forces code for an accident or loss involving military nuclear weapons. A privately funded study of U.S. nuclear weapons accidents reveals:

The first *officially acknowledged [emphasis added]* accident occurred in February 1950 when an American B-36 bomber jettisoned a nuclear bomb into the Pacific Ocean.

By the time SAC stopped the nuclear B-52 flights in the early 1990s, over 35 *known* Broken Arrow accidents had occurred. Yet, one should remain mindful of U.S. Department of Defense nuclear weapons policy, as explained in an independent report:

> The U.S. military's policy is to *neither confirm nor deny* the presence of nuclear weapons in most accidents.

Notwithstanding policy, facts have a way of seeping out. Today it is known that *lost nuclear bombs* lie concealed in at least four locations in or adjacent to the continental United States:

> First, in the Atlantic Ocean off the Delaware coast. On 28 July 1957 after failure of two engines, an Air Force C-124 transport jettisoned two nuclear bombs (carried as cargo) into the sea. The bombs were never recovered.

> Second, in Wassaw Sound off the Georgia coast near the mouth of the Savannah River. On 5 February 1958 a crippled Air Force B-47 bomber jettisoned its nuclear bomb into the sea. It was never recovered.

> Third, in Puget Sound off Whidbey Island, Washington. On 25 September 1959 a Navy P-5M patrol plane carrying a nuclear bomb crashed into the sound. The bomb was never recovered.

> Fourth, in swamps along the Neuse River near Goldsboro, North Carolina. On 24 January 1961 an Air Force B-52 bomber broke apart in flight, and the high explosives in two bombs detonated. The enriched uranium core of one bomb was never recovered.

The Flight of the Valkyrie
Mid-Air Collision near Barstow, California
8 June 1966

North American XB-70 Valkyrie, Serial No. 62-0207
U.S. Air Force and NASA research flight and photo-shoot
2 aboard, 1 killed

Lockheed F-104 Starfighter, Serial No. 813
NASA photo-shoot
1 aboard, 1 killed

Synopsis: Five military aircraft flew in close formation for a photo-shoot. Two of them, a North American XB-70 Valkyrie and a Lockheed F-104 Starfighter, collided in mid-air. The Valkyrie pilot ejected and survived, but the accident killed the Valkyrie copilot and the F-104 pilot.

The North American XB-70 Valkyrie Flight: A sleek futuristic North American XB-70 Valkyrie bomber prototype took off from Edwards AFB, California, at 0715 Hours. This was a research flight conducted jointly by the U.S. Air Force and the National Aeronautics & Space Administration (NASA). The Valkyrie, the second to be built, climbed to 31,000 feet. There the pilots conducted experiments in sonic boom noise reduction technology.

The Valkyrie engine manufacturer, General Electric, had arranged a photo-shoot after completion of the experiments. Four military jets, all powered by General Electric engines, would rendezvous with the Valkyrie and fly in a "V" formation. Cameramen in a trailing LearJet would take photographs to be used on the cover of a General Electric shareholder brochure:

The Valkyrie, then the heaviest aircraft ever to fly, had been conceived in the 1950s as a high altitude nuclear bomber with Mach 3 cruise speed. It would be able to (1) outrun enemy fighters and (2) fly too high to be hit by enemy surface-to-air missiles. Manufactured from titanium and honeycomb stainless steel panels, the Valkyrie featured a forward canard and delta wing. Hinged outer wing panels drooped 65 degrees in super-

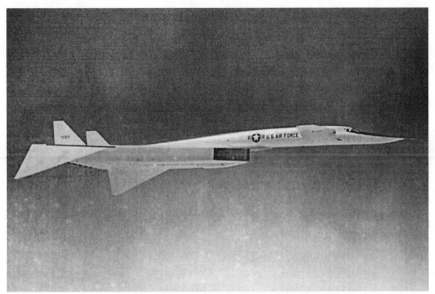

North American XB-70 Valkyrie (official U.S. Air Force photo)

sonic flight to create "compression lift" under the fuselage. The first prototype had flown in 1964, powered by six General Electric J-93 afterburning turbojets producing 180,000 pounds of thrust. By the time of the fatal flight the Valkyrie had exceeded Mach 3 speed and had met all primary flight test objectives.

The Valkyrie prototype was on bailment to North American for flight testing and ultra-high-speed research. The manufacturer's chief test pilot was the pilot-in-command. His copilot was an Air Force major. After completing their "test cards" the pilots flew straight and level at subsonic speed to allow the four other planes to rendezvous for the photo-shoot.

The Lockheed F-104 Starfighter Flight: The NASA test pilot who would fly the Lockheed F-104 Starfighter (owned by NASA) had not been present at the initial formation briefing. He informally had been given the details later. He took off from Edwards and headed for the fateful photo-shoot:

Lockheed's famous Skunk Works had designed the F-104 as a high performance fighter-interceptor. Because of its speed and appearance it was often called the "missile with a man in it."

Powered by an afterburning General Electric J-79 turbojet engine, the plane could exceed Mach 2 in level flight. Its most visible aspects were its razor-thin control surfaces and short wings with negative dihedral. It could be armed with wing-mounted Sidewinder missiles and a six-barrel 20mm Vulcan cannon. More than 1,700 F-104s were built for the armed forces of the United States and allied nations.

Trouble Ahead: In a crystal clear sky the four other aircraft joined the Valkyrie in a tight "V" formation, described below:

Formation leader: . . . . No. American XB-70 Valkyrie
No. 1 position, on the right: . . Lockheed F-104 Starfighter
No. 2 position, on the right: . . Northrop F-5 Freedom Fighter
No. 1 position, on the left: . . . McDonnell F-4 Phantom
No. 2 position, on the left: . . . Northrop T-38 Talon

Photographers in a LearJet took photographs and movie footage for 45 minutes, often asking the formation to "close up." Tragically the NASA test pilot, flying the F-104 on the right wing of the Valkyrie, got too close at 0926 Hours. The high T-tail of his F-104 scraped the drooped right wingtip of the Valkyrie. The F-104 pitched up, and its left wing snagged the bottom of the drooped wingtip.

The radio and ICS transcript, below, has been *reconstructed* from multiple technical reports:

FltDir: . . . . Test Flight Director (flying in the T-38)
Tower: . . . . . . Mission Control at Edwards AFB Tower
ICS-XB-70: . . ICS transmission, Valkyrie pilot

[the small F-104 rolls inverted over the ultra-heavy Valkyrie empennage and tears off the Valkyrie's two vertical stabilizers]

[the crushed F-104 bursts into flame and falls out of control; the big Valkyrie momentarily flies straight and level]

FltDir: Mid-air! Mid-air! Two-oh-seven, you've been hit!

[the Valkyrie pilots are not aware – yet – that they have been involved in a collision]

ICS-XB-70: I wonder who got hit? *[speaking to his copilot]*

FltDir: OK, you're doing fine – he [hit] the verticals *[meaning, the two vertical stabilizers]*, but you're still doing fine.

[after 16 seconds the Valkyrie begins to yaw and roll to the right]

FltDir: Hold on! You may go into a spin!

[the Valkyrie completes a 360 degree roll to the right]

FltDir: Bail out! Bail out!

FltDir: I see no chute! Bail out!

[the Valkyrie pilot ejects; the copilot remains in the aircraft]

FltDir: There's a chute. There's one chute. All aircraft, stay clear. One chute's going down.

[the Valkyrie spins down to the ground and explodes]

Tower: This is Mission Control, give us the crash site.

The Collision: After snagging the Valkyrie's right wing the F-104 had rolled inverted over the rear fuselage of the huge bomber. As it swept over the Valkyrie empennage, the F-104 had sheared off the Valkyrie's two vertical stabilizers. This impact crushed the F-104 cockpit, instantly killing the pilot and rupturing his fuel tanks:

The F-104 burst into flames and fell behind the XB-70.

In the Valkyrie cockpit, over 150 feet forward from the point of impact on their empennage, the pilots had not felt the impact. The big Valkyrie momentarily continued to fly straight and level. Hearing only the "you've been hit" part of the radio message, the Valkyrie pilot had turned to his copilot and asked: "I wonder who got hit?"

Despite a brief period of apparent normalcy the Valkyrie had been mortally wounded. After 16 seconds it began to yaw right, then roll right. The pilots, suddenly aware they were in deep trouble, applied full left rudder input (their rudders had been sheared off), left aileron, and full power on the No. 6 engine. Their efforts had no effect:

Yawing right, [the Valkyrie] made a 360 degree roll to the right, pitched up almost vertically, and entered a flat spin.

The Valkyrie pilot realized his aircraft could not be saved. He fired his ballistic thruster, and it rammed his seat rearward into his ejection pod. The protective clamshell doors slammed shut, the ejection sequence worked perfectly, and the pilot parachuted down to the ground. Unfortunately he suffered serious injuries upon ground impact because the airbag cushion under his ejection pod failed to inflate. A technical report would explain:

> [The pilot] took a 44 G impact – lessened to 33 G's as his seat broke free of its mountings. Amazingly, although banged, battered, and bruised, he suffered no broken bones [and] returned to flight status just three months later.

Meanwhile, the Valkyrie copilot had waited too long to try to eject. Unable to escape from his crippled aircraft, he perished when it spun into the ground and exploded near Barstow, California.

The Investigation: The Air Force convened an Accident Investigation Board. Many people already were asking why the Air Force and NASA had authorized a photo-shoot for the benefit of General Electric, a commercial entity. One critic lamented:

> We lost a four-billion-dollar plane and two [sic] veteran pilots on a mission whose purpose was to obtain a photo for a corporate brochure.

Investigators reconstructed the accident. The Valkyrie's hinged outer wing panels – 20 feet long – had been drooped down into the high speed position. The F-104 pilot had radioed that he experienced turbulence from the resulting airflow vortex. Investigators explained the mechanics of the accident:

> The F-104 horizontal tail contacted the drooped wingtip of the [Valkyrie], causing the F-104 to pitch up, [then] the [F-104] left wingtip struck the right wing tip of the [Valkyrie]. The F-104 rolled into a left bank, going inverted over the [Valkyrie], and it struck both vertical stabilizers . . . .

With regard to the deceased F-104 pilot the board discovered:

> Injuries, helmet damage, and the condition of the ejection seat indicate the pilot was [killed instantly] by the mid-air collision.

Lockheed F-104 Starfighter (photo courtesy of U.S. NASA)

Investigators explored reasons why the Valkyrie copilot had not ejected. They found he had waited too long before firing his ballistic thruster. Under normal conditions the thruster would have slammed his seat back into his ejection pod. Unfortunately the ever-increasing G forces, due to the flat spin, had pinned the seat in the forward position. The copilot tried to eject, but he tried too late. He could not escape from his doomed aircraft:

> The seat/man mass was held forward by G forces in excess of the 3.4 G's for which the thruster was designed.

Probable Cause: The Accident Investigation Board issued its report within weeks after the tragedy. The report pointed out the inadequacy of the Valkyrie seat ballistic thruster in view of "all [G] forces that may be encountered during emergencies." The board also stressed that incomplete data had existed in regard to the powerful airflow vortex generated by the drooped Valkyrie wingtips. The investigators also acknowledged:

> Minimum separation distance between aircraft was not specified during the [formation flight] briefing.

The board discovered that some aircraft in the formation did not have inter-plane communications. Even worse, pilots of two aircraft had not been present at the formation flight and photo-shoot briefing. Nevertheless, the board reverted to elementary facts when stipulating the ***Primary Cause*** of this accident:

> The F-104 aircraft was allowed to attain a hazardous position in proximity to the XB-70.

The board spelled out several ***Contributing Factors*** that included "momentary distraction" of the F-104 pilot. The board dryly noted:

> The aerodynamic properties of the [Valkyrie] wingtip area were sufficiently strong to exceed the [flight] control authority obtainable in the F-104.

Postscript: The Valkyrie never went into production. Its unique invincibility – flying too high and too fast to be shot down – had been militarily sound when the project was conceived in the late 1950s. By the mid-1960s, however, unforseen exponential advances in surface-to-air missile technology made the Valkyrie vulnerable.

The remaining prototype soldiered on alone until 1969, after which it was placed in the National Museum of the U.S. Air Force at Dayton, Ohio. Yet, the technical legacy of "The Great White Bird" survived. Research data obtained during flights of the two prototypes is still used four decades later in the Twenty-First Century. The late great Valkyrie has helped modern-day aeronautical engineers span the gap between supersonic and hypersonic flight.

Helicopter Valley
Mid-Air Collision in the Song Ngan Valley, Vietnam
15 July 1966

<u>Boeing CH-46 Sea Knight, Bureau No. 151930</u> (Rose-Ann 1-17)
U.S. Marine Corps combat assault
16 aboard, 1 killed

<u>Boeing CH-46 Sea Knight, Bureau No. 151936</u> (Rose-Ann 1-18)
U.S. Marine Corps combat assault
16 aboard, 1 killed

<u>Synopsis</u>: Twenty Boeing CH-46 Sea Knight helicopters roared down into a jungle valley on a combat assault. Trying to land downwind and downslope, two collided and crashed. Of the 32 men aboard, only two sustained fatal injuries.

The Boeing CH-46 Sea Knight Flights: Four at a time, 20 Boeing CH-46 Sea Knight helicopters began taking off at 0750 Hours from a staging area at Dong Ha, South Vietnam. The U.S. Marine Corps helicopters, each carrying 4 crewmen and 12 infantrymen, were leading a combat assault into the mountains and jungle 11 miles to the west. A brief historical review is warranted:

North Vietnam had sent its 324-B Division and 341st Division, each reinforced with antiaircraft (AAA) batteries, artillery, and motor transport regiments, across the DMZ into South Vietnam. With over 24,000 troops they staged in the mountains of Quang Tri Province in preparation for a drive southward to capture the city of Hue, the ancient Vietnamese Imperial Capital.

Outnumbered over seven-to-one, U.S. Marines did not have the manpower to defend Hue, and their artillery was no match for the enemy's big siege guns. Out of all other options, the Marines elected to do the unthinkable. They would attack.

Spawning blinding clouds of red laterite dust at Dong Ha, the helicopters lumbered skyward and flew west. Behind the 20 CH-46s (radio call-signs, Rose-Ann 1-1 through Rose-Ann 1-20) came more Marine helicopters, Sikorsky H-34s, Sikorsky H-37s, and nimble Bell UH-1 Hueys. Meanwhile, Douglas A-4 Skyhawks and McDonnell F-4 Phantoms pounded the landing zone in the Song Ngan Valley.

Boeing CH-46 Sea Knight helicopters (official U.S. Marine Corps photo)

The jets dropped napalm and 500 pound bombs into Landing Zone (LZ) Crow and the surrounding jungle:

> *Boeing delivered the first CH-46 to the Marine Corps in 1964. Two General Electric T-58 jet engines powered the state-of-the-art helicopter. It featured a big cabin for cargo, an automated rear ramp for loading troops and small vehicles, protective armor, automatic blade folding, and redundant stabilization systems. The CH-46 carried a crew of four: pilot, copilot, crew chief, and gunner. In combat the crew chief doubled as a second gunner and manned one of the two .50 caliber machineguns.*

Trouble Ahead: The first division (a division is *four helicopters*) thundered down and landed in LZ Crow. Marine infantrymen ran out of the helicopters and began setting up a defensive perimeter:

> Small arms fire was received in [the landing] zone during the troop buildup.

Enemy rifle fire whipped across the LZ, but two more divisions landed and discharged their troops. As the fourth division swooped down into the LZ, Rose-Ann 1-14 overshot the landing area. The pilots tried to abort the landing. Overloaded, they ran out of engine power and slammed into a huge tree. Pieces of shattered rotor blades flew in all directions. The structural imbalance tore the aft pylon and aft rotor system off the fuselage, destroying the helicopter:

> [The] aircraft overshot the landing point and hit a tree, causing strike damage to the aircraft.

Leading the fifth division, Rose-Ann 1-17 overshot the LZ and sank toward the ground. Next, Rose-Ann 1-18 also overshot the landing area. Both helicopters were (1) overloaded, (2) out of ground effect, and (3) out of translational lift. Despite full power on the engines the helicopters mushed toward the ground.

The ICS transcript, below, was recorded aboard Rose-Ann 1-16. This helicopter had flown immediately in front of Rose-Ann 1-17 and Rose-Ann 1-18. The transcript begins as the helicopter rolls onto final approach. The pilot-in-command is flying the CH-46:

> Pilot: . . . . Pilot, Rose-Ann 1-16
> Copilot: . . . . . Copilot, Rose-Ann 1-16
> CrChief: . . . . . Crew Chief, Rose-Ann 1-16

[a preceding helicopter radios that heavy automatic weapons fire is coming from a ridgeline north of the LZ]

Pilot: OK, you hear that?

Copilot: Yeah, [he is] taking fire from that ridge up there.

Pilot: Right, OK, stay with me *[meaning, stay on the flight controls with me]* 'till we're down [on the ground in the LZ].

[Rose-Ann 1-16 lands in the LZ; the copilot activates the ramp, turns, and looks back into the cabin to monitor the debarkation]

Copilot: Get 'em out.

[the 12 infantrymen run out of the helicopter into the LZ]

CrChief: All clear.

Copilot: Ramp comin' up.

[while Rose-Ann 1-16 is still on the ground in the LZ, 2 helicopters (Rose-Ann 1-17 and Rose-Ann 1-18) overshoot the landing area; their rotors intermesh and shatter; the helicopters rip apart and fall to the ground about 80 feet in front of Rose-Ann 1-16]

*[the Rose-Ann 1-16 copilot, still looking back into his cabin, does not see the 2 helicopters **_collide_** and crash]*

Pilot: Look at that! Look at that!

Copilot: Look at what?

Pilot: There! There! There!

[the copilot looks forward and sees the smoking wreckage]

Copilot: Aaahhh, [expletive]!

[both pilots know the LZ must be cleared to allow the remaining helicopters to land and debark their troops]

Pilot: Ready?

Copilot: Yeah, let's go.

[Rose-Ann 1-16 takes off and heads up out of the valley]

The Collision: Beyond their intended landing area, overloaded, out of engine power, the pilots in Rose-Ann 1-18 had spotted Marine infantrymen below their sinking CH-46. They threw their cyclic control to the left to avoid crushing the Marine infantrymen below them. Unfortunately this last-second swerve to the left caused them to ram into Rose-Ann 1-17:

> The [pilot] was unable to land at [an] alternate landing point due to friendly troops within the position. This necessitated selection of a third landing site, which proved to provide insufficient clearance between aircraft. ... [The two helicopters] intermeshed aft rotor blades and suffered strike damage.

The whirling rotors ripped into each other. Tons of centrifugal force imbalance tore both helicopters apart. Meanwhile, west of the LZ on final approach, the pilot in Rose-Ann 1-19 had a bird's-eye view of the accident. His subsequent written report would explain:

> Both aircraft rotor systems contacted each other, creating a large amount of dust and breaking both fuselages in several places . . . Both my copilot and myself were not sure if they had hit each other, or if they had landed on mines, or received mortar fire.

The last CH-46 helicopters, Rose-Ann 1-19 and Rose-Ann 1-20, landed and discharged their troops. Then they waited in the LZ, under fire, to evacuate Marines injured in the crashes. Two had been killed by swirling rotor blades. Five more had critical injuries. With the casualties finally aboard the two CH-46s clawed their way skyward:

> [The two helicopters] received small arms and automatic weapons fire going into and out of the [landing] zone.

The Investigation: Why had many helicopters overshot LZ? Why did one crash into a tree? Why did two others collide? Despite the combat environment the Marine Corps considered the accidents to be *operational* losses. True, the helicopters had been under fire, but that was not deemed a factor. The Marine Corps convened an Accident Investigation Board to find out what had gone wrong.

The pre-assault briefing had called for flying west toward the valley at 3,000 feet, high enough to remain out of the effective range of small arms fire. Then the flight would make a 180 degree turn to the right and approach the LZ from the west. There were two reasons for this plan. First, Marine Intelligence had discovered 12.7mm enemy AAA positions in the valley east of LZ Crow. Helicopters approaching from the west would not overfly the AAA positions:

> [The] approach heading was necessitated by numerous automatic weapons positions located 1200 to 1400 meters east of the LZ.

Also, Marine meteorologists had assumed the morning sun would cause air to warm and rise over the valley. Cooler air offshore, 17 miles to the east, would sweep inland to fill the void. Therefore, the wind was forecast to be whipping through the narrow valley from the east -- a *headwind* for the landing helicopters.

Intelligence had been correct about enemy AAA positions. Two CH-46s that strayed over the enemy guns had been shot down later in the day. An AP photographer had snapped a picture of one of these CH-46s trailing flames and smoke before it crashed into the jungle, killing 13 Marines. The scholarly tome, *Bonnie-Sue: A Marine Corps Helicopter Squadron in Vietnam*, would explain:

> Over the caption, "Hit by Enemy Fire, a U.S. Helicopter Goes Down in Flames," the picture of the doomed CH-46 headed an article in the 1 August 1966 issue of *Newsweek Magazine*.

However, Marine meteorologists had been wrong about the wind direction. Mother Nature had allowed a strong wind to whistle down the valley from the west. The helicopters had landed *downwind*.

On the morning of the assault, Marine Corps infantrymen sit on the grass and wait to board their helicopter (photo by the author, Marion F. Sturkey)

Landing downwind in a heavily loaded helicopter is bad enough, but it got worse. Investigators discovered the terrain under the final approach path sloped steeply *downhill* toward the LZ. Flying over sloping terrain distorts the visual "picture" of the angle of descent. A proper angle of descent – relative to level ground – over terrain that slopes downhill requires a lower than normal height-above-ground. Such a low height above the ground makes it *appear* the angle of descent is too shallow. In the instant case, pilots had flown at what *appeared* to be a proper angle of descent. This resulted in an approach that was far too steep. The board chairman noted:

> The terrain in this area sloped downhill . . . . This downhill gradient gave the pilots the *illusion [emphasis added]* of being at the proper altitude when in fact they were high. *[and also]* An approach over downhill sloping terrain creates the *optical illusion [emphasis added]* of aircraft to [level] ground separation less than the actual aircraft to [level] ground separation.

All pilots know two sure-fire ways to overshoot their intended landing point – too fast an approach, or too steep an approach. In this instance, due to the downwind landing and sloping terrain, the pilots had been suckered into both conditions.

On final approach all ingredients for disaster had surfaced at the same time. The helicopters were (1) landing downwind, (2) landing downslope, (3) approaching too steeply, (4) approaching too fast, (5) out of ground effect, (6) out of translational lift, (7) overloaded, and (8) trying to land in an LZ surrounded by tall trees.

In this hover-out-of-ground-effect (HOGE) condition full power on the engines could not arrest the descents. Too high and too fast, the pilots overshot the LZ. Overloaded, with the sound of decaying rotor RPM in their ears, they had flown into disaster.

Probable Cause: The Accident Investigation Board submitted an undated report to Marine Air Group 16 (MAG-16). The board pointed out that all helicopters had landed downwind and downslope. The board listed several *Conclusions*, which included:

[First], downwind landings in heavy gross weight conditions demand a more precise approach . . . and a slower airspeed. Failure to affect these requirements will result in an overshooting of the intended point of landing.

[Second], an approach over downhill sloping terrain must be recognized early in the approach so that a proper [angle] of descent will be established to prevent being high over the intended point of landing.

Armed with 20/20 hindsight the board pointed out that pilots who overshot the LZ *should* have aborted their landing attempts. Under normal conditions they *could* have waved-off and circled around for another approach to avoid the "high and fast" dilemma. Yet, under the adverse circumstances that had existed, the board admitted that a wave-off had been impossible during the latter stages of the approach.

Postscript: On the first day of the battle the Marines lost five CH-46 helicopters in the Song Ngan Valley (one crashed into a tree, two collided, and enemy AAA fire downed two more). Bitter fighting and carnage raged night and day. Within weeks the wrecks of more helicopters littered the valley floor. This led the international news media to coin a new name for the jungle battleground: "Helicopter Valley." The U.S. Marines and the North Vietnamese Army would remain locked in a bloody slugfest in the mountains and jungle around Helicopter Valley for over two years.

Note: The author has first-hand knowledge of the collision in Helicopter Valley. He was the copilot in Rose-Ann 1-16. Rose-Ann 1-17 and Rose-Ann 1-18 crashed 80 feet in front of him.

The author stands by his CH-46 Sea Knight helicopter at Marble Mountain, South Vietnam, the day before the assault into Helicopter Valley (photo courtesy of Marion F. Sturkey)

Traffic – Twelve O'clock – One Mile
Mid-Air Collision near Urbano, Ohio
9 March 1967

Douglas DC-9, Registration N1063T
Trans World Airlines Flight 553
25 aboard, 25 killed

Beechcraft B-55 Baron, Registration N6127V
Tann Company business flight
1 aboard, 1 killed

Synopsis: Under ATC control an IFR Douglas DC-9 descended to land. In clear weather the airliner overtook a VFR Beechcraft B-55 Baron and plowed into it. The aerial collision destroyed both aircraft and killed the 26 people on board.

The Douglas DC-9 Flight: Trans World Airlines (TWA) Flight 553 climbed skyward from Pittsburgh, Pennsylvania, for the short trip to Dayton, Ohio. The popular Douglas DC-9 twin-jet airliner carried two pilots, two flight attendants, and a slim load of only 21 passengers. Indianapolis Center controlled the IFR flight as the pilots cruised southwest at flight level 200:

Douglas developed the wildly successful DC-9 as a twin engine and medium range airliner. The all-new design featured two turbofan engines mounted at the rear of the fuselage. The aircraft had highly efficient wings and a T-tail atop the vertical stabilizer. It could carry 80 to 135 passengers depending upon the model and seating configuration. The DC-9 entered service with Delta in 1965, and since then over 2,400 have been built.

The captain, age 39, had been a military pilot before he began flying for TWA in 1956. He held an ATP certificate and had 9,832 hours of flight time in his logbook. His brand-new first officer, age 29, also had been a military pilot. He had earned a Commercial Pilot certificate and had recently completed his TWA training. This was his second flight as a first officer. He had logged 1,560 flying hours, which included 1,156 hours as a Naval Aviator prior to his employment with TWA.

Douglas C-9 Skytrain, a military variant of the civil Douglas DC-9 airliner (official U.S. Navy photo)

The Beechcraft B-55 Baron Flight: A twin Beechcraft B-55 Baron took off on a "special VFR" clearance from Detroit, Michigan, at 1101 Hours for a trip to Springfield, Ohio. The pilot did not file a flight plan, and none was required. Two minutes after takeoff he radioed Detroit Tower and reported "on top" of a layer of haze and smoke. That proved to be his last radio contact with ATC.

The pilot, age 54, was the sole occupant of the Baron, and he had logged 4,074 hours of flight time. Although on a business flight for the Tann Company, he held a Private Pilot certificate. He had flown the Detroit-to-Springfield route many times in the past:

Beechcraft delivered the first new Baron in 1961. It usually can carry a pilot plus five passengers, but different models have varying seating configurations. The sophisticated twin engine plane cruises at about 200 knots. Engines range from two 260 horsepower Continentals up to two 380 horsepower Lycomings. The popular Baron has always been near the pinnacle of the light twin aircraft market. The logical "step up" from a Baron is a turbine powered aircraft.

Trouble Ahead: As the DC-9 cruised along airway V-12, Center handed-off the flight to Dayton Approach Control. The Dayton con-

Beechcraft B-55 Baron (photo courtesy of Adrian R. Pingstone)

troller vectored the pilots toward the ILS final approach course and cleared the jetliner to descend to 3,000 feet. Meanwhile, northeast of Dayton, the Baron cruised toward Springfield at 4,500 feet.

The portion of the ATC and CVR transcript, below, includes radio traffic to and from the DC-9. Several radio messages have been *reconstructed* based upon NTSB report content, so some verbiage may not be verbatim. The transcript starts as Center hands-off the flight to Approach Control:

<u>Capt</u>: . . . . Captain, TWA Flight 553
<u>FO</u>: . . . . First Officer, TWA Flight 553
<u>Center</u>: . . . . . . Indianapolis Center (ARTCC)
<u>AppControl</u>: . . Dayton Approach Control
<u>Radio-</u>: . . . . . . (prefix) Radio transmission, TWA Flight 553

[sound of high speed warning clacker at 1151:52]

[sound of landing gear warning hour at 1151:56]

<u>Center</u>: TWA five-five-three, now contact Dayton Approach Control on frequency . . . . (1152:24)

[the Captain makes radio contact with Approach Control]

<u>AppControl</u>: TWA five-fifty-three, radar contact, descend to and maintain five-thousand, fly [heading] two-four-zero for vector, ILS final approach course, report leaving six-thousand. (1152:36)

Radio-Capt: Descending to five-[thousand], left two-four-zero.

Radio-Capt: Five-fifty-three's out of six-thousand. (1153:16)

AppControl: TWA five-fifty-three, descend to and maintain three-thousand, turn left, heading two-three-zero. (1153:22)

Radio-Capt: TWA is descending to three-thousand, left to two-three-zero. (1153:28)

AppControl: TWA five-fifty-three, roger, and [you have] traffic at twelve-thirty, one mile, southbound, slow moving. (1153:31)

Radio-Capt: Roger. (1153:36)

FO: Ready on the checklist, cap'n. (1153:46)

   *[4 seconds later the DC-9 and the Baron **collide** at 1153:50]*

   [the controller does not know the planes have collided]

AppControl: TWA five-fifty-three, you're clear of traffic. (1154:02)

AppControl: TWA five-fifty-three, turn left, heading two-two-zero.

AppControl: TWA five-fifty-three, do you read Dayton?

AppControl: TWA five-five-three, do you read Dayton Approach?

   [there is no response]

The Collision: It was a delightful day. The sun had been shining in a cloudless sky as the Baron had flown south at 4,500 feet. The DC-9 had been flying southwest while descending toward 3,000 feet:

   The [approach control] radar targets of the two aircraft merged, separated, changed shape on the radar screen, and disappeared.

The faster DC-9 had closed on the slower Baron from its left rear quarter. Without taking evasive action the airline pilots had rammed into the hapless Baron 25 miles from the airport at Dayton. The impact ripped the smaller aircraft apart and destroyed the DC-9.

Awed witnesses on the ground watched as a shower of aircraft rubble rained down from the sky. Wreckage fell to earth along a path over two miles long and one-half mile wide. Most of the DC-9 fuselage screamed down nose-first into a wooded area near Urbano,

Ohio, and exploded. The next day the *Dayton Daily News* newspaper headline would exclaim:

Routine Flights Explode into Disaster

The newspaper article related the aftermath of the DC-9 crash:

> A dense column of smoke, climbing into the clear sky, was the airliner's funeral pyre. First witnesses on the scene were stunned by the incredible sight. They knew – instinctively – there were no survivors. [Part of] a woman's red blouse waved as a macabre death flag from a tree. A nylon stocking dangled from another. The sleeve from a man's white shirt flapped in the wind. The scorched earth was littered with personal belongings, an electric shaver, a toothbrush. Officials started marking the [location of] bodies – rather, the pieces of bodies – with [wooden] stakes.

NTSB investigators later would dryly remark that the wreckage was "extensively fragmented." Pieces of the Baron wings were found tangled in the burned DC-9 rubble:

> This was a non-survivable accident. All [26] persons aboard the two aircraft died of traumatic injuries.

The Investigation: The FDR from the DC-9 showed the airliner had been flying at 323 knots and descending at 3,500 feet per minute on a heading of about 230 degrees. The collision had taken place at 4,525 feet in bright daylight:

> Flight checks . . . disclosed that the [Approach Control] radar, navigation aids, and communications systems . . . were operating in a normal manner.

The Baron had no FDR, so NTSB investigators reconstructed its probable flight path. Based on a route from Findlay VOR to the point of impact, the Baron had been flying a heading of about 195 degrees at around 170 knots. Score marks on wreckage from both aircraft confirmed these assumptions.

For the Baron pilot to have seen the onrushing DC-9 he would have had to look backward over his left shoulder. On the other hand, through his front windshield the DC-9 captain would have had a clear unobstructed view of the Baron:

> The [Baron] was in such a position as to be visible to the DC-9 captain in [his] center windshield.

The DC-9 had been controlled by ATC, but the Baron pilot had not filed any type of flight plan. Yet, both aircraft had been flying in VFR conditions. Visibility had been estimated at seven miles:

> The present-day "see and be seen" concept is based on all flight crews maintaining a lookout for other aircraft when they are operating in VFR flight conditions.

Investigators conclusively determined the DC-9 pilots could have seen the Baron in time to avoid a collision. In addition, only 19 seconds before the aerial impact the controller had warned them: "traffic, twelve-thirty, one mile." Nonetheless, the FDR showed that the pilots took no evasive action. CVR verbiage indicated they never saw the Baron, which had been almost directly in front of them.

Probable Cause: Fifteen months after the accident the NTSB adopted the Aircraft Accident Report, File No. 1-0002, on 19 June 1968. Rules governing aerial traffic gave right-of-way to the Baron pilot, for (1) he had been on the right, and (2) the DC-9 had been overtaking him. The DC-9 captain's view of the Baron had not been obstructed, but somehow he obviously had failed to see it:

> CVR transcription [revealed] the DC-9 crew was devoting some of their attention to speed control, clearance response, maneuvering for the approach, and the pre-landing checklist . . . . This activity could [have directed] both pilots' attention inside the cockpit, reducing the effectiveness of any visual search for possibly conflicting traffic.

FARs mandated that arriving aircraft below 10,000 feet fly at a maximum speed of 250 knots within 30 miles of the airport. Yet, the

DC-9 pilots had been zipping along at 323 knots:

> The excess speed . . . reduced the available time for the crew of either aircraft to see and avoid the other, or for the controller to take appropriate action.

The board considered (1) the Baron pilot's absence of radio contact with ATC and (2) the fact that he did not use his transponder. Neither was required. The board pointed out that use of transponders by all aircraft in terminal areas would create a hazard due to "ring-around" and radarscope clutter. Further, the board stressed that a "tremendous overload" on radio frequencies would result if all aircraft in terminal areas maintained radio contact with ATC. The board found a simple and basic *Probable Cause*:

> The Probable Cause of this accident was the failure of the DC-9 crew to see and avoid the Beechcraft [Baron].

The board spelled out a *Contributing Cause*:

> Physiological and environmental conditions and the excessive speed of the DC-9, which reduced visual detection capabilities under an ATC system which was not designed or equipped to separate a mixture of controlled and uncontrolled traffic.

Caution, Fifty-Three Directly Overhead
Mid-Air Collision at MCAF New River, North Carolina
23 June 1967

Sikorsky CH-53 Sea Stallion, Bureau No. 153305
U.S. Marine Corps combat assault training flight
33 aboard, 20 killed

Bell UH-1 Iroquois (Huey), Bureau No. 638572
U.S. Marine Corps pilot training flight
2 aboard, 2 killed

Synopsis: A Sikorsky CH-53 Sea Stallion descended to land as a Bell UH-1 Huey took off. The two military helicopters collided 500 feet above the ground. The impact tore the entire rotorhead off the Huey. It crashed and killed the two pilots. The CH-53 lost its tail rotor, spun to the ground, crashed, and burned, killing 20 of the 33 men on board.

The Sikorsky CH-53 Sea Stallion Flight: A huge Sikorsky CH-53 Sea Stallion helicopter (call-sign, Camel Driver 101) took off from MCAF New River, North Carolina, at 0856 Hours. The U.S. Marine Corps transport helicopter carried 4 crewmen plus 29 troops on a combat assault training mission. The pilots planned to descend parallel to Runway 23 and land in Landing Zone (LZ) Earle beyond the end of the runway. Then the troops would charge out and conduct a mock infantry assault on a simulated bunker complex:

The CH-53 is a heavy-lift helicopter designed to the specifications of the U.S. Marine Corps. With unprecedented weight carrying capability, it was the most powerful helicopter in the Western World. It first flew in 1964 and remains in production over 40 years later. Retractable landing gear and in-flight refueling give it transoceanic range. The twin-engine transport (three engines on later models) first entered combat in Vietnam in 1967. Over four decades later the United States still relies on the CH-53 in combat in the Middle East.

The pilot, a Marine lieutenant colonel, had logged 4,604 hours of flight time. He was the commanding officer of squadron HMH-461.

Sikorsky CH-53 Sea Stallion (official U.S. Marine Corps photo)

The copilot, a captain, had 2,732 hours of flight experience. Their CH-53 had flown a mere 36 hours since the squadron had accepted it from the Sikorsky factory earlier that month.

The Bell UH-1 Iroquois (Huey) Flight: The Bell UH-1 helicopter officially bore the name "Iroquois," but it had become known as the "Huey." One of these wildly successful Marine Corps helicopters (call-sign, Yazoo 4-1) took off from MCAF New River on a pilot training flight at 0850 Hours:

The Huey was the first jet-powered helicopter to go into production in the United States. More than 16,000 have been built. The Huey has served in air assault, cargo, medevac, search and rescue, and ground attack roles. In combat, Huey gunships were armed with rockets, grenade launchers, and machineguns.

The U.S. Army eventually phased out the Huey in favor of the larger Blackhawk. Yet, the Marine Corps twin-turbine Huey has been upgraded and retrofitted with a four-blade rotor system and the latest avionics and electronic countermeasures. The Marine Corps expects the Huey and its lethal variant, the Cobra attack helicopter, to serve until the year 2025 and beyond.

Bell UH-1 Iroquois, the "Huey" (official U.S. Marine Corps photo)

The pilot, a Marine Corps captain in squadron VMO-1, was a Naval Aviator with 2,079 hours in his logbook. The copilot, a first lieutenant, had logged 312 flight hours. The pilot was instructing the inexperienced copilot on tactical maneuvers. Their Huey recently had been transferred to the Marine Corps from the U.S. Army.

Trouble Ahead: At 1,000 feet the troop-carrying CH-53 re-entered the traffic pattern and lined up on the extended centerline of Runway 23. The pilot radioed the tower, got clearance to land in LZ Earle, and began his descent:

> [The pilot] started his approach down the starboard *[right]* side of Runway 23, descending from 1,000 feet. His intended point of landing was just beyond and slightly to the right of the upwind end of Runway 23.

The Huey had been cleared for takeoff *two minutes earlier*, but it remained in a hover over the grass by the runway. Suddenly, without making further radio transmissions, the Huey accelerated skyward. The Huey pilots could not see the big CH-53 descending from above and behind them. The CH-53 pilots could not see the Huey climbing up under them.

The military ATC transcript, below, contains the only recorded radio traffic. It begins as the CH-53 requests clearance to land:

CH-53: .. Sikorsky CH-53 Sea Stallion (Camel Driver 101)
Huey: ... Bell UH-1 Huey (Yazoo 4-1)
Tower: .. MCAF New River Control Tower

[for two minutes the Huey has been cleared for takeoff, but the pilots remain in a low hover over the grass by the runway]

[the CH-53 pilot requests clearance into LZ Earle at 0902]

Tower: Camel Driver one-oh-one, cleared to land, LZ Earle. (0902)

CH-53: Roger, we're cleared to land. (0902)

[the CH-53 pilots begin a straight-in descent from 1,000 feet, paralleling Runway 23]

[below the CH-53, the Huey transitions out of a hover, accelerates, and climbs parallel to the same runway]

Tower: Yazoo four-one, caution, the [CH]-fifty-three [is] directly overhead. (0903)

Huey: Roger. (0903)

[for 24 seconds the Huey climbs straight ahead; the tower issues no further warning to either helicopter]

*[the climbing Huey and descending CH-53 **collide** at 0903]*

The Collision: Landing gear down, the CH-53 had been descending toward LZ Earle beyond the departure end of Runway 23. The Huey had been climbing straight ahead and paralleling the same runway.

The Huey climbed up under the descending CH-53. Five-hundred feet above the ground the Huey main rotor chewed into the bottom of the CH-53 fuselage. The Huey rotorhead ripped off. The mangled airframe then dropped like a manhole cover. It hit the ground and exploded into an orange fireball, instantly killing the two pilots:

The rotorhead separated from the UH-1 .... The main fuselage section struck the ground in an inverted attitude and burned.

The mid-air collision had caused major structural damage to the CH-53. The pilots maintained partial control and descended until the helicopter was about 100 feet above the ground. Then the entire tail pylon, including the tail rotor, ripped off of the airframe. The CH-53 began to spin, and it slammed into the ground. The fuselage broke into two pieces, and JP-4 jet fuel from ruptured tanks ignited:

The tail pylon separated from the CH-53 prior to ground impact, causing the helicopter to become uncontrollable. It struck the ground, rolled on its port *[left]* side, and burned.

One survivor, a battered and burned Marine lance corporal from North Dartmouth, Massachusetts, later would describe the accident in a written statement to JAG investigators:

The [helicopter] floor started shaking. The plane was dropping down . . . and there was more noise, metal shearing and everything. . . . I remember we were still spinning when we hit [the ground]. . . . The rear end of the helicopter must have broken [off] just about where we were sitting, because I remember I was thrown out [of the fuselage]. . . . I was 15 or 20 feet from the plane and lying on my side. My leg [was broken and] all buckled up under me. . . . I started [crawling] to get away from the fire.

Thirteen Marines aboard the transport helicopter suffered serious injuries in the crash and fire, but they were still alive. Ambulances sped them to the Naval Hospital at nearby Camp Lejeune, and they survived. However, the CH-53 crash had killed 20 Marines. An autopsy would show that one had succumbed to the fire. The 19 others had died as a result of impact trauma.

The Marine base at New River had been readying helicopters and training crewmen for the war in Vietnam. This day-and-night routine stopped at 1600 Hours on 26 June, three days after the accident. Flags dropped to half-staff as a memorial service began at the base chapel. Then the commanding general of the Second Marine Air Wing (2nd MAW) placed a wreath at the altar in memory of the 22 men who had been killed. A bugler sounded *Taps*. Within an hour the whine of jet turbines and pop-pop-pop of whirling rotors again filled the air as Marines at MCAF New River prepared for war.

The Investigation: Marine Corps investigators quickly realized this had been an operational accident. There had been no mechanical problem with either helicopter, and weather had not been a factor.

The Huey had been cleared for takeoff. However, it had been on the ground, or in a hover, for two minutes before starting to accelerate and climb. One independent technical report explained:

> Delay between the [takeoff] clearance and the time [the Huey] proceeded on the originally-planned [takeoff] had been too long for [the clearance] to be considered still valid.

Further, the tower controller had not realized where the CH-53 pilot intended to land. The controller *assumed* the helicopter was headed for a landing site a quarter-mile to the right of the runway. If so, lateral separation from the Huey would have been adequate.

Thirty seconds before beginning their descent the CH-53 pilots had seen the Huey hovering by the runway, 1,000 feet below and a mile ahead. Hovering there, it had posed no danger to them. As the CH-53 flew overhead the Huey had slipped out of the CH-53 pilots' sight, hidden from view under their massive fuselage.

Probable Cause: The investigative team completed Naval Aviation Safety Center Investigation 70-67 on 6 September 1967. The board found that air traffic control played the crucial role in this accident:

> The tower controller . . . did not accurately perceive the impending conflict until it was too late.

Pilots in both helicopters had relied on the tower controller. The Huey pilots assumed they were *still* cleared for takeoff. The CH-53 pilots had been cleared to land in LZ Earle.

On final approach the CH-53 pilots had been unable to see the Huey under them. The Huey pilots could not see the CH-53 descending from above. Only the controller could see that the helicopters were on a collision course. Consequently, the board outlined the *Probable Cause* of this accident as follows:

> The most Probable Cause of this accident is personnel factor, in that tower personnel did not continuously monitor the approach

of the CH-53 in order to take more positive action to prevent the mid-air collision.

The board found one ***Contributing Cause***:

A Contributing Cause is pilot factor, in that after being warned by tower personnel of the CH-53 directly overhead, [the Huey pilot] continued to climb his aircraft straight ahead without taking evasive action.

After the accident the burned remains of the Sikorsky CH-53 Sea Stallion, covered with fire-extinguishing foam, lie on the ground at MCAF New River (official U.S. Marine Corps photo)

Death Pours from the Sky
Mid-Air Collision near Hendersonville, North Carolina
19 July 1967

Boeing 727, Registration N68650
Piedmont Airlines Flight 22
79 aboard, 79 killed

Cessna 310, Registration N3121S
Radial Air Inc. business charter flight
3 aboard, 3 killed

Synopsis: A Boeing 727 took off, and a Cessna 310 approached to land. Both planes were on IFR flight plans and controlled by ATC. Yet, they collided head-on in the clouds, killing all 82 people aboard both planes.

The Boeing 727 Flight: At Asheville Municipal Airport in North Carolina the crew and passengers on Piedmont Airlines Flight 22 prepared for their trip to Washington, D.C. At 1158 Hours the pilots released their brakes, and the Boeing 727 tri-jet began rumbling down Runway 16. The airliner accelerated, rose, and powered its way up into hazy and cloudy skies:

The Boeing 727 is a mid-size narrow-body jet airliner, and it first flew in 1963. Its three engines are mounted at the rear of the fuselage. High-lift flaps and slats enable the tri-jet to operate from small airports with modest runway lengths. Versatility, reliability, and low operating cost made the transport attractive for carrying either passengers or freight. For many years it was the world's most popular jetliner with 1,831 deliveries world-wide. In the 1970s it had been called the "DC-3 of the Jet Age."

The captain, age 49, held the required ATP certificate and had accumulated 18,383 hours of flight experience. He had transitioned into the Boeing 727 two months before this final flight. His young first officer, age 30, had earned a Commercial Pilot certificate and had logged 3,364 hours of flight time. A flight engineer and two flight attendants completed the Piedmont crew. The 74 passengers

Boeing C-22, a military variant of the civil Boeing 727 airliner (photo courtesy of U.S. DOD Visual Information Center)

anticipated landing in Washington in 59 minutes. According to a subsequent report from the *Associated Press*:

> John T. McNaughton, who was scheduled to become Secretary of the Navy in about two weeks; his wife, Sarah; and their 11 year old son, Theodore, were aboard the airliner. Theodore had been attending a summer camp [near Asheville], and his parents had come to take him back to Washington.

The Cessna 310 Flight: Two businessmen chartered Radial Air Inc. to fly them from Charlotte, North Carolina, to Asheville. The Cessna 310 took off from Douglas Airport in Charlotte on an IFR flight plan and flew toward Asheville at 8,000 feet. The highly experienced Cessna pilot, age 48, was an employee of Radial Air, and he held a Commercial Pilot certificate. He had over 10,000 hours of flight time documented in his logbook:

> *The Cessna 310 was the first twin engine Cessna to go into production after World War II. Its had modern rakish lines and innovative features such as exhaust thrust augmentor tubes. With four to six seats, depending upon the model, the light twin became a common air-taxi aircraft. A variety of normally aspirated and*

turbocharged piston engines powered the Cessna 310 over the years. The U.S. Air Force purchased 160 of the speedy aircraft as U-3 utility planes. Because of its blue exterior paint, Air Force pilots called the aircraft a "Blue Canoe."

Trouble Ahead: The control tower at Asheville soon cleared the Boeing 727 to climb unrestricted to the VOR. The airline pilots continued their planned ascent toward 21,000 feet.

However, things were not going smoothly for the Cessna pilot. The Center controller had told him to expect an ILS approach at Asheville. However, five minutes later Asheville Approach Control cleared the pilot to the Asheville radio beacon *northwest* of the VOR. This navaid was not depicted on the ILS approach chart. Approach Control failed to specify the type of approach. Consequently, the pilot turned *southwest* toward a different radio beacon in anticipation of the ILS approach. That southwest heading took him into the path of the climbing Boeing 727.

The ATC radio transcript, below, contains transmissions to and from the Boeing 727, the Cessna, and ATC. The transcript begins as the Cessna gets initial clearance to Asheville VOR. The airliner and the Cessna operated on different VHF frequencies, so pilots could not hear transmissions to or from the other aircraft:

> Piedmont 22: . . . Piedmont Airlines Flight 22
> Cessna: . . . . Radial Air Inc. business charter flight
> Center: . . . . Atlanta Center (ARTCC)
> AppControl: . . . . Asheville Approach Control
> Tower: . . . . Asheville Airport Control Tower

Center: [Cessna one-two-one sierra is] cleared to the Asheville VOR, descend and maintain seven-thousand [feet], *expect an ILS approach [emphasis added]* at Asheville. (1151:45)

[the Cessna pilot acknowledges the clearance]

Center: Cessna one-two-one sierra, radar service [is] terminated, contact Asheville Approach on one-two-four-point-three. (1153:00)

[the Cessna pilot contacts Approach Control at 1153:10]

[3 minutes later the controller issues an <u>unexpected clearance</u>]

AppControl: Three-one-two-one sierra, cleared over the VOR to Broad River – correction – make that the Asheville radio beacon, [cleared] over the VOR to the Asheville radio beacon, maintain seven-thousand, report passing the VOR. (1156:28)

Cessna: Two-one sierra, [roger]. (1156:43)

[meanwhile, the Boeing 727 prepares to take off from Asheville]

Tower: Piedmont twenty-two, taxi into position and hold. (1157:33)

Tower: Piedmont twenty-two, maintain runway heading until reaching five-thousand, cleared for takeoff. (1158:01)

Piedmont 22:   Twenty-two [is] rolling. (1158:07)

[the Boeing 727 lifts off and climbs]

Cessna: [Approach Control], two-one sierra just passed over the VOR, we're headed for the – *[a four second pause]* – for – aaahhh – Asheville now. (1158:20)

[the Cessna pilot does not turn northwest toward the Asheville radio beacon; instead, he turns southwest toward the Broad River radio beacon and into the path of the Boeing 727]

AppControl: Two-one sierra, roger, [you are] by the VOR, descend and maintain six-thousand. (1158:41)

Cessna: We're leaving seven-[thousand] now.

[the Cessna has reported passing the VOR, and the controller assumes it is flying northwest toward the Asheville radio beacon, so all airspace between the airport and the VOR should now be clear for the climbing Boeing 727]

Tower: Piedmont twenty-two, climb unrestricted to the VOR, report passing the VOR. (1159:44)

Piedmont 22:   OK, unrestricted to the VOR. (1159:49)

[ATC finally tells the Cessna the type of approach to fly]

AppControl: [Cessna two-one sierra], cleared for an ADF [number] two approach to Runway one-six, report the Asheville radio beacon inbound. (1200:02)

Cessna: Roger. (1200:09)

Piedmont 22: Piedmont twenty-two is – (1201:17)

*[the message is cut off in mid-sentence as the Boeing 727 and the Cessna **collide** at 1201:18]*

The Collision: Zooming through the clouds with a closure rate of about 700 feet per second, the Boeing 727 and the Cessna slammed into each other almost head-on at 6,132 feet. The nose of the Cessna plowed into the left forward fuselage of the airliner:

> [The Cessna] penetrated the Boeing 727 fuselage . . . with parts of the Cessna exiting from the right side of the Boeing 727 forward of the galley doorframe.

The smaller Cessna had rammed completely through the airliner fuselage, tearing it into two main pieces. A witnesses on the ground later would explain: "Bodies fell out of the airliner like confetti." The *Associated Press* later would report:

> The airliner came apart . . . there were two big sections and a thousand little pieces as plane parts, bodies, and luggage plummeted to earth about two miles from Hendersonville.

Large sections of the fuselage fell to the ground 150 feet from Camp Pinewood, a crowded summer camp for young boys and girls. Counselors threw children under overturned canoes to shield them from debris and body parts tumbling down into the camp. The next day a full-page newspaper headline would exclaim:

Death Pours from the Sky at Hendersonville

Wreckage was strewn along a path over a mile long and a half-mile wide. So complete was the destruction of the Cessna that the only identifiable portion was the No. 1 engine. It was embedded inside the forward half of the severed Boeing fuselage:

This was an unsurvivable accident. All [82] persons aboard the two aircraft died of traumatic injuries sustained in the accident.

Later at Washington National Airport, relatives and friends waited at the gate for the Boeing 727 to arrive. It had been due to land at 1257 Hours. It was 40 minutes late, and so far there had been no explanation of the delay. Then the public address system in the terminal announced a change – people waiting to meet passengers on Piedmont Flight 22 were asked to walk to Gate 25.

Col. Robert Hixon, a military aide from The Pentagon who was waiting for the Secretary of the Navy-designee, was among those who hurried toward Gate 25. He worried about the change. He knew he might miss the McNaughton family if there was a mix-up about the proper terminal gate.

A visibly shaken red-eyed Piedmont Airlines hostess met those who arrived at Gate 25. She ushered the colonel and about ninety others into a private lounge. After all had entered, Piedmont's area manager wiped away tears and choked-up as he told them:

Piedmont flight twenty-two has crashed – I am so sorry – There are no survivors.

The Investigation: NTSB investigators began a methodical analysis to determine what had gone awry. The FDR from the airliner and the wreckage locations showed that the Boeing 727 pilots had complied with their clearance. The Cessna pilot had not. He had been cleared to the Asheville radio beacon *northwest* of the VOR:

The Cessna failed to comply with the clearance to proceed from the Asheville VOR to the Asheville RBN *[radio beacon]*.

The Cessna pilot had flown *southwest*. If he had flown toward the Asheville radio beacon, as cleared by ATC, the accident would not have happened.

There were four instrument approaches to the Asheville Airport, two of which were the (1) ILS Runway 34 Approach and the (2) ADF No. 2 Approach. Atlanta Center had told the Cessna to expect an ILS approach. The board deemed it logical that the pilot would have reviewed the ILS approach chart and would have tuned his navigation radios to the frequencies for that approach. Investigators confirmed

Cessna 310 (photo courtesy of Adrian R. Pingstone)

this assumption when they examined the Cessna rubble. One VHF receiver was tuned to the ILS localizer. The ADF receiver was tuned to the Broad River radio beacon:

> The primary approach fix for the ILS approach is the Broad River RBN *[radio beacon]* . . . .

When Approach Control cleared the Cessna to the Asheville radio beacon at 1156:28 Hours the controller failed to specify the type of approach to be flown. Despite his numerous radio messages to the Cessna the controller inexplicably waited until 76 seconds before the collision. Then, finally, he cleared the Cessna for an ADF approach:

> The pilot probably attempted to locate and study the ADF No. 2 approach chart and/or verify the position of the Asheville RBN on the ILS approach chart.

The Asheville radio beacon *was not* depicted on the ILS approach chart. The pilot became confused, as evidenced by his reply to the controller at 1158:20 Hours. Board members theorized the pilot turned southwest for one of the following two reasons:

1. [The pilot was] unable to locate the Asheville RBN on the ILS chart, decided that one of the other facilities depicted on the ILS chart was, in fact, the Asheville RBN, [and flew toward it].

2. [The pilot] believed the Broad River RBN and the Asheville RBN were one and the same facility. A course toward the Broad River RBN, depicted on the ILS chart, was [flown].

Probable Cause: Fourteen months after the accident the NTSB adopted the Aircraft Accident Report, File No. 1-0005, on 5 September 1968. The Cessna pilot's deviation from his clearance resulted in the mid-air collision. However, Approach Control had passed up several golden opportunities to give the pilot advance notice of the ADF approach:

1. Approach Control initially had contacted the pilot at 1153:10 Hours but did not tell him which approach he would fly.

2. Approach Control had cleared the pilot to the Asheville radio beacon at 1156:28 Hours, but the controller again failed to tell the pilot which approach he would fly.

3. Two minutes later the pilot had arrived over the VOR. Center had told him to expect an ILS approach, so *presumably* he was wondering (1) *where* the Asheville radio beacon was, and (2) *why* he had been told to fly to it. Yet, once again the controller failed to specify the type of approach.

4. Approach Control did not specify the ADF approach until 76 seconds before the collision. This gave the pilot scant time to (1) abandon plans for the ILS approach, (2) find the ADF approach chart, (3) tune his VHF and ADF navigation receivers to the proper frequencies, and (4) plan the approach.

Asheville Municipal Airport did not have surveillance radar, and the NTSB emphasized:

In a non-radar environment, radio voice communication . . . becomes the only means by which aircraft separation can be affected. The only safeguard in this system is complete adherence to clearances by pilots . . . .

The board found the following **Probable Cause** of this accident:

The deviation of the Cessna from its IFR clearance, resulting in a flight-path into airspace allocated to the Piedmont Boeing 727. The reason for such deviation cannot be specifically or positively identified.

Notwithstanding the above, the board pointed out that the pilot had been confused. His confusion had been caused by Approach Control's repeated – and needless – delays in specifying the type of approach to be flown. This confusion led to the deviation, and the deviation caused the accident. Consequently, the board found a **Contributing Cause** of the accident to be:

The minimum control procedures utilized by the FAA *[Asheville Approach Control]* in the handling of the Cessna.

What'd Ya Hit Up There?
Mid-Air Collision near Fairland, Indiana
9 September 1969

Douglas DC-9, Registration N988VJ
Allegheny Airlines Flight 853
82 aboard, 82 killed

Piper PA-28 Cherokee, Registration N7374J
Forth Corporation basic training flight
1 aboard, 1 killed

Synopsis: Descending to land in clear weather, a Douglas DC-9 collided with a Piper PA-28 Cherokee. The impact instantly killed the Cherokee pilot. The 82 persons on the DC-9 survived the aerial impact. However, their airliner rolled inverted, dove into the ground, and killed them all.

The Douglas DC-9 Flight: Allegheny Airlines Flight 853, a Douglas DC-9, left Boston, Massachusetts, at noon and made an intermediate stop at Cincinnati, Ohio. During the routine exchange of passengers and baggage, Allegheny got a chance to fill many of its empty seats. In the Cincinnati terminal, 64 unhappy people held tickets for Trans World Airlines (TWA) Flight 69 to Indianapolis, Indiana, but their plane had been delayed. An independent publication explained:

> [Because of the delay], TWA offered its passengers the option of transferring to Allegheny Flight 853, and 38 people took the offer.

At 1516 Hours with 78 passengers aboard – including 38 unlucky souls who could have waited on their TWA plane – Flight 853 took off for the short trip to Indianapolis. The captain, age 47, had amassed 23,818 hours of flight time in his logbook and held an ATP certificate. His young first officer, age 26, had earned a Commercial Pilot certificate. The two experienced hostesses had flown with Allegheny for a total of 17 years.

The Piper PA-28 Cherokee Flight: At Brookside Airpark 20 miles northeast of Indianapolis a student pilot prepared for a solo cross-country flight. He had logged about 51 hours of flight time and had

Piper PA-28 Cherokee (photo courtesy of Adrian R. Pingstone)

passed the written examination for a Private Pilot certificate. On this flight he planned to practice for his upcoming FAA flight test. The general manager at the airpark checked the student pilot's flight plan. He then sent the fledgling pilot on his way in a Piper PA-28 Cherokee rented from the Forth Corporation. At about 1520 Hours the student, a Korean War combat veteran, took off from Runway 36:

> *Piper had introduced the single engine four-seat Cherokee in 1961. The plane featured fixed tricycle landing gear and an all-metal fuselage. Over the years, engines ranged from 140 horsepower up to 180 horsepower. Larger variants had retractable gear, more powerful engines, and seats for six people. The Cherokee offers excellent visibility and is among the most popular trainers used by flight schools around the world.*

Trouble Ahead: As the DC-9 sped along airway V-97, Indianapolis Center cleared the pilots to descend to 6,000 feet at 1522:55 Hours. Five minutes later Approach Control began vectoring the flight for an approach to Runway 31 Left at Indianapolis.

Meanwhile, shortly after takeoff the Cherokee pilot had radioed the local flight service station and had activated his VFR flight plan at 1521 Hours. He climbed to 3,500 feet and cruised southeast. He had not requested flight following, and his aircraft was not equipped with a transponder. The student pilot was not controlled by any ATC agency, nor was he required to be.

The transcript, below, contains radio traffic between ATC and the DC-9, plus content from the CVR. The transcript begins as Center prepares to hand-off the flight to Approach Control:

Capt: . . . . Captain, Allegheny Flight 853
FO: . . . . First Officer, Allegheny Flight 853
Center: . . . . . Indianapolis Center (ARTCC)
AppControl: . Indianapolis Approach Control
Radio-: . . . . . (prefix) Radio transmission, Allegheny Flight 853

Center: Allegheny eight-fifty-three is in radar contact, cross Shelby-ville [VOR] at and maintain six-thousand . . . . (1522:55)

[the DC-9 pilots later contact Indianapolis Approach Control]

AppControl: Allegheny eight-five-three, roger, squawk ident, [fly] heading two-eight-zero for radar vector, visual approach, [Runway] three-one left, descend and maintain two-thousand-five-hundred feet and report level [at that altitude]. (1527:12)

Radio-Capt: Eight-five-three [is] cleared down to two-thousand-five-hundred, and [we'll] report reaching [that altitude]. (1527:29)

FO: Out of thirty-five-[hundred] for twenty-five. (1529:13)

Capt: I'm going down [to twenty-five-hundred feet]. (1529:14)

*[the DC-9 and the Cherokee **collide** at 1529:15]*

[sound of landing gear warning horn and stall buffet at 1529:17]

[the pilots have no pitch or yaw control; the DC-9 rolls inverted; sound of loose items hitting the cockpit ceiling]

Capt or FO: [The] tail [of our DC-9] must be –

[sound of increasing airspeed and vibration]

FO: What'd ya hit up there? *[shouted]*

[sound of impact with the ground at 1529:27]

The Collision: The DC-9 had been descending at 2,460 feet per minute at 256 knots. The Cherokee had been flying straight and level at about 105 knots and moving right-to-left across the path of the big DC-9. The two aircraft collided at 3,550 feet about four miles north of Fairland, Indiana:

> Initial contact between the two aircraft occurred at the forward upper right side of the [DC-9] vertical fin, just below the horizontal stabilizer of the DC-9, and the left forward side of the PA-28 [Cherokee], just forward of the left wing root.

The massive DC-9 vertical stabilizer sliced the Cherokee in half, right through the center of the cockpit. The two halves of the small plane, devoid of any aerodynamic symmetry, fell straight down to the ground. The student pilot was spared the terror of the fall, for the mid-air impact had killed him instantly.

The impact had not damaged the DC-9 forward of the empennage. No one aboard had been materially injured. However, the collision had mortally wounded the airliner. The entire T-tail, including the rudder and elevator, had ripped away. Further controlled flight was not possible. The airplane rolled inverted and dove toward the ground. Twelve seconds after the collision it screamed down into a soybean field. The high speed impact killed everyone on board:

> The DC-9 impact site was approximately 1,300 feet long and 700 feet wide. The aircraft struck the ground in an inverted, almost wings-level, nose-down attitude.

The DC-9 fuselage disintegrated in the field near Shady Acres mobile home park. Flying debris slammed through the sides of several mobile homes. The severed leg of one passenger landed on the front steps of another home. One independent technical account of the crash explained:

> Pieces of everything and everyone [were] scattered everywhere for a distance of over a half-mile. *[and also]* A school bus had stopped to let off some children at the [mobile home] park, and the crash scattered debris and bodies all around it.

The Investigation: The FDR from the DC-9 revealed the airliner had been flying wings-level and descending through 3,550 feet on a heading of 282 degrees when it hit the Cherokee. The absence of aberrant FDR data, plus the CVR-documented monotone voice of the

Douglas C-9 Skytrain, a military variant of the civil Douglas DC-9 airliner (official U.S. Navy photo)

captain one second before impact, indicated the airline pilots never saw the Cherokee:

> The [DC-9] descent [toward] the 2,500-foot assigned altitude was probably made in part while the aircraft was in clouds, and in part under VFR conditions.

The cloud base had been at 4,000 feet, and other planes had reported excellent visibility below the clouds. Based upon the known DC-9 rate of descent it would have broken-out below the cloud base about 14 seconds before the collision. The pilots in both aircraft would have had only those few seconds in which to scan for conflicting traffic. There would have been no relative motion to assist them, for neither plane was turning, and they were on a collision course. NTSB investigators concluded:

> [None of the] pilots would have had sufficient time to "see and avoid" the other aircraft, even if they had devoted virtually their entire attention outside the cockpit, scanning for other aircraft.

Further, conspicuity had been a big problem in the DC-9 cockpit. The Cherokee had approached from the airliner's right side. Seated on the right, the first officer would have had the better opportunity to see it. However, he had been watching his altimeter for the "out of thirty-five" call, which he made two seconds before impact. Thus preoccupied, he would not have been scanning for traffic. Investigators determined that from the left seat the captain's ability to spot the Cherokee had been "virtually nil."

The DC-9 had been under ATC radar control, but neither Center nor Approach Control warned the pilots about converging traffic. Center radar had been operating on low power to reduce the anomalous propagation and clutter caused by a temperature inversion above 7,000 feet. Under this condition the small Cherokee had not been visible on the Center controller's scope. Approach Control radar had been operating at normal power, but the controller had not detected the small Cherokee due to tangential blind speed effect:

> Meteorological circumstances reduced the [radar] safety features below an acceptable level.

Probable Cause: The board adopted the Aircraft Accident Report, File No. 1-0016, on 15 July 1970 about 10 months after the accident. Both airplanes had flown in compliance with all applicable regulations. The transponder-equipped DC-9 had been under radar control. The Cherokee had no transponder, so only a primary radar return could have been displayed. However, due to unique circumstances the Cherokee had been invisible on ATC radar:

> This accident involved an intermix of high-speed aircraft and low speed aircraft under the combined active and passive control of the air traffic control system within a terminal area.

The Cherokee had been flying below the cloud base on a VFR flight plan. The DC-9, on an IFR flight plan, entered VFR conditions when it descended through the cloud base at about 4,000 feet. In the brief seconds thereafter the pilots in both planes had been responsible for maintaining visual separation from other traffic. The board found this visual separation principle was not adequate under the circumstances. In a strongly worded statement the board declared:

> The "see and be seen" concept of collision avoidance, which has been demonstrably deficient in the past, is now totally unacceptable in providing separation between aircraft during descents into

terminal areas where high [speed] and low speed traffic is inter-
mixed under IFR and VFR control.

The board laid full responsibility for the accident upon the FAA
by detailing the following ***Probable Cause***:

The board determines the Probable Cause of this accident to be
deficiencies in the collision avoidance capability of the Air
Traffic Control system of the Federal Aviation Administration in
a terminal area wherein there was mixed Instrument Fight Rules
and Visual Flight Rules traffic. The deficiencies include:

[First], the inadequacy of the "see and avoid" concept under
the circumstances of this case;

[Second], the technical limitations of radar in detecting all
aircraft; and,

[Third], the absence of Federal Aviation Regulations which
would provide a system of adequate separation of mixed VFR
and IFR traffic in terminal areas.

No Problem, We'll File VFR
Mid-Air Collision near Duarte, California
6 June 1971

McDonnell F-4 Phantom, Bureau No. 151458
U.S. Marine Corps cross-country training flight
2 aboard, 1 killed

Douglas DC-9, Registration N9345
Hughes AirWest Flight 706
49 aboard, 49 killed

Synopsis: A McDonnell F-4 Phantom fighter flew toward its home base on a VFR flight plan. An IFR Douglas DC-9 airliner climbed toward cruising altitude and rammed into the Phantom. The accident killed all 49 people on the DC-9 plus the Phantom pilot. The Phantom RIO ejected and survived.

The McDonnell F-4 Phantom Flight: For the last leg of a cross-country flight a U.S. Marine Corps fighter would take off from the Naval Auxiliary Air Station (NAAS) at Fallon, Nevada. Normally this would have been an IFR flight at high altitude. However, the McDonnell F-4 Phantom warplane had developed three problems: (1) an inoperative transponder, (2) an oxygen system leak, and (3) a degraded radar system:

The Phantom was designed as a fleet defense fighter for the U.S. Navy and Marine Corps. It entered service in 1961, and within two years the U.S. Air Force adopted it in the role of fighter-bomber. During 20 years of production the factory built 5,195 Phantoms, more than any other supersonic military aircraft. It could exceed Mach 2 in level flight, and it would remain in service with the United States Armed Forces for 35 years.

Powered by two afterburning General Electric J-79 engines, the Phantom could carry over 18,000 pounds of missiles and bombs on 11 hardpoints under the wings and fuselage. In the 1960s the Phantom set 16 world records. They include the absolute altitude record (98,557 feet), the absolute speed record (1,606 miles per hour), and the highest sustained altitude (66,433 feet).

McDonnell F-4 Phantom (photo courtesy of U.S. DOD Visual Info. Center)

The pilot called his base, the Marine Corps Air Station (MCAS) at El Toro, California, and reported his aircraft problems. He was told to return VFR at low altitude. He filed a VFR flight plan and took off from Fallon at 1716 Hours. After crossing the mountains he zoomed along 1,000 feet above the terrain:

> [He] flew east of the planned course, over Palmdale, to avoid the anticipated heavy traffic over Los Angeles.

The pilot was a Marine Corps first lieutenant, age 27, who had logged 880 flight hours. In addition to his military qualifications he held Commercial Pilot and Flight Instructor certificates. His backseat radar intercept officer (RIO), age 24 and also a first lieutenant, had completed his RIO training five months earlier and had accumulated 195 hours of flight time. The pilot, the RIO, and their Phantom were assigned to Marine Corps squadron VMFA-323 at El Toro.

The Douglas DC-9 Flight: A civil Douglas DC-9, Hughes AirWest Flight 706, left Los Angeles International Airport in California at 1802 Hours and headed for Salt Lake City, Utah. Climbing in VFR conditions, the airliner carried two pilots, three flight attendants, and 44 passengers. The pilots followed radar vectors from Los Angeles Departure Control. They each had earned ATP certificates, and they had logged an impressive total of 32,618 flying hours.

Trouble Ahead: At 1806 Hours, Departure Control handed-off the DC-9 to Los Angeles Center. A journeyman controller began guiding the flight. He was supervising three developmental controllers, so four sets of eyes watched the radarscope transponder return from Flight 706.

Meanwhile, 50 miles from El Toro the Phantom pilot climbed to 15,000 feet for better visibility. After leveling-off from the climb he executed a 360 degree aileron roll at about 420 knots. In the back seat the RIO dutifully peered into his radarscope:

> He was operating the [Phantom] radar in the mapping mode, but due to the extremely degraded air-to-air detection capability no airborne targets were seen.

The short ATC transcript, below, contains the only documented radio messages between Center and the DC-9:

> AirWest 706: . . . Hughes AirWest Flight 706 (Douglas DC-9)
> Center: . . . . Los Angeles Center (ARTCC)

[the DC-9 takes off from Los Angeles at 1802]

[Departure Control hands-off the DC-9 to Center at 1806]

AirWest 706: Center, AirWest seven-zero-six is climbing through twelve-thousand. (1809)

Center: AirWest seven-zero-six, roger, turn left, heading zero-four-zero until receiving Daggett, [then] proceed direct. (1809)

AirWest 706: OK, zero-four-zero direct to Daggett. (1809)

*[2 minutes later the DC-9 and the Phantom **collide** at 1811]*

The Collision: About three seconds before impact the Phantom RIO had glanced up from his radarscope. In his peripheral vision he had seen the DC-9 about 50 degrees to his right and coming straight at him. Simultaneously the Phantom pilot had seen the airliner and had whipped his fighter hard to the left:

> Within the remaining 2 to 4 seconds a left roll was made as an attempt to avoid a collision.

Douglas C-9 Nightingale, a military variant of the civil Douglas DC-9 airliner (photo courtesy of U.S. DOD Visual Information Center)

At 15,150 feet the Phantom wing slashed completely through the DC-9 forward passenger cabin. The collision sheared the cabin into two pieces. An independent report would explain:

> The F-4 right wing cleanly sliced the DC-9 in half longitudinally, separating the cockpit from the rest of the fuselage.

The decapitated DC-9 plummeted down into a canyon in the San Gabriel Mountains near Duarte, California, killing everyone aboard. Pieces of the Phantom were entangled in the twisted DC-9 airframe, which had burst into flames upon ground impact. Meanwhile, the Phantom fuselage had tumbled end-over-end as it fell:

> The mid-air collision was survivable for the [two] occupants of the F-4. The RIO successfully ejected, and he was subsequently rescued uninjured. The pilot was not able to eject.

The Phantom was equipped with Martin-Baker ejection seats, but they were not designed to fire through the canopy. The pilot had pulled his alternate firing handle, but a distortion in the linkage prevented his canopy from jettisoning. Consequently, the ejection sequence terminated, and the pilot had no way to escape from his crippled aircraft. He lived during the long fall, but he died when the Phantom slammed into the ground and exploded.

Ironically the Navy and Marine Corps were modifying the canopy jettison system on their Phantom fighters. Ballistic canopy thrusters were being installed to ensure the canopy would jettison at the start of the ejection sequence. Modifications on Phantoms at El Toro had been scheduled to begin in July, the month *after* the accident.

The Investigation: The NTSB investigation quickly boiled down to two issues: (1) why did ATC not detect the two converging aircraft and radio a warning, and (2) why did the pilots not see each other in time to take evasive action?

The visibility in the area at the time of the accident was good, and there were no clouds between the two aircraft.

Four Los Angeles Center controllers had watched the transponder radar return from the DC-9, which was under their control. The Phantom, which was on a VFR flight plan and not in radio contact with ATC, had an inoperative transponder. No position and altitude "tag" for the warplane would have been displayed on Center radar. In addition the four controllers each testified they did not see any primary target radar return from the Phantom. The FAA conducted radar tests using another Phantom under conditions that simulated the fatal flight. These flight tests showed:

The primary target alone was not of sufficient strength to assure notice by a controller who was unaware of the aircraft presence. *[and also]* Three independent radar systems [had] failed to detect the primary target of [the Phantom] and as a result no warning was given to the crew of [the DC-9].

Investigators learned it would have been extremely difficult for controllers to tell the difference between normal radar clutter and the Phantom primary return "if any target was displayed at all," and:

In this instance, detection of [the Phantom] was hampered by [its small] radar cross-section and a temperature inversion.

Some felt the Phantom pilot had acted irresponsibly by executing an aileron roll. However, the NTSB pointed out that the maneuver had no bearing on the accident. The two aircraft had been separated by 13 miles at the time. Others questioned the pilot's decision to fly

VFR to El Toro at low altitude. Conversely, investigators reported the low-level VFR flight was proper and an "obvious solution" to the oxygen and transponder malfunctions.

The CVR on the DC-9 did not survive the post-crash fire, but the FDR tape was not damaged. It showed the DC-9 pilots had taken no evasive action prior to the collision. They had continued to climb, wings level, toward their cruising altitude. The NTSB concluded:

> It is most likely that the crew of [the DC-9] never saw [the Phantom], or saw it moments before the collision and had no time to initiate an evasive maneuver.

Investigators conducted conspicuity studies. The results indicated that – in theory – the pilots in each aircraft could have seen the other plane in time to take evasive action. Yet, there were many real-world factors which made "see and avoid" highly difficult, including (1) windshield refracture, (2) windshield cleanliness, (3) the background against which the other aircraft would have been viewed, (4) the high closure rate, and (5) the other plane's stationary position on windshields in both planes:

> The lack of relative motion of either target in the peripheral vision of any crew member could have made early detection of the other aircraft highly unlikely.

Probable Cause: The board released the Aircraft Accident Report, NTSB-AAR-72-26, on 22 September 1972, 15 months after the accident. The board concluded the Phantom primary target had not been displayed on ATC radar. The military pilot and his RIO had extensive documented training in heads-up scanning technique, the process of looking for converging aircraft. On the other hand:

> No formal training or evaluation of crew scanning technique and lookout doctrine is accomplished by [Hughes] AirWest.

The board pointed out that the Phantom pilot could have radioed ATC and requested traffic advisories. This would have alerted ATC to the presence of the Phantom and *might* have prompted a controller to detect a faint primary radar return from the fighter. Yet, there was no mandate for the pilot to request advisories. The NTSB outlined the ***Probable Cause*** of this accident as follows:

> The [NTSB] determines that the Probable Cause of this accident was the failure of both crews to see and avoid each other, but [the NTSB] recognizes that [the pilots] had only a marginal capability to detect, assess, and avoid the collision.

The board detailed three ***Contributing Causes***:

> [First], a very high closure rate.

> [Second], comingling of IFR and VFR traffic in an area where the limitation of the ATC system precludes effective separation of such traffic.

> [Third], failure of the crew of [the Phantom] to request radar advisory service [when] they had an inoperable transponder.

Go Around! Go Around!
Mid-Air Collision at NAS Moffett Field, California
12 April 1973

Convair CV-990 Galileo, Registration N711NA
NASA research flight
11 aboard, 11 killed

Lockheed P-3 Orion, Bureau No. 157332
U.S. Navy training flight
6 aboard, 5 killed

Synopsis: A Convair CV-990 Galileo approached the airport as a Lockheed P-3 Orion practiced touch-and-go landings. ATC cleared both aircraft to land on the same runway at the same time. The two planes collided on final approach, crashed, and burned. The accident killed 16 of the 17 men aboard.

The Convair CV-990 Galileo Flight: Thirty miles south of San Francisco, California, NAS Moffett Field served as headquarters for the NASA Ames Research Center. Also, U.S. Navy patrol squadrons for the Pacific Theater were based at the busy military airfield.

A big Convair CV-990 Galileo, one of NASA's research aircraft, took off from Moffett Field and flew over Monterey Bay. Scientists on the Convair tested a new sensing system that tracked migratory sea mammals. In the past they had used the Convair to conduct myriad scientific activities such as chasing a solar eclipse, studying comets, and assisting Soviet Union analysts in surveying the Bering Sea:

The original CV-990 airliner entered production in 1961. The transport was similar to the Boeing 707 and the Douglas DC-8. Four big General Electric CJ-805 turbofans powered the swept-wing jet. It had a maximum takeoff weight of 253,000 pounds and could cruise at 35,000 feet.

Convair modified a CV-990 to meet NASA specifications, and it became the "Galileo" flying research laboratory. NASA used it for aerial mapping and geophysical observations. This aerial laboratory complemented Ames Research Center exploration of life science, space science, and Artificial Intelligence.

Convair CV-990 Galileo (photo courtesy of U.S. NASA)

The Convair pilot, age 28, held an ATP certificate and had logged 11,200 flight hours. His copilot had over 4,000 hours in his logbook, and the flight engineer had 2,300 hours of experience in the Convair. The plane also carried eight NASA scientists and civil contractors.

The Lockheed P-3 Orion Flight: On a pilot training mission a Lockheed P-3 Orion took off from Moffett Field on a planned five hour flight. The U.S. Navy crew practiced high-work over the Big Sur area and then returned to their home airfield. On Runway 32 Left they practiced touch-and-go landings for an hour. The P-3 carried one pilot-in-command, two copilots who were being trained, and three other crewmembers. They and their aircraft were assigned to Navy Patrol Squadron VP-47 at Moffett Field:

The P-3 is a highly modified military variant of the civil Lockheed L-188 Electra airliner. Four Allison T-56 turboprop engines power the aircraft, which has a maximum takeoff weight of 139,760 pounds. The versatile P-3 serves armed forces around the globe in maritime patrol, reconnaissance, and antisubmarine warfare roles. On long missions, one engine – and sometimes two – can be shut down to extend loiter time and range.

A long "stinger" trailing back from the tail houses magnetic anomaly detection (MAD) equipment. For combat the P-3 can carry a variety of air-to-air or air-to-ground missiles on under-wing pylons. An internal bomb-bay holds up to 20,000 pounds of ordnance including torpedoes, mines, and depth charges.

The P-3 pilot, a Navy lieutenant 29 years of age, had accumulated 2,014 flight hours. He was training two young copilots, both of whom had less than 500 hours of flight experience. Their aircraft had been delivered new from the factory in 1970 and had accumulated only 1,232 hours in the air.

Trouble Ahead: Like many huge airports, Moffett Field had two parallel primary runways, Runway 32 Left and Runway 32 Right. On this spring day in VFR conditions the P-3 pilots practiced touch-and-go landings on *Runway 32 Left*. Meanwhile, south of the airport the Convair pilots radioed the tower for a straight-in approach. The controller told them to expect a landing on *Runway 32 Right*.

The portion of the ATC radio transcript, below, includes messages to and from the Convair, the P-3, and the tower. Transmissions to and from other aircraft are excluded in the interest of clarity. The transcript begins as the Convair pilots make initial contact with the tower. The Convair and P-3 use the same UHF radio frequency:

NASA 711: . . Convair CV-990 Galileo (call-sign, NASA 711)
BA 482: . . . . Lockheed P-3 Orion (call-sign, BA 482)
Tower: . . . . . NAS Moffett Field Control Tower
Unk: . . . . Radio transmission from unknown source

NASA 711: Moffett tower, NASA seven-one-one. (1446)

Tower: NASA seven-one-one, Moffett. (1446)

NASA 711: [Moffett tower, we are a] Convair nine-ninety, ten miles south for a straight-in [approach and landing]. (1446)

Tower: [NASA seven-one-one], report a seven mile final for *Runway three-two right [emphasis added]*, altimeter three-zero-zero-one.

NASA 711: OK. (1446)

[meanwhile, the P-3 turns onto base for Runway 32 Left]

BA 482: Tower, BA four-eight-two is turning base, wheels down, touch-and-go. (1448)

Tower: Four-eight-two, continue for the left side. *[meaning, the left runway, Runway 32 Left]* (1448)

BA 482: Four-eight-two. *[an acknowledgment]* (1448)

NASA 711: Seven-one-one [is] seven miles [from the runway].

Tower: Seven-one-one, roger. (1448)

Tower: Seven-one-one, continue for the *right side [emphasis added]*.

NASA 711: [Tower], seven-one-one, gear down and locked. (1449)

[instead of Runway 32 Right, which he has stated twice before, the controller now clears the Convair to land on Runway 32 Left]

Tower: Seven-one-one, wind three-two-zero degrees at ten, cleared to land, *Runway three-two left. [emphasis added]*

NASA 711: Seven-one-one, *three-two left [emphasis added]*, thank you. (1449)

BA 482: [Unintelligible] touch-and-go on the left side. (1449)

Tower: Four-eight-two, continue. (1449)

*[the Convair and the P-3 **collide** on final approach at 1450; within seconds both aircraft crash and explode into flame]*

Unk: Tower, you [see] that? (1450)

Unk: [Garbled transmission]. (1450)

Tower: Go around! Go around! We've – (1450)

Tower: All aircraft in the [traffic] pattern – climb and maintain fifteen-hundred [feet] . . . . (1450)

The Collision: The Navy pilots in the P-3 had turned left onto their base leg, and the controller had told them to "continue for the left side." Unfortunately the controller had cleared the Convair to land on the same runway.

Lockheed P-3 Orion (photo courtesy of U.S. DOD Visual Information Center)

The P-3 had slipped down the final approach path. Above and behind the P-3 the slightly faster Convair had descended toward the same runway. The Convair pilots had been unable to see the P-3 flying below them, and they glided down on top of it:

> NASA 711 struck BA 482 from the rear on top of the vertical fin. BA 482 rotated to a very nose-up attitude, and the nose of NASA 711 [ripped] into the top aft section of [the BA 482] fuselage.

The Convair had struck the vertical stabilizer of the P-3 and crushed it. The impact 200 feet above the ground caused the P-3 to pitch up into the bottom of the Convair fuselage. Interlocked, the two planes slammed down onto the 12th tee at Sunnyvale Municipal Golf Course. They hit the ground one-half mile short of Runway 32 Left and exploded into an orange and red fireball:

> BA 482 crashed entangled with NASA 711. An ensuing fire totally destroyed BA 482 . . . and NASA 711.

Two high school golf teams were playing the 12th hole, and the young golfers ran toward the flaming wreckage. Ignoring the inferno, one teenager used his golf club to pound on the P-3 windshield. He was unable to break through the plexiglass and reach the pilots, who were burning alive in the cockpit. An independent report explained the utter chaos at the crash site:

> The two planes crashed only 200 yards from Bayshore Freeway, and drivers left their cars to scale the fence and rush to the burning wreckage. Golfers and people from nearby office buildings . . . also swarmed around the site.

Two teenage golfers found a survivor, a Navy petty officer who had been thrown from the P-3 wreckage. He lay on the ground 70 feet from the fire, unconscious and horribly burned. The golfers covered him with an open parachute.

Firetrucks arrived within minutes. The driver of one truck saw the parachute on the ground. He did not realize a man was under the chute, and he drove his 13,000 pound firetruck over it. Fortunately the wheels straddled the unconscious victim. This "miracle man" would eventually recover, but the other 16 men died in the crash and fire. One month later the Navy would report:

> There were 6 SOB's *[meaning "souls on board"]* on the P-3 and 11 SOB's on the Convair 990. [The only] survivor . . . remains hospitalized in serious condition.

The Investigation: A Navy-NASA team investigated this accident. The team recovered the Convair CVR. Conversation in the cockpit confirmed the Convair crew never realized the P-3 was below them. Also, conspicuity logic dictated that the P-3 crew would have been unable to see the Convair descending down on top of them from above and behind:

> Runway 32 Right is . . . normally used for final landing traffic when other aircraft are in the touch-and-go pattern on 32 Left.

The P-3 had been in the touch-and-go pattern on Runway 32 Left. The controller twice had told the Convair pilots to expect a landing on the other runway, Runway 32 Right. However, at 1449 Hours the controller cleared the Convair pilots as follows: ". . . cleared to land, Runway three-two Left." The pilots acknowledged the change and turned slightly to the left to line up on the left-hand runway:

> The pilot [accepted] his final clearance to land without question.

Investigators interviewed the controller and confirmed he had made a tragic mistake. He had *intended* for the Convair to use the right-hand runway. Yet, with a simple slip-of-the-tongue he mistakenly cleared the Convair to land on the left-hand runway, which was being used by the P-3 and several other aircraft.

Probable Cause: The U.S. Navy and NASA completed the investigation and forwarded their report to the Naval Safety Center on 11 May 1973, the month following the accident. The weather had been clear and sunny, and visibility had been estimated at seven miles. Neither aircraft had suffered mechanical problems, and both aircrews were properly trained and experienced. The board members logically reached the following *Conclusion*:

> The cause of this accident was an erroneous [radio] transmission by the tower controller, clearing NASA 711 to land on [Runway] 32 Left vice [Runway] 32 Right, as was his intention. This clearance was probably the result of the controller's division of attention between simultaneous landing traffic on parallel runways, coupled with his attention concentrated primarily on traffic clearing Runway 32 Left.

The board pointed out a *Contributing Cause*:

> Failure of the crew of NASA 711 to . . . maintain a proper lookout doctrine for [the P-3], which had transmitted being on base leg and [which had been] cleared to continue for [Runway] 32 Left.

I Just Hit Something!
Mid-Air Collision near Saxis, Virginia
11 October 1974

<u>Convair F-106 Delta Dart, Serial No. 59-0044</u>
New Jersey Air National Guard low-level intercept training
1 aboard, 0 killed

<u>Piper PA-24 Comanche, Registration N6876P</u>
Private business and pleasure flight
3 aboard, 3 killed

<u>Synopsis</u>: NORAD vectored a Convair F-106 Delta Dart toward another F-106 during intercept training at night. The interceptor's radar locked-on the wrong aircraft, a civil Piper PA-24 Comanche. The F-106 overran the Comanche and rammed into it. The impact killed all three people in the civil aircraft. The damaged F-106 landed safely.

The Convair F-106 Delta Dart Flight: Two military New Jersey Air National Guard (ANG) interceptors roared into the night sky from the National Aviation Facilities Experimental Center at Pomona, New Jersey. The North American Air Defense Command (NORAD) guided the Convair F-106 Delta Darts south for low-level intercept practice. NORAD's "Fertile Control" would vector one F-106 toward the other military aircraft. Then the F-106 automated radar guidance system would lock-on and complete the mock shoot-down:

Convair had designed the F-106 as an all-weather interceptor to shoot down Soviet bombers and fighters. The U.S. Air Force had taken delivery of the first new F-106 in 1959. The wasp-waist fuselage, highly swept delta wings, and a Pratt & Whitney J-75 turbojet gave the interceptor blazing speed. The plane had set a world speed record of Mach 2.41.

The F-106 used a Hughes MA-1 electronic guidance and fire control system. This system could control the aircraft and could fly it to the proper altitude and attack position. Then the pilot could fire air-to-air Genie 2A nuclear missiles or conventional Hughes Aim 4 air-to-air Falcon missiles.

A Convair F-106 Delta Dart firing the air-to-air Genie 2A nuclear missile (official U.S. Air Force photo)

Both F-106s were assigned to the 177ᵗʰ Fighter Interceptor Group. One F-106 (call-sign, EL-10) would be the first to play the role of target. NORAD would vector the other interceptor (call-sign, EL-08) for an attack from the rear.

A veteran Air National Guard captain, age 33, piloted EL-08 (the mishap aircraft), and he had logged 1,422 hours of military flight time. In addition he held a civil Commercial Pilot certificate.

The Piper PA-24 Comanche Flight: A civil Piper PA-24 Comanche flew south along airway V-1 at 8,500 feet. The pilot, age 39, had taken off from White Plains, New York, on a flight to Georgetown, South Carolina. He was Vice President of the Bank of New York and had earned a Private Pilot certificate plus an Instrument Rating. His logbook reflected an impressive 1,606 hours of total flight time, 198 hours of night flying, 85 hours of actual instrument time, and 61 hours of simulated instrument time:

The Comanche was a favorite of professional pilots. The low-wing plane had seats for four people, retractable gear, and a big Lycoming engine. The aircraft was a stable IFR platform, and auxiliary fuel tanks gave it a range of well over 1,000 miles.

Trouble Ahead: In the clear night sky the Comanche cruised south on a VFR flight plan. Over Salisbury, Maryland, the pilot radioed

Washington Center and requested flight following.   The civil
controller watched the secondary radar return from the Comanche
transponder as the plane sped along the airway centerline:

> N6876P was on airway V-1 at 8,500 feet, as filed on his VFR
> flight plan, and Washington ARTCC was providing radar
> following and traffic advisory information.

The Center controller knew some type of military aircraft was in
the area because he could see a Mode III, Code 1200, display on his
scope (the secondary return from EL-10). The controller did not see
a primary or secondary return from the other F-106.

Meanwhile, military Fertile Control vectored EL-08 toward its
target, EL-10, which was flying at 3,000 feet. The military controller
saw the secondary return from unidentified traffic (the Comanche) on
his scope and reported it to EL-08 and EL-10.

The combined civil and military radio transcript, below, begins as
the Comanche makes initial contact with Washington Center. The
center controller and the Comanche operated on a civil VHF
frequency. The NORAD controller and the two F-106 interceptors
used a military UHF frequency. The civil controller and Comanche
on the VHF frequency could not hear military radio traffic on the
UHF frequency, and vice versa:

> Comanche: . . . Piper PA-24 Comanche
> EL-08: . . . . . . Convair F-106 Delta Dart (call-sign, EL-08)
> EL-10: . . . . . . Convair F-106 Delta Dart (call-sign, EL-10)
> Center: . . . . . . Washington Center (ARTCC)
> Fertile: . . . . . . Fertile Control (NORAD military controller)

Comanche: Washington Center, Comanche six-eight-seven-six papa
with you, eight-thousand-five-hundred over Salisbury. (2010:29)

[Center acknowledges the contact]

[7 minutes later Fertile Control vectors EL-08 toward EL-10]

Fertile: Zero-eight, short range commit, tight port turn [to] one-five-
zero . . . one target, stern attack, target's bearing is one-three-one at
seven miles, heading one-nine-zero, three-thousand feet, Mach point
four-three, crossing left to right, two-thousand low. (2017:23)

EL-08: [Echo-Lima] zero-eight. *[an acknowledgment]*

Fertile: Zero-eight, target one-three-five at six [miles]. (2017:53)

EL-08: Zero-eight.

Fertile: Zero-eight, target now one-six-two at four miles. (2018:43)

EL-08: Zero-eight.

Fertile: Zero-eight, target one-seven-three at four miles, [turn to] starboard, [heading] one-seven-five. (2018:53)

EL-08: Zero-eight.

Fertile: Zero-eight, target now one-eight-zero at four miles, starboard one-eight – correction – starboard to one-eight-zero. (2019:33)

[Fertile Control gives a traffic advisory to EL-10, the "target"]

Fertile: One-zero, you have traffic one-nine-zero at eight miles, heading south, you should be clear of it. (2019:42)

Fertile: Zero-eight, target now one-seven-five at three [miles], come back to port, one-seven-five. (2019:52)

Fertile: Zero-eight, target one-seven-five at two [miles], minimum range. (2020:27)

["minimum range" means the minimum allowable range astern without radar lock-on by the interceptor]

EL-08: Zero-eight, no Judy *[meaning, no radar lock-on]*

Fertile: Zero-eight, [I] understand no Judy? (2020:32)

EL-08: Affirmative.

Fertile: Roger, break port, one-three-zero, maintain angels five, set speed for four-eight. (2020:35)

[the EL-08 radar suddenly <u>locks-on the Comanche</u>; the EL-08 pilot <u>assumes</u> it has locked-on the "target," EL-10]

EL-08: Zero-eight has a Judy. (2020:49)

Fertile: Roger, you're cleared in on target. (2020:53)

[Fertile Control gives another traffic advisory to EL-10]

Fertile: One-zero, your traffic now one-nine-five at seven miles, heading about one-nine-zero. (2021:04)

EL-10: One-zero, no joy. *[meaning, no visual contact]*

[meanwhile, Center gives traffic advisories to the Comanche]

Center: Seven-six papa, you have traffic in your six o'clock position, five to seven miles, southeast-bound, VFR below ten-[thousand feet], he's slowly overtaking you. (2022:12)

Comanche: [Center, this is] seven-six papa, we'll be looking over our shoulder here. (2022:48)

[Fertile Control gives a traffic advisory to EL-10 and EL-08]

Fertile: Echo Lima one-zero and [Echo Lima] zero-eight, you have traffic paralleling you to port . . . four miles. (2022:48)

EL-10: One-zero, tally ho. *[meaning, I see the traffic]* (2022:57)

EL-08: Zero-eight. (2022:59)

[on radar Center sees only the "target" aircraft, EL-10 (not the rapidly closing EL-08); the Center controller gives a final traffic advisory to the Comanche]

Center: Seven-six papa, he's in your eight o'clock position now, and about five miles . . . gonna be off your left wing. (2023:07)

*[EL-08 **collides** with the Comanche at 2023:09]*

EL-08: [Unintelligible message]. *[spoken under stress]* (2023:11)

Fertile: Zero-eight, say again? (2023:13)

Fertile: Go ahead, zero-eight, what did you say? (2023:17)

Fertile: Echo Lima zero-eight, this is Fertile, say again your first transmission? (2023:22)

EL-08: [This is] Echo Lima zero-eight! [I need] to be [vectored] to an emergency field! (2023:35)

Fertile: Roger, squawk mode three, [code] seventy-seven-hundred, emergency. (2024:04)

Fertile: Zero-eight, what's your problem? (2024:09)

EL-08: . . . I just hit something! (2024:14)

The Collision: The F-106 radar had locked-on the unseen Comanche, a much slower civil aircraft. The airspeed mismatch gave the military F-106 a closure rate of almost 300 knots.

For the F-106 pilot the only warning was a sudden "bright red light." A split-second later his plane slashed through the Comanche empennage and right wing. A piece of the Comanche wing remained wedged in his left air intake. Yet, the sturdy warplane remained flyable. The pilot flew back to his base and landed safely.

However, the Comanche and its occupants would die. Most of their empennage and the majority of their right wing had been sheared off. A witness on the ground watched the Comanche remains fall straight down into Pokomoke Sound, a part of Chesapeake Bay. The mangled fuselage plunged through the shallow water into the mud and muck of the seabed. The three people aboard – the pilot, his wife, and their daughter – perished in the collision and crash.

Witnesses reported that a plane had crashed into the water in the dark. They had no inkling that another aircraft had been involved. The wreckage was submerged and invisible. By the next morning all that was known was that some type of plane had crashed, and that a military plane had hit "something" during intercept practice. The *Richmond Times Dispatch* newspaper headline speculated:

Plane Down in Bay, May Have Hit Fighter

Over the next two days an undersea search located major pieces of the Comanche wreckage. Three days after the mid-air collision the *Associated Press* would explain:

> Part of the wreckage [has] been found imbedded in mud under 15 feet of water, [and] partial remains of three bodies were recovered. The force of the crash reportedly scattered the [human] remains and parts of the plane across a wide area.

The Investigation: NORAD trained F-106 pilots in radar intercepts of low-flying aircraft. This training had to be done at locations where Fertile Control had good radar and radio coverage. The FAA did not require NORAD to coordinate such day or night training with ATC when the flights were conducted in VFR conditions.

Center traffic advisories given to the Comanche just before the collision were based upon the transponder Mode III, Code 1200, secondary radar return from EL-10:

> The [Center] controller did not see a primary or secondary radar return from [the other F-106], probably because the EL-08 transponder was not set for Mode III, Code 1200.

After the collision the secondary return from the Comanche had faded off the center controller's scope. Moments later the controller saw a new secondary return, Mode III, Code 7700. This was the emergency code that EL-08 squawked after the mid-air impact.

Before the collision Fertile Control had seen the transponder return from the Comanche. The military controller had radioed the location of the civil aircraft to EL-08 and EL-10. Yet, neither Fertile Control nor the F-106 pilots suspended the intercept. By the time Fertile Control reported the *real* target (EL-10) at two miles, EL-08 had zoomed to within only one mile of the unsuspecting Comanche:

> [The] EL-08 [pilot] either did not see [the Comanche], or saw it, returned to his intra-cockpit duties, and . . . inadvertently locked-on the civil aircraft rather than EL-10.

Probable Cause: Less than four months after the accident the NTSB adopted an Aircraft Accident Report, NTSB-AAR-75-6, on 29 January 1975. The board pointed out that airway V-1 is heavily traveled by both VFR and IFR traffic. NORAD's intercept practice had been conducted in an area that included not only V-1 but also two more airways, V-139 and V-194. Washington Center had not been aware that night VFR military intercept practice was being conducted in its area of responsibility. In addition NORAD radar could not paint all civil aircraft in the intercept area:

> NORAD is unable to identify all aircraft in the area sufficiently to insure that their interceptors are separated from them.

The NTSB noted that intercepts by military aircraft involve high speed maneuvers and a heavy pilot workload. The F-106 guidance system had locked-on the wrong target. The faster F-106 overran the plodding Comanche and collided with it. The board found that the **Probable Cause** of this accident was:

> Failure of the interceptor pilot to see and avoid a civil aircraft during a high speed, low altitude, intercept training flight conducted in an area which included major north-south airways.

The board took note of a **Contributing Cause**:

> The [present] system which permitted an incompatible mix of traffic in controlled airspace, which resulted in . . . an inadvertent radar lock-on to a civil aircraft.

Radar Lock-On . . . Tally-Ho!
Mid-Air Collision near Kingston, Utah
12 November 1974

<u>General Dynamics F-111 Aardvark, Serial No. 77-055</u>
U.S. Air Force night in-flight refueling training mission
2 aboard, 0 killed

<u>Rockwell 690 Turbo Commander, Registration N40MP</u>
Montana Power Company ferry flight
1 aboard, 1 killed

<u>Synopsis</u>: On an ebony-black night a General Dynamics F-111 Aardvark tried to rendezvous with a tanker for in-flight refueling. The F-111 radar locked-on the wrong aircraft, a civil Rockwell 690 Turbo Commander. In the dark the F-111 closed on the slower airplane and rammed into it. The collision killed the civil pilot, but the two F-111 crewmen ejected and survived.

The General Dynamics F-111 Aardvark Flight: After nightfall two General Dynamics F-111 Aardvarks took off from Nellis AFB near Las Vegas, Nevada. The U.S. Air Force planes were assigned to the 474th Tactical Fighter Squadron at Nellis. On this IFR training mission the F-111 crews planned to meet a Boeing KC-135 Strato-tanker and practice in-flight refueling. They intended to rendezvous on Air Refueling Track 316 (AR-316), which is designated airspace for such operations:

> *The F-111 is one of the more controversial military aircraft ever to fly. In the 1960s the U.S. Air Force needed a long-range strike bomber to penetrate Soviet Union defenses. The U.S. Navy wanted a nimble fighter-interceptor for fleet defense. The new U.S. Secretary of Defense, Robert McNamara, ignorantly claimed that one new aircraft could fill both roles.*

> *The General Dynamics design was adequate for the Air Force mission. Conversely, the proposed Navy variant was far too heavy and grossly underpowered. Studies showed it "could not survive" in a dogfight with Soviet fighters. In testimony before the U.S. Congress, Admiral Tom Connolly was asked if bigger*

General Dynamics F-111 Aardvark (official U.S. Air Force photo)

engines might eliminate the thrust deficiency. Admiral Connolly voiced the now-famous retort:

> *"Gentlemen, there isn't enough thrust in all Christendom to make that airplane a fighter."*

The ill-conceived Navy variant never flew, but the Air Force version became a success. The novel swing-wing design allowed the F-111 to take off at slow speed and accelerate to Mach 2.5 at 60,000 feet. Terrain-following radar and digital electronics allowed low-level and all-weather weapons delivery. Powered by twin Pratt & Whitney turbofans, the F-111 could carry 25,000 pounds of conventional or nuclear munitions.

The mishap F-111 pilot, a captain 32 years of age, had earned an Air Force instructor pilot rating and had logged 2,850 hours of flight time. His weapons system officer (WSO), also a captain, held a navigator rating and had 901 hours of flight experience.

The Rockwell 690 Turbo Commander Flight: A big twin engine Rockwell 690 Turbo Commander left Phoenix, Arizona, on a VFR

Rockwell AC-500 Aero Commander, one of the series of piston-powered planes from which the Rockwell 690 Turbo Commander evolved (photo courtesy of U.S. National Weather Service)

flight plan. The pilot, age 34, held an ATP certificate and had logged 2,754 flight hours. Montana Power Company owned the plane, and the pilot planned this solo night flight to ferry it to Butte, Montana:

> *The Turbo Commander evolved from the Aero Commander series of piston-powered twin engine planes. The turboprop Model 690 first flew in 1968. Two AirResearch turbine engines and full IFR instrumentation made the high-wing Turbo Commander a popular business aircraft for decades.*

Trouble Ahead: The KC-135 Stratotanker arrived over Milford VORTAC, the Air Refueling Control Point (ARCP), at 1739 Hours. Salt Lake City Center cleared the tanker for an altitude block between flight levels 180 and 210. The tanker pilots planned to refuel the two F-111s at flight level 200 within the designated AR-316 area.

Things did not go according to plan. The two F-111s missed their Air Refueling Initial Point (ARIP) because the leader (call-sign, Sigma 72) had developed TACAN problems. Also his vertical steering bar failed to display INS information:

The pilot of Sigma 72 was unable to position his aircraft over the ARIP. He did not know whether he had overflown the ARIP or if he was off to one side.

Because of his navigation problem, Sigma 72 passed the formation lead to his wingman. Thereafter, Sigma 71 led the F-111 flight. Meanwhile, at 1800 Hours the Turbo Commander pilot air-filed an IFR flight plan with Cedar City FSS. He requested flight level 180, the same altitude at which the F-111s were flying. Four minutes later he radioed Center:

> ... OK, we're squawking fourteen-hundred, we're at seventeen-five *[meaning, 17,500 feet]* and would like to go to eighteen, have you got [my] flight plan from flight service?

The portion of the military ATC radio transcript, below, begins as the two F-111s close – in the dark – on what they *think* is the tanker:

> KC-135: ... Boeing KC-135 Stratotanker (call-sign, Toft 51)
> F-111: ..... Gen. Dyn. F-111 Aardvark (call-sign, Sigma 71)
> ICS-WSO: . ICS transmission, Sigma 71 WSO

KC-135: Sigma seven-one ... we're showing forty-seven [miles] and we're going ahead into a left turn at this time, and do you have us on radar yet? (apx. 1801:00)

F-111: Negative. (apx. 1801:10)

> *[at 1801:25, dead ahead, the F-111 pilot sees what be* assumes *is the red anticollision beacon on the tanker, but* actually *he is looking at the anticollision beacon on the Turbo Commander]*

F-111: Toft five-one, I believe I have tally-ho at twelve o'clock *[meaning, I think I see you dead ahead]*, can you turn? (1801:25)

> *[the F-111 radar locks-on the Turbo Commander, but the F-111 pilot and WSO* assume *it has locked-on the tanker]*

F-111: Toft five-one, we have a [radar] lock-on at eight miles, you can maintain your speed [at] this time. (1801:50)

KC-135: Roger, [we] copy. (apx. 1801:56)

[the WSO reports the range has closed to <u>three miles</u>]

[the WSO reports the range has closed to <u>two miles</u>]

[the WSO reports the range has closed to <u>one mile</u>]

<u>ICS-WSO</u>: The range is four-thousand feet, it looks like we have a fast overtake.

<u>ICS-WSO</u>: Range, two-thousand feet – we're closing fast!

*[the F-111 **<u>collides</u>** with the Turbo Commander at apx. 1804:12]*

[the KC-135 crewmen see "fiery streaks" in the night sky]

<u>KC-135</u>: I see flares out there, [Sigma] seven-one! (1804:45)

[there is no response]

The Collision: Without any depth perception in the night sky the lead F-111 pilot, Sigma 71, had closed on the flashing anticollision beacon in the black void ahead. The WSO had peered into his radar scope and called out the range: three miles – two miles – one mile – four-thousand feet – two-thousand feet. Suddenly the outline of an aircraft "blossomed" out of the darkness ahead.

At 17,900 feet with a closure rate of 158 knots the F-111 overran the Turbo Commander and rammed through it. The collision ripped the civil aircraft apart. The pilot literally never knew what hit him. The impact tore him out of his cockpit, and his body eventually would be found two miles from the main wreckage site:

> The Turbo Commander broke into several large and many small pieces. The pieces fell to earth . . . along a 9-mile wide path which ended about 2 miles from the wreckage of the F-111.

The F-111 pilot and WSO *assumed* they had hit the tanker. They ejected, the parachute opened, and the escape module floated down into the mountains. The module struck the ground on an incline, rolled over twice, and stopped. Glad to be alive, the flyers crawled out of the module with only minor injuries. The next day the *Las Vegas Review Journal* would carry a front page story that began:

> Two Nellis Air Force Base pilots were in good condition Wednesday after their F-111 collided with a twin-engine plane in

flight over southern Utah on Tuesday night in an explosive "ball of fire" seen for 50 miles.

The Investigation: The NTSB began digging into the accident and looking for answers. The tanker and the two F-111s had been on IFR flight plans under positive radar control. Center radar data showed the tanker had been at flight level 200 inside the AR-316 refueling track, which was protected airspace.

The F-111 radar had locked-on the Turbo Commander, which was *outside* the refueling track. Score marks on the wreckage confirmed the F-111 had overtaken the Turbo Commander from the rear.

Investigators found that trouble had begun after the first F-111 flight leader encountered TACAN and INS problems. By the time he realized his navigation difficulty and passed the lead to his wingman (the mishap aircraft, Sigma 71), the two F-111s had missed their ARIP by 15 to 17 miles:

> This position error, which was caused by navigation equipment errors . . . placed them *outside [emphasis added]* the protected airspace of the refueling track.

The new leader, Sigma 71, had not noticed the navigation error. Based on a *faulty assumption* that they were at the ARIP, the pilot and WSO had flown directly toward where they expected the tanker to be. The pilot had spotted an expected anticollision beacon ahead in the darkness. His radar had locked-on a target at the expected distance and heading. However, because of the initial navigation error, Sigma 71 had been homing on the Turbo Commander, not the tanker.

The Turbo Commander had been flying at 170 knots, so the F-111 rapidly closed the range. The WSO warned his pilot about the "fast overtake" and called out the decreasing range. Yet, the pilot had visual contact with the flashing beacon in the black abyss ahead. He *expected* to rendezvous at about 305 knots. With (1) no other visual cues, (2) no depth perception, and (3) no relative motion, he had homed-in on the only thing he could see, the flashing beacon:

> The tanker's beacon is the only recognizable visual cue during a night rendezvous. The tanker's underbelly lighting and engine nacelle lighting cannot be seen until the [rendezvousing aircraft] is close to the tanker.

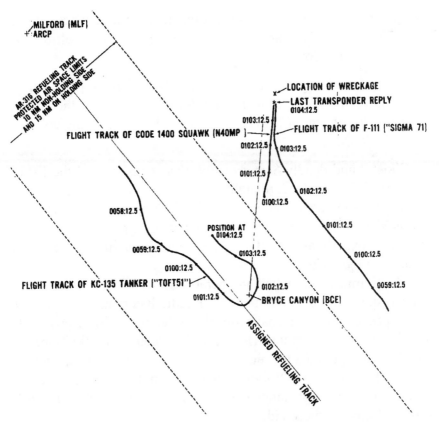

Diagram showing (1) track of the KC-135 tanker within protected airspace of AR-316 and (2) tracks of the F-111 and the Turbo Commander outside of AR-316 (diagram courtesy of U.S. NTSB)

The tanker had been flying at flight level 200, and the F-111 had been at flight level 180 as planned. If the F-111 pilot had been flying toward the tanker, the tanker beacon would have appeared to move up on his windshield as he closed the range. Yet, the pilot's visual picture had never changed. The beacon had remained stationary on his windshield. Homing on the light, he had not sensed the visual anomaly. He continued to close on the light and collided with the much slower Turbo Commander.

Probable Cause: Nine months after the accident the NTSB adopted the Aircraft Accident Report, NTSB-AAR-75-12, on 1 August 1975. The Turbo Commander had been flying at 17,900 feet. The Salt Lake City Center controller had been watching transponder displays from the tanker and both F-111s. The last recorded altitude for Sigma 71 was 17,900 feet. By chance that was precisely the same altitude at which the Turbo Commander was flying:

> The pilot of Sigma 71 mistook the [anticollision] beacon of N40MP *[the Turbo Commander]* for that of Toft 51, the tanker. The pilot of Sigma 71 closed on N40MP's beacon and collided with N40MP.

Sinister and deadly *Scenario Fulfillment* had snared the F-111 pilot and WSO. In the dark night sky with no other visual cues, they *expected* to spot the tanker beacon. The flashing red beacon visually appeared exactly where they expected it to appear. Their radar locked-on exactly as expected. They logically assumed they were homing on the tanker.

The Sigma 71 crew had an operable TACAN with an air-to-air range display option. Also their UHF radio had homing capability. The pilot and WSO could have used these electronic aids to positively identify the tanker location. However, their radar had locked-on, and the pilot had spotted a flashing beacon. During the investigation he testified that he simply "followed that beacon the rest of the way."

In brief and simple language the board spelled out the ***Probable Cause*** of this unique accident:

> The F-111 pilot's misidentification of the Turbo Commander as a refueling tanker with which he intended to rendevous.

The board noted a ***Contributing Cause***:

> Contributing to the misidentification was [the pilot's] failure to use prescribed procedures and techniques during rendezvous with a tanker for refueling.

Unavoidable Collision?
Mid-Air Collision over Whittier, California
9 January 1975

de Havilland Canada DHC-6 Twin Otter, Registration N6383
Golden West Airlines Flight 261
12 aboard, 12 killed

Cessna 150, Registration N11421
CessnAir Aviation Inc. pilot training flight
2 aboard, 2 killed

Synopsis: In clear weather a de Havilland Canada DHC-6 Twin Otter airliner, in radio and radar contact with ATC, approached to land. A Cessna 150 cruised toward the Twin Otter approach course. Because of extraordinarily rare circumstances (1) the Cessna was invisible on radar, and (2) it was impossible for any of the pilots to see the other aircraft. The planes collided, killing all 14 people aboard.

The de Havilland Canada DHC-6 Twin Otter Flight: On a sunny afternoon Golden West Airlines Flight 261, a de Havilland Canada DHC-6 Twin Otter, climbed skyward from Ontario, California. The regional airliner carried 2 pilots and 10 passengers on a scheduled short flight toward Los Angeles International Airport. At 1604 Hours the pilots radioed Los Angeles Approach Control as they flew over Rose Hills, a local landmark:

> The arrival controller . . . assigned the flight a new transponder code and cleared [the pilots] for a Terminal Control Area (TCA) No. 2 arrival to Runway 24 Left.

The captain, age 47, held an ATP certificate and had been flying for Golden West since 1967. His logbook reflected 9,366 hours of flight experience. The first officer, age 27, had been flying for the airline for four years. He had earned a Commercial Pilot certificate and had logged 2,555 flying hours:

> *The Twin Otter has been the most successful aircraft program in Canadian history. The plane has carved out a huge niche in the*

de Havilland Canada DHC-6 Twin Otter (photo courtesy of U.S. DOD Visual Information Center)

commuter airline and freight business. It can carry up to 20 passengers, and STOL capability allows it to fly from short and primitive runways. Two Pratt & Whitney turboprop engines power the Twin Otter. Military variants have served in the armed forces of the United States and 19 allied countries.

The Cessna 150 Flight: At 1546 Hours a flight instructor and a student pilot took off from Long Beach Airport in the sprawling Los Angeles metropolitan area. The Cessna 150 flight had been scheduled as the student's last instruction flight before his FAA check-ride for a Private Pilot certificate. No flight plan was needed for this VFR training mission:

The Cessna 150 is a two seat plane designed for flight training and personal use. It has fixed gear and is powered by either a Lycoming or a Continental piston engine. The small aircraft is arguably the most popular trainer in the world. An astounding 23,949 have been built: 22,138 in the United States, 1,764 in France, and 47 in Argentina.

The instructor, age 47, was the chief pilot for CessnAir Aviation Inc., which owned the Cessna. Although a corporate pilot, he often

Cessna 150 (photo courtesy of Adrian R. Pingstone)

flew training flights in the flight school department of his company. He held an ATP certificate and had amassed 22,010 hours of flight time. His student, age 25, had logged 42 hours in the air. CessnAir Aviation used the Cessna 150 as a VFR primary trainer, so it was not equipped with a transponder.

Trouble Ahead: Approach Control issued a traffic advisory to the Twin Otter pilots at 1605:50 Hours. The controller reported a police helicopter five miles away at twelve o'clock. The Twin Otter pilots looked for the traffic but did not spot it. Meanwhile, undetected by ATC and the two airline pilots, the Cessna flew straight toward the approach path of the Twin Otter.

The portion of the ATC radio transcript, below, contains only the radio messages between Approach Control and the Twin Otter pilots. The transcript begins as they make initial contact with ATC:

Golden West 261: ... Golden West Airlines Flight 261
AppControl: .... Los Angeles Approach Control

Golden West 261: [Approach Control, this is] Golden West two-six-one at Rose Hills .... (1604:45)

AppControl: Golden West two-sixty-one, squawk zero-seven-two-two and ident – radar contact twenty-three miles east of the airport, TCA number two to Runway two-four left. (1604:50)

Golden West 261: Two-four left it is, two-sixty-one. (1604:55)

AppControl: Golden West two-six-one, verify leaving two-thousand-six-hundred [feet]. (1605:45)

Golden West 261: [We are passing through] six-[thousand now].

AppControl: Roger, at twelve o'clock and five-and-a-half miles is a police helicopter climbing out of one-thousand-five-hundred, [unintelligible] three-thousand VFR, I'll point him out again when he's a little closer, let me know when you have him in sight. (1605:50)

Golden West 261: Two-sixty-one, we'll do it. (1605:55)

*[63 seconds later the airliner and Cessna **collide** at 1606:58; the controller does not know the planes have collided]*

AppControl: Golden West two-sixty-one, that helicopter is at eleven-thirty and three miles now, looks like he's northbound . . . . (1607:35)

[there is no response, so the controller repeats the advisory]

AppControl: Golden West two-sixty-one, that helicopter is now at eleven-thirty and three miles, northbound . . . . (1607:55)

[the controller sees that the primary and secondary radar returns from the Twin Otter have dropped off his scope]

AppControl: Golden West two-sixty-one. *[a query]* (1608:05)

AppControl: Golden West two-six-one, [this is] Los Angeles, if you hear me, ident. (1608:10)

AppControl: Golden West two-six-one, Los Angeles Approach Control, how do you hear? One, two, three . . . . (1608:25)

AppControl: Golden West two-six-one, radar contact [has been] lost, last position observed [was] one-seven miles east of the Los Angeles Airport, if you hear me attempt [to] contact the tower on one-two-zero-point-eight. (1608:35)

[there is no response]

The Collision: The Twin Otter had flown toward the airport at 150 knots while descending at 300 feet per minute. The pilots had been heading toward the ILS localizer for Runway 24 Left. Off to their left the Cessna had flown north across their approach path. At 2,200 feet over Whittier, California, the small plane had rammed through the airliner fuselage in front of the wing. A witness on the ground heard a noise "like a thunderclap" and looked up. She explained:

> I saw the airliner twisting down out of the sky without its wings, with people falling out [of it].

Parts of the Twin Otter fell onto the athletic field at Katherine Edwards Intermediate School, where about 200 young children were watching an outdoor basketball game. Miraculously none of them were injured. A body thrown from the Twin Otter splattered onto the basketball court, missing the players. A chunk of wreckage plunged through the schoolhouse roof and crushed the basketball coach's desk and chair. Fortunately he was outside with his team.

Another body fell into a front yard on Reichling Lane across the street from the school. Mangled wreckage fell onto nearby homes. The next day the *Los Angeles Times* newspaper headline would read:

Air Crash Over Whittier Kills 14

A newspaper photograph showed a priest giving the Last Rites to small scraps and pieces of human remains. The newspaper quoted a shocked Whittier resident:

> There's a wing and an engine sitting right in my baby's bedroom. I have a three-year-old daughter, but she wasn't here when it happened. She was [with] a babysitter. Thank God!

The collision and crashes had killed all 14 people on the two planes. Pieces of human remains were strewn throughout Whittier.

The Los Angeles Sheriff's Office appealed to the public for help in finding parts of bodies. The NTSB later would note:

> Wreckage of both aircraft [and human remains] were scattered over an 8 to 10 city block area.

The Investigation: The NTSB found that this was an operational accident. No mechanical problems, electronic failures, or communications issues had played a role in the collision. Visibility had been estimated at 40 miles. Both planes had been outside the Los Angeles TCA in a seen-and-be-seen environment.

Investigators focused on why the pilots in each aircraft had not seen the other plane in time to avoid a collision. Also, radar had painted the police helicopter. Why had radar not painted the Cessna?

Investigators unmasked an electronic gremlin that had concealed the Cessna. The ASR-4 radar used by Approach Control featured a Moving Target Indicator (MTI) electronic gate to eliminate radar ground clutter caused by buildings, antenna towers, hills, etc:

> The MTI circuitry electronically cancels primary target returns that are stationary, or appear to be stationary, with respect to their distance from the [radar] antenna [and clears the controller's scope] of ground clutter.

The NTSB acknowledged a down-side to this unique capability:

> A negative characteristic of the MTI [circuitry] is that a non-transponder-equipped aircraft flying a course that is tangential to the radar antenna produces an apparently stationary target . . . . This phenomenon is called tangential effect.

Investigators conducted a flight test. They flew a plane without a transponder along the flight path of the Cessna. On radar a primary return was not displayed. The Cessna had been *invisible* on radar:

> The Cessna's tangential course [had] produced a return which was canceled out by the MTI circuitry so that there was no video return [from the Cessna] on the controllers's scope.

Despite the Cessna's invisibility on radar, investigators initially assumed the pilots in both planes could have seen the other aircraft. Ironically that turned out to be untrue. At the time of the accident the

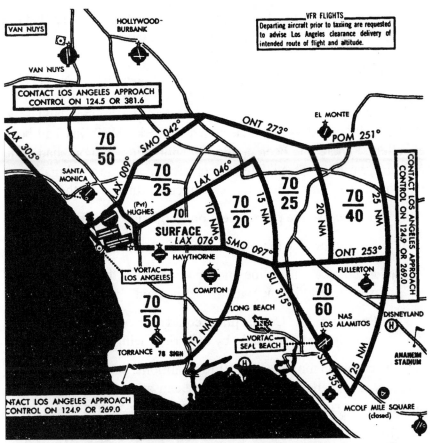

Diagram depicting the Los Angeles Terminal Control Area (diagram courtesy of U.S. NTSB)

sun had been setting in the western sky nine degrees above the horizon. The Cessna flight path had placed it exactly between the blazing setting sun and the Twin Otter:

> It would have been . . . impossible to see the Cessna against the backdrop of the setting sun. *[and also]* Because they were on a collision course there would have been no change in the relative sighting angles *[meaning, no relative motion]* as the [two] aircraft approached the collision point.

Tragically, it got worse. The Cessna pilots also had been blinded. The Twin Otter had been above them, 90 degrees to their right. The high right wing of the Cessna had totally blocked its pilots' view of the Twin Otter, which was descending toward them:

The angle of closure between the [two] aircraft was such that the Twin Otter was masked by the Cessna's wing.

Probable Cause: Seven months after the accident the NTSB adopted the Aircraft Accident Report, NTSB-AAR-75-14, on 7 August 1975. ATC had made no errors. The pilots had performed flawlessly. Radar had worked as it was designed to work. Visibility had been excellent. There had been no aircraft anomalies. No one had done anything wrong. Instead, fate had contrived a unique combination of extraordinarily rare circumstances that constituted a death sentence:

1. *By chance*, the planes were on a collision course.

2. *By chance*, due to its track, the Cessna was invisible on radar.

3. *By chance*, the airline pilots could not see the Cessna, which was directly in line with the setting sun.

4. *By chance*, the Cessna pilots could not see the airliner, which was blocked from sight by their right wing.

The collision took place in VFR conditions, so the board did not consider radar tangential effect to be a causative factor. The board detailed the *Probable Cause* of this unfortunate accident:

The board believes that [it was impossible for] both crews to detect the other aircraft in time to avoid a collision . . . because of the position of the sun [and] the closure angle of the [two] aircraft.

Clear Sky and Unrestricted Visibility
Mid-Air Collision at Newport News, Virginia
9 January 1975

<u>Convair VT-29 (C-131), Serial No. 52-5826</u>
U.S. Air Force administrative flight
7 aboard, 7 killed

<u>Cessna 150, Registration N50430</u>
Cavalier Flyers Inc. local pleasure flight
2 aboard, 2 killed

Synopsis: In VMC weather at night a military GCA controller vectored a Convair VT-29 toward the precision approach path. A civil Cessna 150 flew across the GCA approach course. The planes collided, killing all nine people on both aircraft.

The Convair VT-29 (C-131) Flight: U.S. Air Force pilots planned an administrative flight from Langley AFB in Virginia to Shaw AFB in South Carolina, then to Key Field in Mississippi, and then back to Langley. They flew the first two legs of the mission as planned in their Convair VT-29 transport. At 1520 Hours they took off from Key Field on an IFR flight plan for the return trip to Langley. In addition to two pilots and a flight mechanic the plane carried two military "flight stewards" and two military VIP passengers:

The VT-29 (also known as the C-131) is a military variant of the Convair CV-340 airliner. The civil CV-340 can seat 52 passengers, and it had been the backbone of medium-range airline service in the United States in the 1950s. The VT-29 would serve in the United States Armed Forces for over 25 years.

The pressurized VT-29 cruised at around 240 knots, powered by two 2,400 horsepower Pratt & Whitney 18 cylinder Double Wasp radial engines. The U.S. Air Force slated the VT-29 for medical evacuation, cargo, and VIP transport. Many civil CV-340s would eventually be converted to turboprop power.

The Convair pilot-in-command, an Air Force lieutenant colonel 47 years of age, was a command pilot and flight examiner who had

Convair VT-29 (photo courtesy of U.S. DOD Visual Information Center)

logged 6,840 flight hours. His copilot, age 33, was a major who had 2,206 hours documented in his logbook.

The Cessna 150 Flight: A Cessna 150 took off from Runway 5 at Norfolk Regional Airport about 20 miles southeast of Langley AFB. The pilot had not filed a flight plan, and none was required for his proposed night VFR pleasure flight. At 1803:22 Hours the tower cleared him for a downwind departure from the airport traffic area. Thereafter ATC had no further contact with the Cessna.

The pilot, age 19, was a Seaman E-1 in the U.S. Navy. He was stationed aboard the *USS Pensacola* at Little Creek amphibious base. He held a Private Pilot certificate and had logged about 195 hours of flight time. One of his Navy shipmates had tagged along as a passenger. Cavalier Flyers Inc. owned the Cessna and had rented it to the teenage pilot several times. The single engine aircraft was not equipped with a transponder.

Trouble Ahead: Washington Center handed-off the IFR Convair to Norfolk Approach Control five miles south of Cofield VORTAC.

The new controller cleared the pilots to descend to 1,500 feet and vectored them for a final hand-off to Ground Control Approach (GCA) at Langley AFB. At 1832 Hours the military GCA controller took over when the Convair was 12 miles from the airfield. The pilots and controller anticipated the plane would intercept the GCA glideslope five miles from the Runway 7 threshold.

Meanwhile, the Cessna pilot had flown northwest to the James River bridge. He then flew up the river and followed the brightly lit shoreline of Newport News, Virginia, off to his right.

The ATC transcript, below, documents radio messages between controllers and the Convair pilots. Parts of the transcript have been *reconstructed* from the subsequent NTSB report, so some verbiage may not be verbatim. The transcript begins as the pilots get clearance to descend to 1,500 feet. The pilot-in-command is flying the aircraft, and the copilot makes all radio transmissions:

Convair: . . . . Convair VT-29 (call-sign, Motel 32)
AppControl: . . . . Norfolk Approach Control
GCA: . . . . GCA military controller at Langley AFB
Tower: . . . . Langley AFB Control Tower
Inter-: . . . . (prefix) Langley AFB interphone call

AppControl: Motel three-two, Norfolk Approach Control, descend and maintain one-thousand-five-hundred feet.

Convair: Motel three-two, we're going down to one-five.

AppControl: Motel three-two, contact Langley [GCA] . . . .

[the pilots switch to the GCA frequency and radio the military precision controller]

GCA: Roger, Motel three-two, radar contact, you're twelve miles west of Langley [AFB], maintain one-thousand-five-hundred feet for GCA to Runway seven . . . . (1832)

[the pilots acknowledge the instructions]

GCA: Motel three-two, you are ten miles from the runway, do not acknowledge further transmissions . . . . (1834:20)

GCA: Maintain heading zero-seven-three, traffic at one o'clock, two miles, northwest-bound. (1835:09)

Convair: Roger. (1835:14)

GCA: . . . eight miles from the runway, traffic [is] now slightly higher than you on precision radar. (1835:25)

Convair: Roger. (1835:31)

GCA: Now seven miles from the runway . . . .

*[the Convair and the Cessna **collide** at 1835:51]*

[the two targets merge on the GCA scope, then vanish]

GCA: Motel three-two, radar contact lost.

GCA: Motel three-two, Langley GCA?

GCA: Motel three-two, Langley GCA, how do you hear?

[there is no response]

Inter-GCA: Tower, GCA, I've lost all radio and radar contact with Motel three-two – uuuhhh – he was about seven miles out.

Inter-Tower: Aaahhh – I don't see him, stand by.

Tower: Motel three-two, Langley Tower?

[there is no response]

The Collision: The Convair had tracked east on a heading of 073 degrees. The Cessna had flown west on a heading of about 298 degrees. Flying toward each other, the planes had collided over the James River just offshore from Newport News:

> The collision occurred . . . during hours of darkness at an altitude of 1,500 feet and at a point about 7 miles west of the threshold of Runway 7 at Langley AFB.

The Cessna had slashed into the right rear side of the Convair fuselage. The aerial impact tore open the Cessna cockpit and ripped off the right wing. The entire Convair empennage and its control surfaces broke away. Both aircraft instantly tumbled down into shallow water that covers tidal mud flats along the James River:

Both aircraft were destroyed as the result of the in-flight collision and impact with the water.

Wreckage of both planes lay entombed in mud under a few feet of water two miles north of the James River bridge near the Mariner's Museum. A crowd of spectators quickly gathered at the museum to watch rescue efforts. Three Coast Guard helicopters and a crash boat from the naval base at Norfolk searched for survivors. By the next morning four bodies had been found, and the local *Virginian-Pilot* newspaper headline would reference the ongoing search:

----

4 Killed, 5 Sought as Planes Collide

No survivors were ever found. The seven men on the Convair and the two occupants of the Cessna had been killed. The body of the Cessna passenger was recovered on 25 February almost two months after the accident. The body of the Cessna pilot was never found.

The Armed Forces Institute of Pathology and the Virginia State Medical Examiner conducted autopsies on remains of the military men aboard the Convair. These examinations did not reveal any preexisting disease that might have played a role in the accident. Toxicological tests proved negative.

The Investigation: NTSB analysts determined the night sky had been clear with no restrictions to visibility. There was no moonlight. The Convair pilots had extensive experience in night-flying, but the young Cessna pilot had only about 12 hours of flight time at night.

Operational checks of the military GCA radar indicated it was working properly. The GCA audio tape revealed the controller had given the pilots proper guidance.

The Convair pilot had replied "roger" in response to two GCA traffic advisories about the Cessna . The meaning of "roger" was not clear, for the pilot did not say he had the traffic in sight. Investigators *theorized* the pilot (1) might merely have been acknowledging receipt of the advisory, or (2) might have mistaken certain lights on the ground for the Cessna and believed that he had the traffic in sight:

> The collision tracks of the two aircraft were reconstructed for the last 180 seconds of flight using radar plots . . . .

Both planes had red, green, and white position lights plus 150 candlepower flashing red anticollision lights. For more than 30 seconds before the collision the Cessna pilot and his passenger should have been able to see the Convair lights:

> The Convair lights would not have been obstructed . . . . They would have appeared [through the front windshield of the Cessna] slightly to the left of the [Cessna] aircraft's centerline.

On the other hand, pilots in the Convair would have had trouble spotting the Cessna lights. Binocular photographs showed the small plane was likely hidden by the Convair cockpit structure. Also, the Cessna had been directly between the Convair and bright lights on the ground in Newport News:

> Even without any masking of the Cessna by the [Convair] cockpit structure, the Cessna lights would have been difficult to detect against the ground lights. *[and also]* Ground lights had a mixture of hues from incandescent lights, sodium lights, and mercury lights, as well as hues from multicolored [neon] lights.

The Cessna lights would have blended into a bright kaleidoscope of background lights. Adding to this conspicuity problem, the two planes had been flying straight and level on a collision course. There had been no visible relative motion to alert the pilots to their peril.

Probable Cause: Five months after the accident the NTSB adopted the Aircraft Accident Report, NTSB-AAR-75-10, on 18 June 1975. Weather and visibility had not been a factor. The Convair had been on an IFR flight plan under GCA control. Nonetheless, both aircraft had been operating in a VMC see-and-be-seen environment. Pilots in both planes had been responsible for avoiding other traffic:

> The accident is another example of the problem created by a heterogenous mix of controlled and uncontrolled traffic in a high-

density terminal area where regulations place the burden on both crews to see and avoid the other aircraft. The effectiveness of the see-and-avoid concept is governed by the capability and reliability of the human element, [and] therein lies its inherent limitation.

Both aircraft had been flying in a clear night sky. Simply stated, the pilots did not see each other, and the planes collided. The board explained the *Probable Cause* of the accident as follows:

The Probable Cause of this accident was the *human limitation [emphasis added]* inherent in the see-and-avoid concept, which can be critical in a terminal area with a combination of controlled and uncontrolled traffic.

The board pointed out a *Contributing Cause*:

The reduced nighttime conspicuity of the Cessna against a background of city lights.

One-Eighty-Two, Descend Immediately!
Near-Collision near Carleton, Michigan
26 November 1975

Douglas DC-10, Registration N124
American Airlines Flight 182
205 aboard, 0 killed

Lockheed L-1011 TriStar, Registration N11002
Trans World Airlines Flight 37
114 aboard, 0 killed

Synopsis: Two big jumbo-jets, a Douglas DC-10 and a Lockheed L-1011 TriStar, flew toward each other (1) on the same jet route (2) at the same altitude. ATC failed to notice the conflict. The DC-10 captain spotted the TriStar, made a violent last-second maneuver, and missed the TriStar by 50 to 100 feet. No one was killed, but the evasive maneuver required hospitalization of 24 seriously injured people from the DC-10.

The Douglas DC-10 Flight: American Airlines Flight 182, a big Douglas DC-10, took off from Chicago, Illinois, on a scheduled trip to Newark, New Jersey. The widebody airliner carried a crew of 13 plus 192 passengers. Under the control of Cleveland Center the IFR tri-jet climbed east on jet route J-584 toward its cruising altitude:

The DC-10 first flew in 1970. The medium-range to long-range widebody airliner is powered by three engines. Two are mounted on underwing pylons, and the third is embedded in the vertical stabilizer. In the 1970s the DC-10 experienced a string of catastrophic crashes, most due to cargo door design or hydraulic system problems. After ironing out these maladies the plane enjoyed great success. Some models had a range of over 5,000 miles and a maximum takeoff weight up to 555,000 pounds. The factory built 386 DC-10s as passenger or freight transports and another 60 as KC-10 military tankers.

The captain, age 47, held an ATP certificate and had logged 21,600 flight hours. His first officer, age 43, also held an ATP certificate and had 7,500 hours documented in his logbook. As the

Douglas KC-10 Extender, a military tanker variant of the civil Douglas DC-10 airliner (official U.S. Air Force photo)

plane neared its cruising altitude the flight attendants began serving refreshments to their passengers.

The Lockheed L-1011 TriStar Flight: A Lockheed L-1011 TriStar, Trans World Airlines (TWA) Flight 37, cruised west on a scheduled IFR flight from Philadelphia, Pennsylvania, to Los Angeles, California. The TriStar carried a crew of 11 and a light load of only 103 passengers. The tri-jet tracked the centerline of jet route J-584 and skimmed through the top of a cloud deck:

> *The TriStar was the third widebody passenger jet to enter airline service. The huge transport had an engine layout similar to that of the DC-10. Three Rolls-Royce turbofans allowed the aircraft to cruise at Mach .9 with takeoff weights up to 496,000 pounds. The plane earned an excellent safety record and featured a twin-isle cabin, low noise emissions, cost-efficient operation, and seating for almost 400 passengers. Delta Airlines operated more TriStars than any other carrier. From 1968 through 1984, Lockheed built 250 of the big tri-jets for carrying passengers and freight, and for military use as tankers.*

Trouble Ahead: The DC-10 climbed toward its intended cruising altitude, flight level 370. Meanwhile, flying in the *opposite direction*, the TriStar cruised at flight level 350. The controller at Cleveland Center became distracted with other traffic and other duties, and he did not recognize the danger.

The portion of the ATC radio transcript, below, includes transmissions from the DC-10, the TriStar, and two other aircraft. It begins as the DC-10 pilots check-in with Cleveland Center:

> American 182: .. American Airlines Flight 182 (DC-10)
> TWA 37: ...... TWA Flight 37 (TriStar)
> United 680: ..... United Airlines Flight 680
> American 26: ... American Airlines Flight 26
> Center: .... Cleveland Center (ARTCC)

American 182: Cleveland Center, American flight one-eighty-two heavy [is] with you out of [flight level] two-eight-zero for [flight level] three-seven-zero. (1916:24)

Center: American one-eighty-two, roger, squawk three-two-zero-two and ident. (1916:31)

> *[the pilots of a different airliner on a different jet route ask the controller about the height of the clouds]*

United 680: Center, United six-eight-zero, any idea of the [height of the cloud] tops? (1922:05)

Center: Well, they were at thirty-five-[thousand feet] earlier – just a minute, let me check. (1922:08)

Center: TWA thirty-seven, Cleveland, what are the tops? (1922:13)

TWA 37: They are higher than we are, it's hard to say – you can see through [the tops of the clouds], I'd say [the tops] must be at least thirty-seven-[thousand feet]. (1922:17)

Center: OK, TWA thirty-seven, thank you. (1922:25)

Center: [United] six-eighty, did you copy? (1922:29)

United 680: Yes, thank you. (1922:31)

American 26: Center, American twenty-six *[cruising at flight level 370]* is just skimming the tops [of the clouds]. (1922:38)

Center: OK, American twenty-six, thank you – United six-eighty, that aircraft *[American Flight 26]* is at three-seven-zero. (1922:42)

> *[the controller suddenly notices the DC-10 is about to climb through the altitude of the TriStar, and the planes are approaching each other head-on; in <u>disbelief</u> he radios the DC-10 pilots]*

Center: American one-eighty-two, Cleveland! What is your altitude?

American 182: American one-eighty-two is passing through three-four-seven *[meaning, 34,700 feet]* at this time, and we can see the stars above us, but we're still in the area of the clouds. (1922:55)

> *[the DC-10 pilots suddenly see the TriStar ahead; they <u>violently jam</u> their control columns forward]*

> *[the DC-10 **misses** the TriStar by 50 to 100 feet]*

> *[simultaneously the controller <u>frantically</u> radios the pilots]*

Center: American one-eighty-two! Descend immediately to [flight level] three-three zero! (1923:03)

American 182: Descending to three-three-zero at this time! (1923:06)

> *[the pilots stop the descent at flight level 330]*

American 182: American one-eighty-two is at three-three-zero.

Center: American one-eighty-two, thank you. (1923:46)

American 182: What altitude was that other aircraft at? (1923:52)

Center: He was at three-five *[flight level 350]*, sir. (1923:57)

American 182: Check on that! (1924:02)

Center: Yes, sir, will do. (1924:07)

> *[flight attendants tell the pilots that many people in the cabin have severe lacerations and broken bones]*

> *[the pilots declare an emergency; they request vectors to the nearest suitable airport]*

Lockheed L-1011 TriStar (photo courtesy of U.S. NASA)

The Near-Collision: The TriStar had cruised west at flight level 350. The DC-10 had flown east on the same jet route while climbing through the TriStar's altitude. The two jumbo-jets had approached head-on over Carleton, Michigan:

> The two aircraft were on reciprocal courses and were closing at a [combined] speed of about 850 knots.

The DC-10 captain had spotted the lights of the onrushing TriStar and had shouted: "There he is!" Both pilots jammed their control columns forward. Pitch attitude dropped to minus 11 degrees:

> Flight attendants and service carts were thrown against the cabin ceiling by negative G forces. Passengers who did not have their seatbelts fastened . . . also were thrown against the ceiling.

The DC-10 screamed under the TriStar. The captain then pulled back on his control column to stop the descent at flight level 330. Vertical acceleration forces in the cabin instantly changed from negative G's to positive G's:

All unrestrained persons, service carts, and other objects which had been momentarily pinned to the [ceiling] came [crashing] down and hit the floor . . . 10 flight attendants and 14 passengers [later would be hospitalized with] abrasions, contusions, lacerations, and . . . fractured bones.

The DC-10 captain diverted toward Wayne County Metropolitan Airport at Detroit, Michigan. He safely landed there at 1950 Hours. Ambulances rushed 24 injured flight attendants and passengers to Wayne County General Hospital.

Ironically the TriStar pilots had not seen the DC-10 until too late to take evasive action. They barely glimpsed the DC-10 lights as the huge Douglas airliner flashed under them. After the near-miss they continued their flight to Los Angeles and landed without incident.

The Investigation: The planes had missed each other. However, 24 people from the DC-10 had been hospitalized, so the near-miss was categorized as an "accident."

NTSB investigators soon had aircraft FDR data and Cleveland Center radar data in hand. Pilots in both aircraft had followed all ATC instructions. The near-collision had resulted because Center had issued clearances that let the DC-10 climb – unchecked – through the cruising altitude of the TriStar. Immediately before the controller realized the danger the following condition existed on the radarscope at Cleveland Center:

American 182's data block showed the [DC-10] to be at FL-345 and climbing to FL-370. TWA 37's data block showed that the [TriStar] was maintaining FL-350.

The NTSB reconstructed events leading to the conflict. The Wayne Sector radar controller at Cleveland Center had been handling 11 aircraft, including the DC-10 and the TriStar. The radar secondary (transponder) return for each aircraft consisted of a symbol for the plane's position, plus an alphanumeric data block. This data block included the flight number and assigned altitude:

In the case of [the DC-10], which was climbing, the alphanumeric data also included actual altitude. The display on the [radarscope] was updated every 12 seconds.

FDR diagram depicting the <u>TWA</u> TriStar cruising at flight level 350, and depicting the <u>AAL</u> DC-10 climbing to flight level 350 (where it missed the TriStar by 50 to 100 feet), and then rapidly descending to flight level 330 (diagram courtesy of U.S. NTSB)

According to the controller, when he accepted the hand-off of the DC-10 he realized it *might* conflict with the TriStar. Yet, he thought the climbing DC-10 would reach flight level 370 before passing the TriStar, which was flying in the opposite direction at flight level 350:

> He *assumed [emphasis added]* that by keeping an eye on the situation he would be able to take [further] steps if the anticipated [vertical] separation did not materialize.

Unfortunately the controller became distracted with other duties. He failed to notice the two aircraft drawing closer together on his scope. At 1922 Hours another controller returned from lunch and relieved him. The original controller briefed the new controller, but he forgot to mention the climbing DC-10.

About 50 seconds later the new controller spotted the conflict. The DC-10 was at flight level 345 (500 feet below the TriStar) and climbing. In disbelief the new controller radioed the DC-10 and asked the pilots to verify their altitude. By the time they responded they had almost reached the TriStar's altitude:

> Both flights were on the same jet route and approaching each other head-on.

Fortunately the DC-10 captain had spotted the TriStar's lights and had dived out of its path. Had he not seen the other plane hurtling toward him the jumbo-jets would have collided.

Probable Cause: Two months after the near-collision the NTSB adopted NTSB-AAR-76-3 on 28 January 1976. The board emphasized that controller "intent" to separate traffic is never a substitute for "positive action at the first opportunity" to do so.

The original controller had *assumed* the DC-10 would climb above the cruising altitude of the other plane before they passed each other. If that did not happen he had *planned* to hold the DC-10 at flight level 330 until the TriStar passed overhead:

> The automated altitude read-outs [for the DC-10] induced [the controller] to rely solely on his own observations of the [radar] data. He did not consider the possibility that he might become distracted.

The NTSB explained the ***Probable Cause*** of this near-mid-air collision as follows:

> [First], failure of the [original] radar controller to apply prescribed separation criteria when he first became aware of a potential traffic conflict.

> [Second, the controller] allowed secondary duties to interfere with the timely detection of the impending traffic conflict [which] was displayed clearly on his radarscope.

The board pointed out a ***Contributing Cause***:

> An incomplete sector briefing during the change of controller personnel about 1 minute before the accident.

We Are Finished! Goodbye! Goodbye!
Mid-Air Collision over Vrbovec, Croatia
10 September 1976

Hawker Siddeley HS-121 Trident, Registration G-AWZT
British European Airways (BEA) Flight 476
63 aboard, 63 killed

Douglas DC-9, Registration YU-AJR
Inex-Adria Aviopromet (Adria) Flight 550
113 aboard, 113 killed

Synopsis: A Hawker Siddeley HS-121 Trident and a Douglas DC-9, both on IFR flight plans, flew toward the same VOR. Following their ATC clearances, the two airliners reached the VOR at the (1) same altitude and (2) same time. They collided, killing all 176 people aboard both planes.

The Hawker Siddeley HS-121 Trident Flight: At busy Heathrow Airport in London, England, British European Airways (BEA) Flight 476 roared down the runway and climbed skyward. ATC vectored the Hawker Siddeley HS-121 Trident out over the English Channel. The plane flew southeast over continental Europe and crossed into Austrian airspace. The pilots continued southeast on airway Upper Blue 5 (UB-5) on their way toward Istanbul, Turkey:

The Trident entered airline service with BEA in 1964. It was the first tri-jet design and the first airliner to use a FDR. The new jetliner replaced turboprop Vickers Viscounts on BEA's longer European routes. Similar to the later Boeing 727, the Trident could carry about 115 passengers, depending upon the model and seating configuration. With three Rolls-Royce engines the plane had a maximum takeoff weight of 143,500 pounds.

The captain, age 44, had logged 10,781 flying hours and held the requisite ATP certificate. His first officer was backed up by an "acting" first officer. Two stewards and four stewardesses took charge of their 54 passengers. Soon Austrian ATC handed-off the Trident to Zagreb Center in Croatia.

Douglas C-9 Nightingale, a military variant of the civil Douglas DC-9 airliner (photo courtesy of U.S. DOD Visual Information Center)

The Douglas DC-9 Flight: At 1048 Hours a Douglas DC-9 took off from Split, a seaside resort on the Adriatic Coast. The chartered airliner, Inex-Adria Aviopromet (Adria) Flight 550, headed northeast on a two hour trip toward Cologne, Germany. The plane carried 2 pilots, 3 flight attendants, and 108 passengers. Most passengers were German citizens returning from a vacation at the seashore.

The captain, with 10,157 flight hours in his logbook, had planned to cruise at flight level 310. However, conflicting traffic required him to level-off at flight level 260 on airway Upper Blue 9 (UB-9). He told the Zagreb Center controller he wanted a higher altitude as soon as traffic would permit.

Trouble Ahead: The Trident cruised southeast on UB-5 at flight level 330. The DC-9 cruised northeast on UB-9 at flight level 260. These airways intersected at Zagreb VOR. Both airliners were under the control of Zagreb Center. An ATC Guild publication reported:

> In 1976, Zagreb Air Traffic Control was one of the busiest [ATC] centers in Europe. ... To look after all this traffic was a group of understaffed and overworked air traffic controllers using a radar system which was "under test" and radio transmitters that often failed to work.

The middle-sector controller, responsible for airspace up to flight level 310, blindly cleared the DC-9 to climb to flight level 350 in the upper-sector. Both airliners were estimating their arrival over Zagreb VOR at the same time, 1114 Hours.

The portion of the ATC transcript, below, includes post-collision verbiage from the CVR in the DC-9. ATC uses the radio call-sign "Bealine" for the Trident:

Adria 550: . . . . . Adria Flight 550 (DC-9)
Bealine 476: . . . . BEA Flight 476 (Trident)
Lufthansa 360: . . Lufthansa Airlines Flight 360
Center-Mid: . . . . Zagreb Center, middle-sector controller
Center-Up: . . . . . Zagreb Center, upper-sector controller
CVR-: . . . . (prefix) CVR, Adria Flight 550 (DC-9)

[the Trident checks-in with the Center upper-sector controller]

Bealine 476: Four-seven-six . . . [is at flight level] three-three-zero, estimating Zagreb at one-four *[meaning, 1114 Hours]*. (1104:19)

Center-Up: Bealine four-seven-six, roger, call me passing Zagreb [VOR], flight level three-three-zero, squawk alpha two-three-one-two. (1104:27)

[the DC-9 checks-in with the Center middle-sector controller]

Adria 550: [Center], Adria five-five-zero leveling [at flight level] two-six-zero, standing by for [a] higher [altitude]. (1105:57)

Center-Mid: Five-five-zero, sorry, aaahhh, [flight level] three-one-zero is not available, two-eight-zero also. Are you able to climb to, maybe, [flight level] three-five-zero? (1106:03)

Adria 550: Affirmative, affirmative, with pleasure. (1106:11)

Center-Mid: Roger, [stand by, I'll] call you back. (1106:13)

[the middle-sector controller (who knows nothing about traffic in the upper-sector) fills out a flight progress strip and hands it to the upper-sector controller; the middle-sector controller then clears the DC-9 to climb into the upper-sector]

Center-Mid: Adria five-five-zero, re-cleared [to] flight level three-five-zero. (1107:40)

Center-Mid: Adria five-five-zero, [contact] Zagreb [upper-sector controller] on one-three-four-decimal-four-five, squawk standby and good-day, sir. (1112:06)

[the upper-sector frequency is constantly busy, the controller is talking with many aircraft, and the DC-9 pilots can not check-in for almost 2 minutes]

[while they wait to check-in the DC-9 pilots have not been given a transponder code; ATC radar does not display their secondary radar return, their identification, or their altitude]

[the <u>DC-9 pilots continue to climb</u> into the upper-sector; they <u>finally check-in</u> as they reach Zagreb VOR]

Adria 550:   Good morning, Adria five-five-zero crossing Zagreb [VOR] at one-four *[meaning, 1114 Hours]*. (1114:10)

[seeing no secondary radar return, the controller has to ask the DC-9 pilots what flight level they are maintaining]

Center-Up: What is your present level? (1114:14)

Adria 550:   [Flight level] three-two-seven. *[32,700 feet]* (1114:17)

[the DC-9 is only 300 feet below the altitude of the Trident and <u>still climbing</u>; the upper-sector controller has learned of this conflict only 31 seconds before the coming collision]

[the controller <u>stammers</u>, then <u>desperately pleads</u> with the DC-9 pilot to stop his climb]

Center-Up: Aaa! Aaahhh! Hold yourself at that height! (1114:22)

Adria 550:   What height? (1114:27)

Center-Up: The height you are climbing through!   You have an aircraft in front of you at [unintelligible]! (1114:29)

Adria 550:   OK, we'll remain precisely at three-three-zero. (1114:38)

*[10 seconds later the DC-9 and the Trident **collide** at 1114:48]*

[the Trident disintegrates; the DC-9 tumbles out of control]

CVR-Adria 550:   [A scream of despair]. (1114:56)

CVR-Adria 550: [Unintelligible].

CVR-Adria 550: We are finished! Goodbye! Goodbye! Goodbye!

CVR-Adria 550: [A final scream of despair].

[sounds of final structural break-up begin at 1115:00]

[end of the CVR tape at 1115:13]

Center-Up: Adria five-five-zero, Zagreb [Center]. (1115:50)

Center-Up: Adria five-five-zero, Zagreb. (1116:00)

Center-Up: Adria five-five-zero, Zagreb. (1116:14)

[there is no response]

[Lufthansa Flight 360 is near the collision site; the Captain radios the middle-sector controller]

Lufthansa 360: Zagreb! Zagreb! It is possible we have a mid-air collision in sight! We [see] two aircraft going down!

Center-Mid: Yes, two aircraft are below you, I don't understand.

Lufthansa 360: I think there's been a mid-air collision! Two aircraft are going down with a very fast rate of descent!

Center-Mid: I'm sorry, sir, I don't understand you.

Center-Mid: Lufthansa three-six-zero, this is Zagreb, will you be so kind as to say again? Do you have any problem? (11:18:02)

Lufthansa 360: We don't have any problem. But in front of us . . . I think we did see a mid-air collision. Over Zagreb . . . two aircraft going down . . . smoke coming out [of them]. (1118:12)

The Collision: At a closure rate approaching 800 knots the two airliners had torn into and through each other at flight level 330. The outboard 15 feet of the DC-9 wing had sliced through the Trident cockpit. Like a scythe the wing ripped the Trident apart, tearing rearward through the cabin, then sawing through the empennage. Demolished, the remains of the Trident fell earthward.

The collision had torn off most of the DC-9 left wing. The resulting violent right yaw at over 400 knots caused the whole tail section to break away. Although the DC-9 tumbled toward the ground the CVR operated for 25 seconds after the collision. It documented the pitiful cries of the doomed first officer.

The planes ripped apart in the air. Passengers and flight attendants were tossed out into the sky. A woman walking toward her home in the village of Vrbovec later would explain:

> The body of a young girl crashed to the ground five meters in front of me. I screamed in horror. Next, suitcases started falling around me, and I began to run. I almost fell over the body of a man who fell from the skies. . . . [Soon there were] arms, legs, and heads everywhere.

The largest pieces of the Trident screamed down into a cornfield. Wreckage of the DC-9 fell into woods near the town of Dvoriste. There had been 176 people on the two planes. None survived.

The Investigation: A skilled Yugoslav accident investigation team recovered the FDR and CVR from the rubble of each plane. The Trident had been flying at 479 knots and 32,960 feet with the autopilot engaged. The DC-9 had been climbing at 430 knots:

> The tragic oversight here was that the DC-9 [had been] climbing through 33,000 feet, directly in the path of the Trident.

This sensational accident and great loss of life prompted international scrutiny of the Croatian and Yugoslavian ATC system. The mercenary news media rabble, always thirsting to peddle tales of misfortune and scandal for profit, demanded that "justice" be served. Someone, somewhere, had to be at fault.

Blame rolls downhill. In this case the "fall guy" would be the upper-sector controller. He was a quiet man, married with one child, the youngest controller at Zagreb Center. He was arrested, jailed, and indicted under the Penal Code of Yugoslavia. In part the multiple criminal charges against him read:

> He was acting contrary to the rules, Part 25 of Air Traffic Control, which demand a minimum of separation as a priority; and contrary to the rules of Chapter 4 of the same rules he accepted coordination from the middle-sector . . . and did not ensure that minimum separation was available between the planes.

Hawker Siddeley HS-121 Trident (photo courtesy of Adrian R. Pingstone)

The Aircraft Accident Investigation Commission of the Yugoslav Federal Civil Aviation Administration completed its investigation in December 1976. The commission had compiled a thorough technical report. Nonetheless, with a highly publicized criminal trial pending the government made a decision. "Higher Interest" would be served by withholding the report from the public.

The trial before the Grand Council of the Zagreb District Court began in April 1977, seven months after the accident. There was no testimony or evidence indicative of criminal conduct. There was no evidence of wilful negligence. Yet, the outcome was obvious from the start. The German press saw through the charade:

> [The upper-sector controller], as has become clear in only a few days in court, has been psychologically isolated in this Zagreb trial and has been marked out as the scapegoat.

The court found the young controller guilty. He was sentenced to serve "seven years rigorous imprisonment." The controller alone, the court decreed, was to blame for the accident. However, the German news magazine *Der Spiegel* pointed out:

> It was clear that the [controller] would pay for the incompetence and confusion of the Yugoslav airspace administration.

Probable Cause: The plight of the imprisoned controller infuriated aviation industry professionals. Knowledgeable parties knew the young man had done nothing wrong. In fact, with the resources and time constraints involved he had done all within his power to prevent the aerial collision. The ATC *system* had failed, and an independent technical report explained:

Unfortunately [the controller] had been made a scapegoat for an air traffic control system that was allowed to run down to a [condition] which beckoned for a disaster to happen.

In 1973, three years before the accident, a modern radar system had been installed at Zagreb. However, it had never been calibrated. It remained unreliable and was used only as a back-up for manual flight progress strips. There had been an astounding 32 known "near misses" over the five years leading up to the collision between the Trident and DC-9. An aviation authority wrote:

Zagreb ARTCC . . . [had] a staff of 30 [controllers] desperately coping with traffic which required *at least [emphasis added]* double their number.

Three controllers had been scheduled to work the upper-sector on the morning of the accident. However, one had not shown up for work, and another left his post to look for him. That had left the "guilty" controller alone and juggling 11 aircraft. The supervisor had been busy in his office, writing the work schedule for the next day:

The Zagreb ARTCC was working on a knife-edge of nerves.

The middle-sector controller blindly had cleared the DC-9 into the upper-sector. Routine procedure required upper-sector *consent* before a plane could be cleared to climb into the upper-sector. On the morning of the accident, consent never had been given. The lone upper-sector controller had been talking nonstop into his microphone and telephone land-lines, taking position reports, issuing directives, and coordinating clearances with other ATC centers:

The upper-sector controller was heavily involved with traffic in his sector, without an assistant; he also had to arrange clearance for a number of aircraft to enter Belgrade airspace.

The harried controller had learned of the DC-9 in his sector only 31 seconds before the collision. It was climbing toward Zagreb VOR, merging with the Trident at the speed of a rifle bullet. Stammering, lapsing into a Serbo-Croatian dialect, the controller had *pleaded* with the DC-9 pilots to stop their climb. His warning came too late. Time had run out.

Postscript: After the trial the International Federation of Air Traffic Controller's Associations (IFATCA) mounted a legal appeal of the court verdict. Without any question ATC had failed. The appeal maintained that the controller merely had been the visible symbol of the *failed system.*

Without addressing the merit of the appeal the Supreme Court of the Socialist Republic of Croatia announced the controller's sentence would be cut in half. He would serve three and one-half years.

Months later Marshall Tito, the Yugoslav head of state who had become President-for-Life in 1974, flew toward North America in his personal jetliner. He had a scheduled meeting with Jimmy Carter, President of the United States. Irish controllers learned the dictator's plane would refuel at Shannon Airport prior to flying west across the Atlantic Ocean:

> The controllers [agreed] that until [Marshall Tito] *personally accepted [emphasis added]* a copy of their Petition, his aircraft would not get a takeoff clearance.

Takeoff clearance was indeed withheld. Then, on the tarmac at Shannon Airport an IFATCA attorney put a Petition in the Yugoslav leader's hands. Soon thereafter on 29 November 1978 the controller became a free man. Pardoned, he walked out of Dom Zabela prison near Povarevac, Yugoslavia, to rejoin his wife and child and begin life anew – a life that would not include air traffic control.

My Wingman Hit Something!
Mid-Air Collision near Brighton, Florida
13 September 1976

McDonnell F-4 Phantom, Serial No. 67-0255
U.S. Air Force gunnery training mission
2 aboard, 1 killed

Cessna 414, Registration N8PR
Ruel Insurance Corporation business flight
4 aboard, 4 killed

Synopsis: Three McDonnell F-4 Phantoms cancelled their IFR clearance and descended toward an aerial gunnery range. The No. 3 Phantom ripped through a Cessna 414 that was cruising VFR. The Phantom instructor ejected and survived. The other Phantom pilot plus four people aboard the Cessna died in the collision and crashes.

The McDonnell F-4 Phantom Flight: With afterburners glowing, three McDonnell F-4 Phantoms roared into the sunny morning sky from Homestead AFB near Miami, Florida. The formation flight of three U.S. Air Force fighters from the 31st Tactical Fighter Command flew north on an IFR flight plan under the control of Miami Center. Cleared for a block altitude between flight levels 180 and 210, they headed toward Avon Park military gunnery range 135 miles north-west of Miami:

> The flight had [been cleared] toward the 34 mile [DME] fix on the 322 degree radial of the Pahokee VORTAC, the entry point for the Avon Park gunnery range.

The No. 3 Phantom, the mishap aircraft (call-sign, Reed 13), was flown by an instructor pilot, age 35, in the rear seat. He was an Air Force major and had logged 2,591 hours of flight time. His "student" in the front seat was a relatively new first lieutenant, age 23, who had accumulated a mere 293 flying hours. He recently had completed his Air Force basic flight training and was transitioning into the supersonic Phantom.

McDonnell F-4 Phantom (official U.S. Air Force photo)

The Cessna 414 Flight: In a twin engine Cessna 414, a pilot and three passengers took off from Executive Airport at Fort Lauderdale, Florida, on a business trip to Tallahassee, Florida. The pilot did not file a flight plan for this trip in VMC weather conditions. He climbed to 12,500 feet and squawked Code 1200 as he cruised northwest toward Florida's capital city:

> *The Cessna 414 is a twin engine piston-powered aircraft that was built from 1968 through the mid-1980s. Two 310 horsepower engines give the plane a range of over 1,000 miles when cruising at 170 knots. It has a maximum takeoff weight of 6,750 pounds and can seat two pilots and five to eight passengers.*

The pilot, age 49, was a principal at Ruel Insurance Corporation, which owned the Cessna. He held a Private Pilot certificate plus an Instrument Rating and had logged 798 hours of flight time. His three passengers were Fort Lauderdale city officials. They were going to Tallahassee to meet with state legislators and discuss insurance for police officers.

Trouble Ahead: As the Phantom flight leader, Reed 11, descended through 17,000 feet he cancelled the IFR flight plan for the flight-of-three. Reed 12 flew off his right wing, and Reed 13 flew on his left. Reed 11 squawked Code 1200 and switched to the Avon Park Control frequency. He continued his descent and turned left to a heading of 320 degrees. Meanwhile, ahead of the three Phantoms the Cessna cruised toward Tallahassee on a heading of 327 degrees.

The portion of the ATC radio transcript, below, contains messages between Miami Center, Avon Park Control, and Reed 11. Parts of the transcript have been *reconstructed* from a subsequent NTSB report, so some verbiage may not be verbatim. The transcript begins as Center clears the Phantoms to descend to 14,000 feet:

Reed 11: . . . McDonnell F-4 Phantom (call-sign, Reed 11)
Center: . . . . Miami Center (ARTCC)
Avon: . . . . . Avon Park Control (gunnery range)

Reed 11: Center, Reed one-one, we're approaching the Avon range, we need to start down. (0948:52)

Center: Reed one-one, roger, cleared to descend to and maintain one-four-thousand, altimeter three-zero-zero-two. (0948:59)

[3 minutes and 21 seconds later Reed 11 cancels his flight plan]

Reed 11: Center, Reed one-one, passing one-seven-thousand, cancel our IFR clearance. (0952:20)

Center: Reed one-one, roger [the IFR] cancellation, [I'll] see you on [your] return. (0952:26)

[the Cessna is 5.5 miles ahead and flying at a lower altitude]

[Reed 11 descends and <u>turns toward the unseen Cessna</u>]

*[the left wingman, Reed 13, and the Cessna **<u>collide</u>** at 0954:06]*

Reed 11: Mayday! Mayday! Avon, Reed one-one on Guard!

Avon: Reed one-one, go ahead.

Reed 11: Avon, this is Reed one-one, my wingman hit something, and he's gone down out of control! We see one chute!

The Collision: As the military formation flight descended through 12,500 feet it had overtaken the much slower Cessna near Brighton, Florida. The flight leader later would write:

> Just as Avon [Control] finished reading the 3,000 foot winds, I noticed a flash past the left side of my radome [and] it collided with No. 3, who was on my left wing at the time.

Reed 13 had sliced off the Cessna's left horizontal stabilizer and left wing. The Cessna fell earthward, dooming the four men aboard. The fuselage slammed into a marshy field on the Seminole Indian Reservation near Lake Okeechobee, 10 miles south of the gunnery range. Other pieces of the aircraft lay scattered in a bog three-fourths of a mile from the main wreckage.

After the collision Reed 13 had rolled left and had spiraled toward the ground. The instructor ejected and parachuted to safety. The lieutenant in the front seat repeatedly – *desperately* – tried to eject but could not do so. Aware he was going to die, the lieutenant rode his mortally wounded Phantom all the way into the ground:

> The pilot in the front seat was killed when his ejection system failed because of damage received in the collision.

The Phantom buried itself nose-down in the muck at the bottom of a pond a mile from the Cessna forward fuselage. Potential rescuers used swamp buggies to reach the crash sites. The Phantom front-seat pilot and the Cessna occupants were found dead in the wreckage.

The Investigation: Radar at Miami Center had tracked Reed 11, transponder Code 4556, before the pilot cancelled his IFR clearance. The controller's scope had depicted a full alphanumeric data block, including the Phantom's altitude, as the fighter descended.

The Cessna pilot had not filed a flight plan, but his aircraft had a Mode C transponder. Radar had identified the Cessna shortly after takeoff. The data block showed the pilot had squawked Code 1200, climbed to 12,500 feet, and had cruised straight toward Tallahassee:

> A VFR aircraft transmitting on Mode C . . . normally displays a limited data block on the [radarscope], including the aircraft's altitude information.

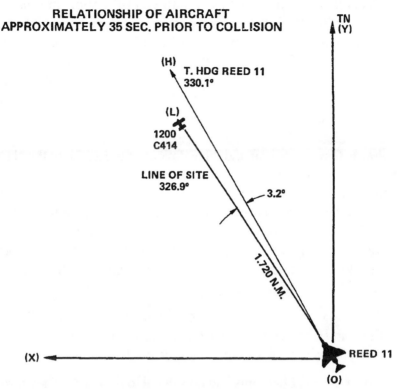

**RELATIONSHIP OF AIRCRAFT
APPROXIMATELY 35 SEC. PRIOR TO COLLISION**

TN
(Y)

(H)

T. HDG REED 11
330.1°

(L)

1200
C414

LINE OF SITE
326.9°

3.2°

1,720 N.M.

(X)

REED 11

(O)

Diagram depicting headings of the aircraft 32 seconds before the collision, as the Phantoms closed on the Cessna (diagram courtesy of U.S. NTSB)

Radar data showed Reed 11 had been 22 miles behind the Cessna when the controller cleared the Phantoms to descend to 14,000 feet. The controller had no reason to issue a traffic advisory regarding the Cessna, which was then an uncorrelated VFR target at 12,500 feet.

However, after cancelling his IFR clearance Reed 11 had turned left, directly behind the Cessna, and had continued to descend VFR. In this see-and-be-seen environment the lieutenant in the front seat had been responsible for keeping the flight-of-three away from all other traffic. Visibility had been estimated at 10 miles:

> Because the F-4 flight was overtaking [the Cessna] from the rear and from above, it would have been impossible for the pilot of [the Cessna] to see and avoid the F-4s. *[and also]* No reason could be found for the inability of the pilot in the F-4 front seat to see [the Cessna].

Radar had continued to track the Reed 11 flight-of-three and the Cessna. As the Air Force fighters descended through 12,500 feet the two targets had merged on the controller's scope. Reed 13 had rammed into the Cessna. Radar showed that the Phantom leader, Reed 11, then entered a tight 360 degree turn around the collision and crash sites. By then the radar target that had been the Cessna had vanished from the controller's scope.

Probable Cause: A year and three days after the accident the NTSB adopted the Aircraft Accident Report, NTSB-AAR-77-5, on 16 September 1977. Before Reed 11 had cancelled his IFR flight plan the controller had been responsible for providing separation from all other IFR traffic. The controller also would have provided traffic advisories for any VFR aircraft that conflicted with the Phantom's flight path. However, no such conflict had existed when Reed 11 cancelled his IFR flight plan:

> About 5.5 miles and 4,500 feet of altitude separated [the] Reed 11 flight and [the Cessna] when the [military] flight cancelled its IFR clearance. Their respective tracks were diverging.

However, the Phantoms had continued to descend. They turned left to a position directly behind the Cessna. It should have been visible, dead ahead, through the Reed 11 windshield as the fighters swooped down toward it. At a closure rate of 186 knots the left wingman had ripped into the hapless civil aircraft. Consequently, the board found the *Probable Cause* of this accident to be basic:

> The Probable Cause of this accident was the failure of the pilots of [the] Reed 11 flight to maintain adequate vigilance in order to see and avoid the [Cessna].

The board took note of a *Contributing Cause*:

> Inconsistency between Homestead AFB Supplement 1 to Tactical Air Command Manual 55-4, which allowed a VFR descent into the Avon Park gunnery range, and Air Force Regulation 60-16, which required all flights to be conducted under IFR [clearances] unless to do so would result in unacceptable mission derogation.

We Are Now "at" Takeoff
Runway Collision at Tenerife, Canary Islands
27 March 1977

Boeing 747, Registration N736PA
Pan American World Airways (Pan Am) Flight 1736
396 aboard, 335 killed

Boeing 747, Registration PH-BUF
KLM Royal Dutch Airlines (KLM) Flight 4805
248 aboard, 248 killed

Synopsis: In blinding fog a Boeing 747 began its takeoff roll without a takeoff clearance. At 140 knots it ripped into another Boeing 747 that was taxiing on the same runway. Of the 644 people aboard the two jumbo-jets, 583 perished in the collision and raging fires that followed.

The Boeing 747 Flight (Pan Am): Pan American World Airways (Pan Am) Flight 1736, a big Boeing 747, left New York and headed for Las Palmas Airport in the Canary Islands. In antiquity this chain of islands had been known as the "Fortunate Isles" because of warm and fair weather that lasted most of the year. Spain's volcanic archipelago, stretching for 250 miles off the northwest coast of Africa, was a popular vacation spot. Throughout Europe for decades, airlines and cruise ships had zeroed-in on this tourist Mecca:

Following their successful Boeing 707, Boeing engineers vied for the contract to build a U.S. Air Force ultra-large strategic transport. Lockheed won that contract with the big C-5 Galaxy. Undeterred, Boeing used its plans for the huge military transport and built the Boeing 747 commercial airliner.

The new plane first flew in 1969, and Pan Am began line service the next year. The Boeing 747 was larger and more complex than any other airliner. It revolutionized intercontinental air travel. On long routes the seat cost per mile was lower than anything else the market could offer. Later models could seat over 500 passengers, and maximum takeoff weights gradually climbed to over 900,000 pounds. The massive Boeing 747 remained the

*world's largest commercial airliner for 35 years. It relinquished
that title to the super-jumbo Airbus A-380 prototype in 2005.*

The captain and first officer held ATP certificates, and each had
logged over 21,000 hours of flight time. Their flight engineer had
logged 15,210 hours in the air. In the cabin the six flight attendants
had their hands full with 387 passengers, over 60 per flight attendant.

The Boeing 747 Flight (KLM): Another Boeing 747 jumbo-jet,
KLM Royal Dutch Airlines (KLM) Flight 4805, left Schiphol Airport
in Amsterdam with 248 people aboard. This KLM flight also was
headed for the airport at Las Palmas. The plane had been chartered
by Holland International Travel Group.

The captain was featured in the airline's in-flight magazine. He
was KLM's chief flying instructor, a senior pilot of great prestige. He
had been flying for 30 years, and now he specialized in transitioning
veteran pilots into Boeing 747 airliners. Behind the cockpit, 11 flight
attendants catered to the needs of their passengers.

Trouble Ahead: While both airliners were crossing the Atlantic a
small bomb exploded in a florist shop in the terminal at Las Palmas.
A terrorist group claimed responsibility, and Spanish police closed
the Las Palmas airport. ATC diverted inbound flights to one of the
archipelago's other airports on the island of Tenerife.

The KLM and Pan Am flights landed at Tenerife and found the
airport packed with other diverted flights. However, the airport at Las
Palmas soon reopened, and aircraft waiting at Tenerife began asking
for clearance to taxi to the single runway for takeoff. Las Palmas was
only 25 minutes away; however:

Weather conditions [at] the airport were getting rapidly worse.

Thick fog began descending on Tenerife. Three aircraft managed
to taxi out to the runway and take off. Next the KLM flight began to
lumber out toward Runway 30. To get the Boeing 747 to the proper
takeoff position, ATC told the pilots to taxi onto the runway and
back-taxi down to the approach end.

The Pan Am flight also was ready to go. ATC told the Pan Am
pilots to follow the KLM airliner, taxi onto the runway, and back-taxi
toward the approach end. ATC planned to have the Pan Am airliner

Two Boeing 747 airliners (photo courtesy of U.S. DOD Visual Info. Center)

later turn off onto a vacant taxiway. That would clear the runway for the KLM airliner to take off. Because of the thick fog the KLM pilots and Pan Am pilots could not see each other. The tower controller could not see either airplane. He relied on their radio reports to determine their position on the airport.

The portion of the CVR and ATC transcript, below, contains verbiage from both the Pan Am and KLM cockpits and the tower. The transcript begins as the KLM flight is back-taxiing down the runway. Far behind, out of sight in the fog, the Pan Am aircraft also is back-taxiing down the same runway. The captains handle the controls in each plane. In compliance with international practice the Pan Am radio call-sign is "Clipper":

PanAm-Capt: . . . Captain, Pan Am Flight 1736
PanAm-FO: . . . . First Officer, Pan Am Flight 1736
KLM-Capt: . . . . . Captain, KLM Flight 4805
KLM-FO: . . . . . . First Officer, KLM Flight 4805
KLM-FltEngr: . . Flight Engineer, KLM Flight 4805
Tower: . . . . Tenerife Airport Control Tower
Radio-: . . . . (prefix) Aircraft radio transmissions

Radio-KLM-FO: [Tower,] you want us to turn left at Charlie one, Taxiway Charlie one? (1659:28)

Tower: Negative, negative, taxi straight ahead – aaahhh – up to the end of the runway and make back-track. (1659:32)

Radio-KLM-FO: OK, sir. (1659:39)

[2 minutes later the controller radios the Pan Am pilots]

Tower: Clipper one-seven-three-six, Tenerife. (1702:01)

Radio-PanAm-FO: [Tower,] ah, we were instructed to contact you and also to taxi down the runway, is that correct? (1702:03)

Tower: Affirmative, taxi onto the runway and – aaahhh – leave the runway [at the] third [taxiway] to your left. (1702:08)

Radio-PanAm-FO: Third [taxiway] to the left, OK. (1702:16)

Tower: KLM four-eight-zero-five, how many taxiways – aaahhh – did you pass? (1702:49)

Radio-KLM-FO: I think we passed Charlie four [taxiway] now.

Tower: OK – at the end of the runway make [a] one-eighty [degree turn] and report, aaahhh, for ATC clearance. (1702:55)

Radio-PanAm-FO: Would you confirm that you want Clipper one-seven-three-six to turn left at the third intersection? (1703:35)

Tower: Third one, sir – one, two, three – third one. (1703:38)

PanAm-Capt: Good, that's what we need, the third one. (1703:39)

Tower: Clipper one-seven-three-six, report leaving the runway.

Radio-PanAm-FO: [Roger], Clipper one-seven-three-six. (1703:56)

[the KLM flight reaches the approach end of the runway, turns around into takeoff position, and requests its ATC clearance]

Radio-KLM-FO: [Tower], KLM four-eight-zero-five is now ready for takeoff, and we are waiting for our ATC clearance. (1705:50)

Tower: KLM four-eight-zero-five, you are cleared to the Papa Beacon, climb to and maintain flight level nine-zero, right turn after takeoff, proceed [on] heading zero-four-zero until intercepting the three-two-five radial from Las Palmas [VOR]. (1705:53)

[the KLM flight has en route clearance but <u>does not have takeoff clearance</u>; yet, the Captain releases his brakes]

KLM-FO: Wait a minute, we don't have an ATC [takeoff] clearance!

[the KLM Captain ignores his First Officer's warning]

KLM-Capt: Let's go, check thrust. (1706:12)

[sound of the KLM engines spooling up at 1706:14]

[the KLM First Officer does not challenge his Captain further; he "reads back" the en route clearance during the takeoff roll]

Radio-KLM-FO: Ah, roger, sir, we're cleared to the Papa Beacon, flight level nine-zero, right turn out [to heading] zero-four-zero until intercepting the three-two-five [radial of the Las Palmas VOR], and *we are now at takeoff. [emphasis added]* (1706:17)

[the tower does not know the KLM plane is on its takeoff roll]

Tower: OK, stand-by for takeoff, I will call you. (1706:21)

[the Pan Am pilots radio that they are still on the runway]

Radio-PanAm-FO: And, we're still taxiing down the runway, [this is] the Clipper one-seven-three-six. (1706:22)

Tower: Papa Alpha *[phonetic for "Pan Am"]* one-seven-three-six, report [when] clear [of the runway]. (1706:25)

Radio-PanAm-FO: OK, we'll report when we're clear. (1706:29)

Tower: Thank you.

[the Pan Am Captain is not sure of the KLM pilots' intent, so he heads for the nearest taxiway to clear the runway]

PanAm-Capt: Let's get the [expletive] out of here! (1706:30)

[as the KLM airliner accelerates down the runway, the nervous KLM Flight Engineer queries his Captain]

KLM-FltEngr: Is he not clear [of the runway] then? (1706:32)

KLM-Capt: What do you say? (1706:34)

KLM-FltEngr: Is he not clear, that Pan American [plane]? (1706:34)

KLM-Capt: Oh, yes! *[spoken emphatically]* (1706:35)

[the Pan Am Captain sees the landing lights of the KLM aircraft materialize out of the fog, <u>headed straight toward him</u>]

<u>PanAm-Capt</u>: There he is! Look at him coming! [Expletive], that [expletive] is coming! Get off [the runway]! Get off [the runway]! Get off [the runway]! (1706:40)

[the KLM Captain suddenly sees the Pan Am airliner on the runway ahead of him]

<u>KLM-Capt</u>: [Unintelligible scream]! (1706:47)

[the KLM aircraft lifts off and <u>almost</u> clears the top of the Pan Am aircraft below on the runway]

*[sound of the <u>**collision**</u> at 1706:50]*

The Collision: Speeding down the runway at 140 knots, the KLM captain saw a ghostly apparition materialize dead ahead in the fog. In a moment he recognized the outline of the Pan Am airliner. In desperation he hauled back on his control column and tried to take off. He *almost* made it.

The KLM airliner lifted from the runway. Its fuselage skimmed over the Pan Am aircraft, but its main landing gear and engines tore through the top of the Pan Am cabin. The KLM airliner slammed back down onto the runway 450 feet beyond the collision point. It skidded over 1,000 feet and exploded into flames. Fire consumed the KLM airliner and killed all 248 people on board:

> An immediate and raging fire must have prevented emergency [evacuation] because all of the aircraft's doors remained shut.

While beginning to turn off the runway the Pan Am pilots had spotted the KLM plane rushing out of the fog toward them. They had jammed their throttles forward to evade the monster hurtling toward them. Unfortunately they were too late:

> In the Pan Am aircraft the first class lounge disappeared as a result of the impact [by the KLM airliner], as well as nearly the whole top of the fuselage.

The impact instantly killed many Pan Am passengers. Others fell prey to flames and smoke. A fortunate few managed to leap out of

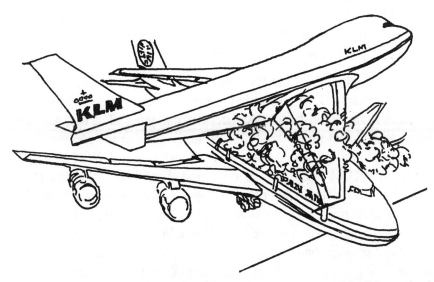

An illustration depicting the accelerating KLM airliner, which had mushed into the air, striking the top of the Pan Am airliner, which had been taxiing on the runway (illustration courtesy of Spain's "Commission of Investigation of Accidents and Civil Aircraft Incidents")

the mangled remains of the Boeing 747. There were 396 people on the Pan Am airliner, and the collision and fire killed 335 of them.

Because of thick fog the two burning aircraft were not visible from the control tower. An aircraft on the parking apron radioed that it saw a red glow in the mist. Alerted, the fire brigade slowly drove around the airport and eventually stumbled upon the two flaming wrecks. Despite the best efforts of firefighters the two airliners burned until 0330 Hours the following morning.

The Investigation: Spanish, Dutch, and American analysts began searching for answers. They recovered the CVR and FDR from each aircraft and found the tapes to be in good condition. This was indeed fortunate, for the keys to the puzzle lay in the tapes.

When the KLM first officer had "read back" the en route ATC clearance he had added the phrase: ". . . we are now at takeoff." What did that mean? What was "at" takeoff? When investigators first heard those words on the tape, they had no idea they had been spoken while the KLM plane was accelerating down the runway:

> The tower, which was not expecting the aircraft to take off, as it had not given [a takeoff] clearance, interpreted the sentence as: "We are now at takeoff *position*." *[emphasis added]*

Neither the tower nor the Pan Am pilots had been sure what the phrase meant. Both responded out of caution. The controller had replied: "OK, stand-by for takeoff, I will call you." The Pan Am crew had radioed that they were ". . . still taxiing down the runway."

Investigators learned there was confusion in the KLM cockpit as well. The KLM first officer had warned his captain: "We don't have an ATC [takeoff] clearance!" The captain ignored the warning. The KLM flight engineer had realized the Pan Am aircraft might be on the runway ahead, hidden in the fog. Twice he had asked his captain if the Pan Am flight had cleared the runway ahead. The captain brushed aside both questions and continued the takeoff.

From the CVR tapes, investigators realized the alarmed Pan Am pilots had been heading toward a taxiway to clear the runway. Then, through the fog the Pan Am captain spotted the landing lights on the onrushing KLM airliner. He had firewalled his throttles, but there was not enough time to get out of the way:

> The Pan Am crew saw the KLM plane approximately 8.5 seconds before the impact. Amidst logical exclamations of alarm they accelerated in order to try to get off the runway, but the collision was already inevitable.

The KLM captain apparently had spotted the Pan Am airliner six seconds before the collision. He pulled back on his control column to try to take off. His aircraft had mushed into the air, but his engines and landing gear failed to clear the top of the Pan Am airliner.

Probable Cause: The Spanish government's Commission of Investigation of Accidents and Civil Aircraft Incidents published a lengthy "Collision Aeronaves" report over a year later on 7 December 1978. The board found the KLM captain had begun his takeoff roll *without a takeoff clearance*. In thick fog the KLM aircraft had collided with the Pan Am aircraft, which had been taxiing on the same runway per instructions from the controller. The board found the ***Fundamental***

Cause of the accident to be four-fold, in that the KLM captain:

[First], took off without [a takeoff] clearance.

[Second], did not obey the "stand-by for takeoff" [instructions from the tower].

[Third], did not interrupt [his] takeoff when [the] Pan Am [pilots] reported that they were still on the runway.

[Fourth], in reply to [his] flight engineer's query as to whether the Pan Am [aircraft] had left the runway, replied emphatically in the affirmative.

The report listed several *Contributing Factors*:

[First, the KLM captain was in a hurry.] If he did not take off within a relatively short space of time he might have to [postpone] the flight [because of "duty time" restrictions].

[Second], layers of low-lying [fog restricted] visibility.

[Third], two radio transmissions took place at the same time. The "stand-by for takeoff, I will call you" [transmission] from the tower coincided with Pan Am's "we're still taxiing down the runway" [transmission]. *[the KLM pilots may not have clearly heard either message, both of which were crucial]*

[Fourth], when the KLM [first officer stated], "we are now at takeoff," the controller, who had not been asked for a takeoff clearance, and who consequently had not granted it, did not understand that [the KLM aircraft] was taking off.

Ma! I Love Ya!
Mid-Air Collision over San Diego, California
25 September 1978

Boeing 727, Registration N533PS
Pacific Southwest Airlines (PSA) Flight 182
135 aboard, 135 killed (plus 7 killed on the ground)

Cessna 172 Skyhawk, Registration N7711G
Gibbs Flite Center instrument training flight
2 aboard, 2 killed

Synopsis: A Boeing 727 was cleared to land in VFR conditions. The airline pilots overran a Cessna 172 and slammed into it. All 135 people on the airliner, the 2 Cessna pilot-occupants, and 7 people on the ground died in the collision and crashes.

The Boeing 727 Flight: A Boeing 727, Pacific Southwest Airlines (PSA) Flight 182, left Los Angeles, California, for a short trip down to San Diego. Those aboard assumed this would be a quick flight. Unfortunately it would end three minutes sooner than planned.

The captain, age 42, held an ATP certificate and had 14,382 hours of experience in his logbook. His first officer, equally rated, had over 10,000 hours of experience. Backing them up in the cockpit was a veteran flight engineer. An off-duty PSA captain was riding in the cockpit jump-seat, but he was not part of the crew. Three flight attendants were responsible for 128 passengers in the cabin.

The Cessna 172 Skyhawk Flight: A Cessna 172 owned by Gibbs Flite Center took off from Montgomery Field, six miles northeast of the main San Diego airport, Lindbergh Field. Aboard the Cessna were a flight instructor and his experienced student. They flew to Lindbergh Field where they practiced ILS approaches:

The Cessna 172 evolved from the Cessna 170 tail-dragger. The first new "one-seventy-two" was delivered in 1956, and the plane remains in production over 50 years later. The high wing aircraft seats four, and most models are powered by a Lycoming

Cessna 172 (photo courtesy of Adrian R. Pingstone)

160 horsepower piston engine. Over 40,000 have been built. The Cessna 172 is considered to be the most popular mass-produced light aircraft in the world.

The instructor, age 32, was an employee of Gibbs Flite Center, and he held Commercial Pilot and Flight Instructor certificates. His student also was a Commercial Pilot, and on this flight he practiced instrument approach procedures. The pilots had not filed a flight plan and none was required, for the weather was clear with a reported visibility of 10 miles. The Cessna pilots had the benefit of Stage II terminal radar services in the busy sky over San Diego.

Trouble Ahead: The Boeing pilots radioed San Diego Approach Control and descended toward the coastal city. They soon reported the airport in sight. Approach Control cleared the airliner for a visual approach to Runway 27 at Lindbergh Field.

Meanwhile, the two Cessna pilots completed their second ILS approach and began a climbout to the northeast. Lindbergh Tower instructed them to contact Approach Control:

> Approach Control told [the Cessna pilot] that he was in radar contact and instructed him to maintain VFR conditions . . . .

As the Cessna climbs out from the airport, Flight 182 is descending to land. The first officer is flying the Boeing 727. The portion of the radio and CVR transcript, below, begins after Flight 182 contacts

Boeing C-22, a military variant of the civil Boeing 727 airliner (official U.S. Air Force photo)

Approach Control. The airspace over San Diego is crowded, and ATC is in contact with several aircraft. In the interest of clarity, radio contacts with these other planes are omitted:

Capt: .... Captain, PSA Flight 182
FO: .... First Officer, PSA Flight 182
FltEngr: ...... Flight Engineer, PSA Flight 182
UnidCrew: .... Unidentified crewmember, PSA Flight 182
OffDutyCapt: .. Off-duty Captain, PSA Flight 182
Radio-: .... (prefix) Radio transmission, PSA Flight 182
Tower: .... Lindbergh Field Control Tower
AppControl: ... San Diego Approach Control

Radio-FO: Approach [Control], PSA one-eighty-two's out of nine-five, descending to seven-thousand, the airport's in sight. (0857:01)

AppControl: PSA one-eighty-two's cleared [for a] visual approach, Runway two-seven. (0857:06)

Radio-FO: Thank you, cleared [for a] visual approach, [Runway] two-seven. (0857:09)

[Flight 182 receives its <u>first traffic advisory</u>]

AppControl:: PSA one-eighty-two, traffic [at] twelve o'clock, one mile, northbound. (0859:30)

Radio-Capt: We're looking [for the traffic]. (0859:35)

[Flight 182 receives a second traffic advisory]

AppControl: PSA one-eighty-two, additional traffic's -- ah -- twelve o'clock, three miles, just north of the field, northeast-bound, a Cessna one-seventy-two climbing VFR out of one-thousand-four-hundred [feet]. (0859:39)

Radio-FO: OK, we've got that other [traffic at] twelve. (0859:50)

AppControl: Cessna seven-seven-one-one-golf, [this is] San Diego Departure, radar contact, maintain VFR conditions at or below three-thousand-five-hundred [feet], fly heading zero-seven-zero, vector [to the ILS] final approach course. (0859:57)

[Flight 182 receives a third traffic advisory]

AppControl: PSA one-eighty-two, traffic's at twelve o'clock, three miles, out of one-thousand-seven-hundred. (0900:15)

FO: Got 'em [in sight]. (0900:21)

Radio-Capt: Traffic in sight. (0900:22)

AppControl: [PSA one-eighty-two], *maintain visual separation, [emphasis added]*, contact Lindbergh Tower [on] one-three-three-point-three, have a nice day now. (0900:23)

Radio-Capt: OK. (0900:28)

Radio-Capt: . . . PSA one-eighty-two, downwind. (0900:34)

[Flight 182 receives a fourth traffic advisory]

Tower: PSA one-eighty-two, Lindbergh Tower -- ah -- traffic [at] twelve o'clock, one mile, a Cessna. (0900:38)

[ATC has reported the Cessna dead ahead and only one mile away; the airline pilots look for the Cessna but do not see it; they discuss where it might be]

Capt: Is that the one we were looking at? (0900:42)

FO: Yeah, but I don't see him now. (0900:43)

Radio-Capt: OK, we had it there a minute ago. (0900:44)

Tower: [PSA] one-eighty-two, roger. (0900:47)

Radio-Capt: I think he passed off to our right. (0900:50)

Tower: Yeah. (0900:51)

Capt: He was right over here a minute ago. (0900:52)

FO: Yeah. (0900:53)

Tower: How far are you going to take your downwind, one-eighty-two? Company traffic is waiting for departure. (0900:53)

Radio-Capt: Ah, probably about three to four miles. (0900:57)

Tower: PSA one-eighty-two, [you are] cleared to land. (0901:07)

Radio-Capt: One-eighty-two's cleared to land. (0901:08)

FO: Are we clear of that Cessna? (0901:11)

FltEngr: Supposed to be. (0901:13)

Capt: I guess. (0901:14)

OffDutyCapt: I hope. (0901:20)

Capt: Oh, yeah, before we turned downwind I saw him [at] about one o'clock, [he is] probably behind us now. (0901:21)

FO: Gear down. (0901:31)

  [the pilots suddenly see the Cessna dead ahead]

Capt: Whoop! (0901:45)

FO: Aaaggghhh! (0901:46)

  *[the Boeing 727 and the Cessna **collide** at 0901:47]*

OffDutyCapt: Oh, [expletive]! (0901:47)

Capt: Easy, baby! Easy, baby! What have we got? (0901:49)

FO: It's bad. (0901:52)

<u>Capt</u>: Huh? (0901:52)

<u>FO</u>: We're hit, man! We are hit! (0901:53)

   [the airliner, right wing low, begins to <u>spiral earthward</u>]

<u>Radio-Capt</u>: Tower, we're going down, this is PSA! (0901:55)

<u>Tower</u>: OK, we'll call the [crash] equipment for you. (0901:57)

<u>UnidCrew</u>: Whhhooo! (0901:58)

   [sound of stall warning at 0901:58]

<u>Radio-Capt</u>: This is it, baby! (0901:59)

<u>UnidCrew</u>: Bob! (0901:59)

<u>Capt</u>: Brace yourself! (0902:03)

<u>UnidCrew</u>: Ma! I love ya! (0902:04)

   [3 seconds later, <u>impact with the ground</u> at 0902:07]

The Collision: Both aircraft had been flying in the same direction, with the big Boeing 727 rapidly overtaking the small Cessna 172.

From behind, the airliner had caught up with the hapless Cessna and had slammed into it 2,600 feet above the ground.

Later, enhancement of two in-flight post-collision photographs would show massive damage to the Boeing right wing and control surfaces. Several leading edge devices had been ripped away. Part of the leading edge, back to the spar, had been crushed:

> Because of fire which covered the aft section of the wing in the area of the inboard aileron, it was not possible to ascertain [from the photographs] whether any of the [trailing edge control] surfaces in that area were missing.

After the aerial impact the Boeing 727 had begun a right turn. The fire had increased in intensity as the doomed airliner arced toward the ground. Nose down, banked 50 degrees to the right, the airliner dove into a residential area of San Diego three miles north of the airport. Seismological instruments recorded the massive ground impact, and all 135 people aboard died instantly. The crash and intense fire that followed killed seven people on the ground and destroyed or damaged 22 homes:

> The Cessna 172 was damaged extensively by the collision and fell to the ground in pieces.

Unlike the airliner, the Cessna did not *arc* down to the ground. Instead, the small aircraft "broke up immediately and exploded," instantly killing both pilots.

The Investigation: ATC had maintained radar and radio contact with both aircraft. Both were receiving active Stage II radar services. How could this accident have happened?

The key lay in the ATC terminology, *maintain VFR conditions* and *visual approach*. In the clear sky over San Diego those terms placed the burden to "see and avoid" other aircraft on the pilots. After issuing "visual" clearances, conflicting traffic warnings from ATC became mere *advisories* to aid the pilots:

> After Flight 182 was cleared for the visual approach . . . federal regulations required the crew to "see and avoid" other aircraft.

Investigators documented three traffic advisories from Approach Control and another from Lindbergh Tower. The controllers had pointed out potentially conflicting traffic to the pilots of Flight 182. However, between 0858 Hours and 0905 Hours, Lindbergh Tower was in radio contact with six aircraft. Consequently, investigators could not be certain which aircraft the PSA pilots had been looking at when they had replied, "traffic in sight."

Analysis suggested, however, that the PSA pilots were looking at a *different aircraft*. The controller had no way to know this. In any event the PSA pilots did not have, or did not keep, the Cessna 172 in sight and did not convey this information to the controller.

ATC had issued its last advisory to Flight 182 at 0900:38 Hours, 69 seconds before the collision. ATC had reported the Cessna was one mile ahead and flying essentially the same heading:

> [A cockpit] visibility study showed that when the [last two] advisories were issued, the Cessna would have been almost centered on both [PSA] pilots' windshields.

An overtaking aircraft is responsible for avoiding a slower aircraft under visual separation procedures. Yet, CVR verbiage in the Boeing cockpit showed that the pilots thought the Cessna had passed behind or beneath them. Instead, it was dead ahead. The automated "conflict alert" had sounded in Approach Control at 0901:28 Hours, a mere 19 seconds before the collision:

> . . . indicating to the controller that the predicted flight paths of Flight 182 and the Cessna would enter the computer's prescribed warning parameters.

The controller had radioed an advisory to the Cessna pilots:

> . . . traffic in your vicinity, a PSA jet, [he] has you in sight, he's descending for [landing at] Lindbergh.

ATC did not radio a similar advisory to Flight 182, for its pilots had reported (erroneously) that they had the Cessna in sight. The final crucial seconds ticked away, and the two aircraft collided.

Probable Cause: Seven months after the accident the NTSB adopted the Aircraft Accident Report, NTSB-AAR-79-5, on 20 April 1979.

The board made it clear the pilots of both aircraft had been responsible for maintaining *visual* clearance from other aircraft in flight. They also were required to comply with ATC instructions in the controlled airspace around Lindbergh Field:

> Flight 182 was an IFR aircraft; the Cessna was a participating VFR aircraft. Proper Stage II services were afforded [to] both aircraft.

The board concluded that when ATC told the Boeing pilots the Cessna was straight ahead, one mile away, the Boeing pilots did not see it. The Cessna pilots could not have seen the huge airliner bearing down on them from behind. Therefore, the NTSB found the **Probable Cause** of this accident to be:

> . . . the failure of the flight crew of Flight 182 *[the Boeing 727]* to comply with the provisions of a maintain-visual-separation clearance, including the requirement to inform the controller when they no longer had the other aircraft in sight.

The board found a **Contributing Cause**, the ATC procedure in use. This procedure authorized controllers to "use visual separation procedures to separate two aircraft on potentially conflicting tracks" in controlled airspace around the airport. ATC controllers had the capability to "provide either lateral or vertical radar separation to either aircraft" but did not do so.

Plain and Simple Bad Luck
Mid-Air Collision near Cannon AFB, New Mexico
6 February 1980

General Dynamics F-111 Aardvark, Serial No. 68-119
U.S. Air Force cross-country training flight
2 aboard, 2 killed

Cessna TU-206 Turbo Stationair, Registration N7393N
Building Contractors Inc. business flight
2 aboard, 2 killed

Synopsis: During an IFR approach a General Dynamics F-111 Aardvark collided with a Cessna TU-206 Turbo Stationair. The Cessna disintegrated, killing the pilot and his passenger. The two F-111 crewmen ejected outside the ejection envelope, and impact with the ground killed both of them.

The General Dynamics F-111 Aardvark Flight: A swing-wing General Dynamics F-111 Aardvark roared into the sky from Cannon AFB, New Mexico, on an IFR cross-country training flight. The U.S. Air Force plane carried a pilot and a student weapons systems officer (WSO). After completing their planned training the two crewmen headed back toward their base. The pilot planned to execute a Hi-TACAN penetration to Runway 21. Albuquerque Center handed-off the flight to Cannon AFB Approach Control.

The crew and their supersonic fighter-bomber were assigned to the 27th Tactical Fighter Wing. The pilot, a captain 34 years of age, had been in the Air Force for 11 years and had accumulated 2,505 hours of flight time. He was rated as a senior pilot, and he had 516 hours of experience in the F-111. The WSO, a second lieutenant 23 years of age, had logged a mere 126 hours of flight time.

The Cessna TU-206 Turbo Stationair Flight: A civil pilot and his passenger climbed into a Cessna TU-206 Turbo Stationair and took off from Albuquerque, New Mexico. They landed at Tucumcari to refuel. The pilot filed a VFR flight plan to Clovis Municipal Airport in eastern New Mexico. He got a weather briefing and information about military jet traffic in the vicinity of Cannon AFB:

Cessna 206 Stationair (photo courtesy of Adrian R. Pingstone)

True to Cessna tradition the TU-206 features a high wing with excellent visibility on all sides. A 285 horsepower Continental engine and a McCauley constant speed propeller give the plane a cruising speed of about 160 knots.

The Cessna left Tucumcari and flew toward Clovis at 7,500 feet. The pilot, age 43, was president of Building Contractors Inc., and he held a Private Pilot certificate. He had logged about 220 hours in the air. He and his passenger, a project manager, were going to Clovis to inspect a site where they planned to build two group homes for retarded children.

Trouble Ahead: With its speed brake extended the F-111 warplane descended at about 340 knots and about 3,600 feet per minute. On a heading of 208 degrees the pilot tracked inbound toward the TACAN. He passed the 10 mile DME fix at 5,800 feet, and Approach Control cleared him for a low approach to Runway 21. The F-111 primary and secondary returns were clearly visible on the controller's scope. Meanwhile, the Cessna motored toward nearby Clovis Municipal Airport, descending to 5,800 feet on a heading of 111 degrees.

The portion of the ATC radio transcript, below, begins as the military F-111 pilot is handed-off to Cannon Approach Control. Parts of the transcript have been *reconstructed* from the subsequent NTSB report, so some verbiage may not be verbatim:

<u>F-111</u>: . . . . Gen. Dyn. F-111 Aardvark (call-sign, Leggs 45)
<u>AppControl</u>: . . Cannon AFB Approach Control
<u>Center</u>: . . . . . . Albuquerque Center (ARTCC)

General Dynamics F-111 Aardvark (photo courtesy of U.S. DOD Visual Information Center)

<u>Center</u>: Leggs forty-five, descend to and maintain one-six-thousand feet, contact Cannon Approach Control . . . . (1006)

<u>F-111</u>: Center, Leggs four-five, roger, I'm descending to one-six-thousand and contacting Cannon Approach . . . . (1006)

[the F-111 pilot radios Cannon Approach Control]

<u>AppControl</u>: Leggs four-five, radar contact, you're four-zero miles northwest, hold northwest of Curry Intersection . . . . (1007)

[the F-111 enters the holding pattern]

<u>AppControl</u>: Leggs four-five . . . descend to and maintain one-four-thousand feet. (1010)

<u>F-111</u>: Approach, Leggs forty-five, descending to one-four-thousand in the holding pattern. (1010)

<u>AppControl</u>: . . . descend to one-two-thousand. (1017)

<u>AppControl</u>: Leggs four-five, cleared for a high-TACAN penetration and approach to Runway two-one, expect radar vectors . . . . (1019)

*[the F-111 **<u>collides</u>** with the Cessna at 5,800 feet at 1026:34]*

*[the F-111 secondary return instantly drops off the radarscope;
10 seconds later the primary return vanishes]*

[the controller hears an ELT on the tower watch frequency]

AppControl: Leggs four-five – uuuhhh – radar contact lost at this
time, say your altitude. (1026)

AppControl: Leggs four-five, say your altitude. (1026)

AppControl: Leggs, four-five, this is Cannon Approach transmitting
on Guard, say your altitude. (1027)

AppControl: Leggs four-five, Cannon Approach, over. (1027)

AppControl: Leggs four-five, Cannon Approach, over. (1028)

[there is no response]

The Collision: The F-111 pilot and WSO had leveled-off at 5,800
feet msl (only 1,505 feet agl) and had expected to get radar vectors at
the 6.5 mile DME fix. Descending through the same altitude, the
Cessna had crossed the F-111 approach path while flying toward the
civil airport 13 miles away:

> None of the [Cannon AFB] controllers had visually detected the
> Cessna or observed . . . the transponder signal of the [Cessna].

Eleven miles from Cannon AFB the two planes had rammed into
each other at a closure rate of about 370 knots. The Cessna literally
disintegrated, and pieces of the plane rained down from the sky.

Engulfed in a fireball, the F-111 rolled right and pitched down.
The pilot and WSO ejected as their plane rolled past 90 degrees:

> An altitude of about 2,000 feet is required for a successful module
> ejection. [However], evidence indicates that module ejection
> occurred at an altitude of about 1,300 feet above ground level.

The small stabilization chute deployed properly, but the crew had
ejected too close to the ground. The main parachute did not have
time to deploy before the module slammed into the earth:

> The crew ejection module impacted the ground nose-down on its
> left side, . . . bounced 30 feet, and came to rest in an inverted

position. [It] sustained severe structural damage . . . . The crew of the F-111 died as a result of the . . . impact with the ground.

The Investigation: The NTSB investigated this accident. The F-111 had been on an IFR flight plan and in radio and radar contact with ATC. Its transponder had Mode C capability, so Approach Control could monitor its altitude. Radar contacts confirmed the F-111 pilot was complying with ATC clearances and the mandates found on the Hi-TACAN penetration chart.

The Cessna pilot had been on a VFR flight plan and had not been in contact with any ATC facility. In fact, there was no control tower or flight service station at Clovis. The Cessna transponder had been set to Code 1200, the usual VFR code, but it lacked Mode C capability. The secondary return from the Cessna, if seen by ATC, would not have displayed the altitude of the aircraft.

Controllers at Cannon AFB told investigators they never saw a primary or secondary radar return from the Cessna. Yet, based upon (1) checks of Approach Control radar and (2) flight tests, the NTSB found the Cessna primary and secondary radar returns *would* have been displayed. The investigative staff concluded:

> The [primary return] and [transponder secondary] return of the Cessna *were [emphasis added]* displayed on the [Approach Control] radarscope. However, the Safety Board was unable to determine why the controllers did not observe the radar returns.

Although Approach Control never detected the Cessna, the plane *was* detected and tracked by radar at Albuquerque Center. This radar data was used to conduct a visibility study, and the results revealed a tragic coincidence. The weather had been clear with an estimated visibility of at least 30 miles, but neither pilot could have seen the other plane. The Cessna had been to the extreme right of the F-111, and the F-111 had been to the extreme left of the Cessna:

> During the 105 seconds before the collision . . . the F-111 would have been above the [Cessna's] left wing and completely hidden from the Cessna pilot. During the 90 seconds before the collision the vision of the F-111 pilot *[who was seated on the left side of the cockpit]* was obscured by his cockpit structure.

The study revealed the F-111 student WSO *might* have been able to see the Cessna if he had been scanning. However, the approach required him to constantly monitor his cockpit radar. Consequently, neither the two pilots nor the WSO had a chance to see the other plane barreling toward them – plain and simple bad luck.

Probable Cause: The NTSB completed a two year investigation and finally adopted the Aircraft Accident Report, NTSB-AAR-82-10, on 24 August 1982. On request, Cannon Approach Control had offered radar advisories to civil and military aircraft operating near the base. Advisories were available to both VFR and IFR traffic. However, the F-111 pilot had not requested radar advisories, and the Cessna pilot never made radio contact with Approach Control.

The weather had been clear with excellent visibility. Yet, in a cruel coincidence the position, altitude, speed, and track of the two planes had combined to hide each of them from the pilot of the other aircraft. Neither pilot had been aware of the presence or proximity of the other plane. For an undetermined reason the military controllers never saw radar returns from the Cessna. The NTSB found that the *Probable Cause* of this accident was threefold:

[First], failure of both aircraft to request radar advisories.

[Second], failure of the F-111 flightcrew to see and avoid the Cessna TU-206.

[Third], failure of the [Cannon AFB] controllers to observe the Cessna radar target and to issue traffic advisories to the F-111.

The board detailed a *Contributing Cause* as follows:

Contributing to the accident were the limitations of the see-and-avoid concept in a terminal area with [both] low speed [and] high speed traffic.

Where Are the Skydivers?
Mid-Air Collision near Loveland, Colorado
17 April 1981

Handley Page HP-137 Jetstream, Registration N11360
Air-US Flight 716
13 aboard, 13 killed

Cessna TU-206 Turbo Stationair, Registration N4862F
Sky's West Parachute Center skydiving flight
6 aboard, 2 killed

Synopsis: A Handley Page HP-137 Jetstream airliner, cruising at 13,000 feet, collided with a Cessna TU-206 Turbo Stationair that was carrying skydivers. The collision and crash killed all 13 persons on the Jetstream. The aerial impact killed two Cessna occupants, but the pilot and three skydivers parachuted to safety.

The Handley Page HP-137 Jetstream Flight: Visibility high over Colorado was estimated at 60 miles on Good Friday afternoon, two days before Easter Sunday. Air-US Flight 716, a sleek Handley Page HP-137 Jetstream, took off from Stapleton International Airport in Denver on a regularly scheduled flight to Gillette, Wyoming. The IFR Jetstream carried a captain, first officer, flight attendant, and 10 revenue passengers as it cruised through the crystal clear sky:

Handley Page had designed the HP-137 in the 1960s. The British pressurized twin turboprop was aimed at the regional airline market and first flew in 1968. British Aerospace (BAE) acquired the high-tech design in 1978 and turned out more Jetstreams for European and American markets.

Powered by two American-built Garrett AirResearch turboprop engines and Hartzell propellers, the Jetstream was ideally suited for regional airlines. It offered operators a rare blend of speed and economy. The aircraft evolved from the original 12 seat design into an efficient 18 seat commuter. Executive transport versions became popular, and militarized variants were pressed into service as multiengine trainers and VIP transports.

Handley Page HP-137 Jetstream (official U.S. Air Force photo)

The young captain, age 27, had been employed by Air-US since July of the previous year, and he held an ATP certificate. He had logged 4,784 hours of flight time. His younger first officer, age 23, had earned a Commercial Pilot certificate, and his logbook reflected 2,280 hours of flight experience.

The Cessna TU-206 Turbo Stationair Flight: Sky's West Parachute Center, based at the Fort Collins-Loveland Municipal Airport, had been in business since November 1979. The company specialized in parachute sales and skydiving instruction. During the previous 12 months its two Cessna TU-206 Turbo Stationair aircraft had facilitated over 10,000 individual jumps near the airport:

> *The TU-206 made an ideal skydiving platform. Only the pilot's seat remained in the plane. All other seats had been removed to allow skydivers to sit on the floor. The standard door on the right side of the cabin had been removed and replaced with a sliding wooden door. It would remain closed in climb and cruise, but skydivers would slide it open to make their jumps.*

The Cessna carried a pilot and five skydivers. The pilot squawked transponder Code 1234 and began climbing toward the jump altitude at 15,500 feet. He held an ATP certificate and flew part-time for Sky's West, and his logbook showed 4,600 hours of flight time.

Trouble Ahead: The Jetstream cruised on a direct route from the Denver VOR toward the Douglas, Wyoming, VOR. This route would take the aircraft almost directly over the airport at Loveland. The regional airliner was on an IFR flight plan and under the control of Denver Center. The controller monitored the Jetstream secondary transponder display as the plane sped northwest.

Meanwhile, the Cessna climbed up toward the intended jump location near the airport. The pilot had not filed a flight plan. He was taking five skydivers up for their second jump of the day:

> The Cessna began climbing in a left racetrack pattern over the airport toward an altitude of 15,500 feet.

So far on this *second flight* the pilot had not made radio contact with ATC. Yet, two hours earlier on a previous flight he had radioed his intentions to Center. The ATC transcript, below, reflects radio transmissions on that *previous flight*:

> <u>Cessna</u>: . . . . Cessna TU-206 Turbo Stationair
> <u>Center</u>: . . . . Denver Center (ARTCC)

<u>Cessna</u>: Denver Center, . . . six-two fox will be skydiving [at] eight-thousand-five-hundred feet a mile and a half southeast [of the] Fort Collins-Loveland Airport [in] approximately one minute, then we'll be climbing to fifteen-thousand-five-hundred [feet]. (1410:32)

<u>Center</u>: [Cessna] six-two fox, roger. (apx. 1410:48)

<u>Cessna</u>: Denver Center, Cessna six-two fox [will be] skydiving [at] fifteen-thousand-five-hundred feet [in] one minute [at] Fort Collins-Loveland Airport. (1421:07)

<u>Center</u>: [Cessna] six-two fox, roger. (1421:14)

> *[the skydivers make their jumps, and the Cessna returns to the airport and lands]*
>
> *[the Cessna begins the second flight at apx. 1530]*
>
> *[the Cessna and the Jetstream **collide** at 1601:17]*

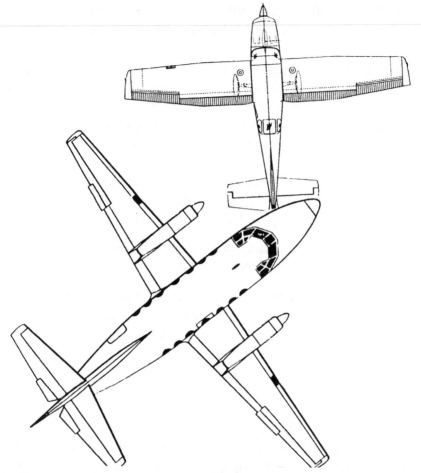

An illustration showing the Jetstream, lower left, and the Cessna, upper right, at the moment of collision (illustration courtesy of U.S. NTSB)

The Collision: On his *second flight* the pilot had never contacted Denver Center. The Cessna had climbed in a constant left turn two miles east of the airport.

The Jetstream flew straight and level at 13,000 feet, heading northwest. From the left rear quarter the Jetstream overtook the Cessna and rammed into it. First, the Jetstream nose tore off the Cessna empennage, and next:

> . . . the No. 1 propeller of the Jetstream cut through the aft fuselage section of the Cessna, resulting in an immediate loss of control to both aircraft.

The Jetstream propeller literally sawed off the rear of the Cessna fuselage. With what later would be described as enormous "chop-slash" injuries, the spinning propeller instantly killed two skydivers sitting in the rear of the Cessna cabin. Fortunately the Cessna pilot was wearing a parachute. He and three skydivers were thrown out of the plane through the gaping hole where the rear of the cabin should have been. Each sustained severe lacerations, and one got his right foot chopped off. Yet, they parachuted to safety:

> The pilot and three parachutists fell free of the aircraft and parachuted to the ground.

Everyone aboard the Jetstream airliner would die. The turboprop dove over two miles straight down. So severe was the high speed vertical ground impact that the fuselage and its priceless human cargo totally disappeared. The following day the *Reporter-Herald* newspaper in Loveland would explain the vanishing act:

> The commuter plane . . . nosedived into a wheat field, drilling . . . . the wreckage and most of the bodies 10 to 20 feet underground.

A deep smouldering pit marked the grave of 13 poor souls who had been flying toward Wyoming in the Jetstream moments before. Witnesses told of the "thunderous fiery explosion" when the plane dove into the field. Manual digging proved fruitless. Law Enforcement used a mechanized backhoe to excavate the smoking pit. The Loveland newspaper would report:

> By midnight sheriff's officers and Loveland firemen using . . . a backhoe had recovered six of the 13 bodies from the [pit]. They expected to keep digging [through the night].

The Investigation: Using recorded radar data, investigators found the Jetstream had been on its assigned course and altitude. Radio contacts between Center and the Jetstream pilots had been routine.

When investigators delved into the Cessna flight they uncovered a host of problems. FARs prohibited flight above 12,500 feet without transponder Mode C altitude encoding capability. Neither of Sky's West two aircraft had such capability, but they had routinely flown skydivers above 12,500 feet. Sky's West had no FAA waiver to permit such flights, and it rapidly got worse:

Denver Center controllers, on a routine basis, allowed the flights to operate at altitudes above 12,500 feet. *[and also]* The permissiveness of Denver Center created a situation wherein Sky's West [pilots] believed they had a standing waiver . . . .

Denver Center management maintained it had *not given permission* for such flights, but instead had simply acknowledged they were being made – a dubious distinction.

Controllers had sometimes assigned transponder Code 1234 to Sky's West flights. The pilots gradually *assumed* this was their permanent code and always would allow ATC to track them. Such was not the case. On the fatal flight, and on many other flights, the controller had not seen the Cessna on radar:

[The controller's] non-Mode C filter button was not activated, thereby eliminating radar target display of non-Mode C aircraft.

FARs require that skydiving flights be reported to ATC one hour in advance so that a NOTAM can be issued. In the instant case Sky's West notified the Denver FSS only 23 minutes before the first flight. There was no notification before the fatal flight. The FSS supervisor prepared a NOTAM and posted it at his local facility. He never made any further distribution, although the FAA required such distribution.

Notwithstanding all the above, the collision had taken place in clear weather with a visibility of about 60 miles. Why had the pilots not seen each other? The Cessna pilot admitted he had not seen the Jetstream because he had been looking at the ground and concentrating on the drop zone. He testified that he thought ATC was protecting him from collisions with other aircraft, even though he had not radioed ATC during the fatal flight. He said he had been "about to call" Center when the accident occurred.

The Jetstream pilots should have had a perfect unobstructed view of the Cessna. However, they were under radar control and relied upon Denver Center for traffic advisories. Investigators concluded:

The Jetstream flight crew had not been advised by ATC of any air traffic, and therefore probably were not scanning.

Probable Cause: Eight months after the accident the NTSB adopted the Aircraft Accident Report, NTSB-AAR-81-18, on 17 December

1981. The board found that the Cessna pilot did not establish and maintain contact with ATC, as required by Sky's West procedures and by FARs. He flew above 12,500 feet without Mode C capability, which is required for such flights by FARs:

> Had the Cessna been equipped with Mode C the resultant target, with an indication of [its] altitude, would have been presented clearly on the controller's radar display. *[and also]* A primary target was [weakly] presented on the controller's display for about 75 percent of the Cessna's flight path, but [it] was not noticed by the controller.

FAA management had known of the flights above 12,500 feet without Mode C transponders. They had done nothing to stop it. The FSS had not distributed the NOTAM as required. The planes had collided within two miles of three airways (V-101, V-4, and V-19) and in airspace normally reserved for planes departing from Stapleton International Airport. In consideration of the foregoing the board detailed the *Probable Cause* of this accident as follows:

> [First], failure of the Cessna pilot to establish communications with the Denver Center.
>
> [Second, the Cessna pilot's] climb into controlled airspace above 12,500 feet without an authorized deviation from the altitude encoding transponder requirement.
>
> [Third], the practice of Denver Center routinely condoning Sky's West parachute jump operations above 12,500 feet without a Mode C transponder.
>
> [Fourth], failure of the pilots of both aircraft to "see and avoid" each other.

The board also explained a *Contributing Cause*:

> Contributing to the accident was the fact that existing regulations do not prohibit parachute jumping in, or immediately adjacent to, federal airways.

You've Got F-4s on Your Tail
Mid-Air Collision over the Atlantic Ocean
9 January 1983

McDonnell F-4 Phantom, Serial No. 63-7536
U.S. Air Force and NORAD intercept mission
2 aboard, 0 killed

Beechcraft D-55 Baron, Registration N7142N
U.S. Atlantic Coastal ADIZ unauthorized penetration
7 aboard, 7 killed

Synopsis: *NORAD scrambled two supersonic McDonnell F-4 Phantom fighters when an unidentified bogey penetrated the U.S. Atlantic Coastal ADIZ. In solid clouds one of the F-4s collided with the bogey, a Beechcraft D-55 Baron. The collision ripped the Baron apart and killed its seven occupants. The damaged F-4 landed without injury to its two crewmen.*

The Beechcraft D-55 Baron Flight: A Beechcraft D-55 Baron pilot walked into the flight service station at Nassau, Bahama Islands. He tried to file a four hour VFR flight plan direct to Norfolk, Virginia. Bahamian authorities told him that U.S. Customs required entry into the United States through Florida. Hearing this, the pilot filed a flight plan to Fort Pierce, Florida, then direct to Norfolk. Nassau Tower cleared the Baron for takeoff at 1349 Hours:

> The flight . . . was instructed to contact en route VFR advisory service for activation of its VFR flight plan.

The pilot, age 47 with 4,455 hours of flight time and a Private Pilot certificate, ignored these instructions. He neither activated his flight plan nor flew toward Fort Pierce. In his twin engine Baron he flew north over the Atlantic Ocean for almost three hours without any type of flight plan. Headed toward Norfolk, the Baron penetrated the U.S. Atlantic Coastal Air Defense Identification Zone (ADIZ).

The McDonnell F-4 Phantom Flight: Long range radar operated by the U.S. Air Force, in conjunction with the North American Air Defense Command (NORAD), detected the inbound bogey. The

McDonnell F-4 Phantom (photo courtesy of U.S. DOD Visual Info. Center)

bogey could not be correlated with any known flight. The Air Force called Washington Center which, at that time, had no record of the flight and no contact with it. Consequently, the Air Force issued a "scramble" order to intercept and identify the bogey.

Two McDonnell F-4 Phantoms took off from Seymour Johnson AFB in North Carolina. Vectored by NORAD "Fertile Control," the supersonic fighters headed out over the Atlantic Ocean toward the bogey. The Phantoms and their pilots, call-signs JL-25 and JL-26, belonged to the 171st Fighter Interceptor Squadron of the Michigan Air National Guard. They were assigned to the Air Force in support of NORAD. The wingman, JL-26, would conduct the intercept:

> Four radar-guided Sparrow [air-to-air] missiles were mounted on the . . . F-4 [Phantom].

Trouble Ahead: As Fertile Control vectored the Phantoms toward the unsuspecting Baron, the civil pilot was having trouble with bad weather. At 1631 Hours, 13 minutes after the Phantoms took off, he made initial radio contact with Washington Center and asked for radar vectors around thunderstorms.

Washington Center controllers faced a dilemma. They dealt with (1) an unidentified voice-on-the-radio and (2) an unidentified radar target. There was no record of the flight, no flight plan, no way to identify the plane, and no way to identify the pilot. Meanwhile, the Air Force told Center that two Phantoms were closing on a bogey at that same location.

The radio and internal communications transcript, below, begins as the Baron pilot makes initial contact with Washington Center. The Baron and Washington Center operate on a civil VHF frequency. The Phantoms and Fertile Control use a military UHF frequency. The Baron pilot and Washington Center can not hear the Phantom pilots and Fertile Control, and vice versa:

Baron: . . . . Beechcraft D-55 Baron
Phantom: . McDonnell F-4 Phantom wingman (call-sign, JL-26)
Center: . . . Washington Center (ARTCC)
Fertile: . . . Fertile Control (NORAD military controller)
Inter-: . . . . (prefix) Fertile Control internal communications

[the Baron pilot radios Washington Center]

Center: Seven-one-four-two-November, this is Washington Center, go ahead. (1631:26)

Baron: . . . we're about fifty miles south of New Bern, squawking one-two-zero-zero, and would like radar advisories . . . . (1631:28)

[the Baron pilot says he is at 9,500 feet and heading for Norfolk; he continues to talk with Center as he flies north]

[Center informs Fertile Control that an unidentified pilot claims he is flying a civil Beechcraft D-55 Baron]

[Fertile Control keeps the Phantom pilots informed as it vectors them toward the bogey]

Fertile: We talked to Center, [they] reported that the aircraft you're [going to intercept] is at approximately nine-thousand feet, climb [to] angels eleven *[11,000 feet]* and call when level. (1639)

Inter-Fertile: [We have a] Center report, they think this [airplane] is a D-fifty-five *[Baron D-55]* that's lost. (1640:12)

Inter-Fertile: Be advised Washington [Center] has informed us of a possible ID, a D-fifty-five, call-sign November-seven-one-four-two-November.

Inter-Fertile: A *possible [emphasis added]* ID is not close enough, we have to get a [positive visual] ID.

Inter-Fertile: The type of aircraft [supposedly] is a D-fifty-five, and he's supposed to be landing at Norfolk, but this is all speculation.

[the Phantom onboard radar locks-on the bogey]

Phantom: Twenty-six has a [radar] contact, zero-two-zero for six.

[Fertile Control confirms the Phantom has the right target]

Phantom: Roger, [I'm] going to move on in, [flight leader], you drop back two or three miles.

Phantom: Judy. *[meaning, I have radar contact and will complete the intercept without further help from Fertile Control]* (1641:48)

[the Baron pilot is not aware a Phantom is closing behind him]

Baron: [Center], say, you have us in radar contact? (1642:33)

Center: Yes sir, [and] I want to advise you of something, sir. You are in a warning area, and they [sent] out military aircraft to scramble on your flight . . . OK, sir, well, you got some F-4s on your tail, sir. I want you to be aware of that . . . . (1642:35)

Baron: . . . we're going direct to New Bern right now, aren't we?

Center: I don't know. I asked you before if you were going direct to Norfolk, and you said yes. (1643:25)

Baron: . . . you want us to go direct [to] New Bern? (1643:50)

Center: Seven-one-four-two-November, yes sir, head toward New Bern, and I'll give you vectors to Norfolk [later], but right now head toward New Bern. (1644:04)

[the Baron <u>turns</u> toward New Bern, <u>into the path of the Phantom</u>]

[in solid clouds, the Phantom pilot can not see the Baron]

Phantom: [I think] we went by him . . . No! (1644:46)

*[the Phantom **collides** with the Baron at 1644:46]*

[the Baron disintegrates, but the Phantom remains flyable]

[3 seconds later Fertile Control decides to abandon the intercept; the military controller does not know about the collision]

Fertile: Juliet Lima flight, knock it off, post attack, left [to a heading of] two-five-zero. (1644:49)

Fertile: Juliet Lima flight, did you copy? Knock it off.

Phantom: [Fertile Control, this is Juliet Lima] twenty-six, [I] had a mid-air with the target, [I'm] climbing now past eleven-thousand, leaking fuel out of the left wing! (1644:57)

[the Phantom flies to Seymour Johnson AFB and lands safely]

The Collision: The Phantom had closed on the unseen Baron, which had been concealed in solid clouds. Closing to within 1,800 feet, the Phantom pilot still had been unable to see his target. He had turned slightly left to abort the intercept attempt. Unfortunately, at that moment the Baron pilot had turned hard to the left to fly toward New Bern. He turned directly into the path of the Phantom:

> The F-4 [left wing] came in contact with the tail of the Beech Baron and continued in a forward descending direction through the fuselage and cockpit of the Baron.

The Phantom had overrun the Baron and had sliced through it over the ocean 30 miles south of Cherry Point, North Carolina. The impact heavily damaged the Phantom left wing, but the pilot was able to nurse the crippled fighter back to Seymour Johnson AFB and land safely. Propeller slashes covered the wing, and:

> . . . a section of a Beech window frame was found embedded in the leading edge of the F-4 wing.

Also, parts of the Baron instrument panel were found jammed into the wing, which had literally sawed the Baron and its occupants in half. Coast Guard and Marine Corps search and rescue helicopters combed the ocean. They found no trace of the Baron or its occupants. Later, contacts with relatives and friends would confirm that seven people had been aboard the Baron. They took off from Nassau and were never seen again:

> The only evidence relevant to the [Baron] occupants was blood and hair found embedded in the leading edge of the F-4 left wing.

Beechcraft T-42 Cochise, a military variant similar to the Beechcraft D-55 Baron (official U.S. Army photo)

The Investigation: The NTSB relied upon radio transcripts from Washington Center and the Air Force. Combined with radar data and input from Bahamian authorities, the transcripts enabled the NTSB to reconstruct the flights and the collision.

Radar data consistently placed the Baron on a northerly heading at 180 knots. The Phantom had been angling in from the left rear quarter at a closure rate of 127 knots. Unable to see the target in the IMC weather, the Phantom pilot had started a left breakaway about 10 seconds before the collision. Under normal circumstances the Phantom would have passed harmlessly to the left of the target. However, the Baron pilot had turned left toward New Bern:

> The turn . . . placed the Beech Baron in the path of the F-4.

The NTSB noted that the Phantom pilot had closed on the Baron at a closure rate in excess of the recommended 50 knots. On the other hand, investigators realized the conduct of the Baron pilot had made an intercept attempt necessary:

> The need for airborne identification of unidentified aircraft as a matter of national security is clear.

The Baron pilot had failed to follow instructions from Bahamian aviation authorities. He had not activated his VFR flight plan. He had ignored the U.S. Customs mandate to enter the United States through Florida. He had penetrated the ADIZ, presumably *knowing* he would do so, and *knowing* it would prompt a military intercept. Also, he had been flying without any type of flight plan in IMC weather. In view of this conduct, investigators were not surprised to learn the Baron somehow had carried seven persons, although it had seats for only six.

Probable Cause: The NTSB released its Aircraft Accident Report, NTSB-AAR-84-07, on 19 June 1984. The Baron pilot's failure to activate a flight plan made it impossible for Fertile Control to correlate its radar contact with a known aircraft. Therefore, the scramble and intercept order had been appropriate. The intercept attempt had been conducted in IMC weather:

> During [the previous] year the 20th NORAD Region intercepted 10 Soviet Block aircraft operating off the east coast of the United States between the Virginia Capes and Florida.

Board members proposed three broad cause scenarios, the last of which was titled, *Statement on Probable Causation.* It included the following five causes of the accident:

> [First], the unauthorized penetration of an ADIZ [by the Baron].

> [Second], deviation by the Baron pilot from a requirement to activate a previously filed VFR flight plan and thus declare [his intended] ADIZ penetration.

> [Third], failure of the Baron pilot to file and activate an instrument flight plan before operating in IMC.

> [Fourth], inadequate tracking sensitivity for both ground and airborne radar for the intended mission.

> [Fifth], use of an excessive closure rate between [the] unidentified aircraft and F-4 while depending on visual identification in IMC.

Head-On Collision on the ILS
Mid-Air Collision near San Luis Obispo, California
24 August 1984

Beechcraft C-99 Commuter, Registration N6399U
Wings West Airlines Flight 628
15 aboard, 15 killed

Rockwell 112-TC Alpine Commander, Registration N112SM
Aesthetec Corporation check-out flight
2 aboard, 2 killed

Synopsis: A Beechcraft C-99 Commuter airliner followed a SID and flew outbound on the ILS localizer. A Rockwell 112-TC Alpine Commander was flying inbound on the same localizer. The two aircraft rammed head-on into each other, killing all 17 people on board both planes.

The Beechcraft C-99 Commuter Flight: Late on a sunny Sunday morning Wings West Airlines Flight 628 landed at San Luis Obispo Airport, California. After a quick exchange of baggage and passengers the Beechcraft C-99 Commuter airliner left the gate at 1110 Hours. The plane carried 2 pilots and 13 passengers for the planned IFR flight to San Francisco International Airport, California:

The Beechcraft C-99 Commuter is a larger model of the highly successful Beechcraft King Air. The airliner/freighter twin turboprop is powered by Pratt & Whitney PT-6 turbine engines and cruises at about 250 knots.

The young captain, age 28, had 4,110 hours of flying experience and held an ATP certificate. His older first officer, age 46, was a former U.S. Air Force B-52 bomber pilot who had logged 6,194 hours. He held a Commercial Pilot certificate and had been employed by the airline for slightly over a month. He had passed his initial Wings West proficiency check only 10 days before this final flight.

The Rockwell 112-TC Alpine Commander Flight: Two experienced pilots took off from Paso Robles Airport, California, in a sleek Rockwell 112-TC Alpine Commander. One pilot was an instructor.

The other pilot was getting checked-out in the Rockwell plane. They filed no flight plan, and none was required:

The Rockwell 112-TC was developed in 1976. The plane featured retractable landing gear and a turbocharged Lycoming IO-360 engine. The speedy aircraft became wildly popular as a sophisticated cross-country cruiser.

Aesthetec Corporation owned and operated the Rockwell. The pilot, age 58, held Commercial Pilot and Flight Instructor certificates and had logged 4,857 hours of flight time. The copilot, age 42, had been a military pilot in the U.S. Navy and had 2,450 hours of flight experience documented in his logbook. He held a Commercial Pilot certificate with an airplane multiengine land rating. On this flight he planned to get checked-out in the Rockwell in preparation for a new single engine land rating.

Trouble Ahead: There was no control tower at San Luis Obispo Airport. The weather was clear with visibility estimated at 15 miles. The airline pilots took off from Runway 29 in VFR conditions and radioed Los Angeles Center to activate their IFR flight plan. They followed the preplanned Standard Instrument Departure (SID) and climbed outbound on the ILS localizer while in radio and radar contact with Center.

Meanwhile, the VFR Rockwell pilots were practicing an ILS approach to Runway 11. They intercepted the localizer and flew inbound toward the climbing airliner.

The portion of the ATC transcript, below, contains radio transmissions between Center and the airline pilots. The transcript begins as the pilots make initial contact with the controller:

Wings West 628: . . . Wings West Airlines Flight 628
Center: . . . . Los Angeles Center (ARTCC)

Wings West 628: Center, Wings West six-twenty-eight. (1116:00)

Center: Wings West . . . Los Angeles Center. (1116:40)

Wings West 628: [Center, this is] Wings West six-two-eight with a request. (1116:46)

Center: Wings West six-two-eight, go ahead. (1116:49)

<u>Wings West 628</u>: [Unintelligible] intersection VFR, two-point-seven, climbing IFR to San Francisco. (1116:51)

<u>Center</u>: Wings West six-twenty-eight, roger, squawk [transponder code] six-seven-two-one. (1116:56)

<u>Wings West 628</u>: Wings West six-two-eight, [we are squawking code] six-seven-two-one. (1117:00)

<u>Center</u>: Wings West six-two-eight, radar contact six [miles] north-west of San Luis Obispo Airport, say [your] altitude. (1117:16)

<u>Wings West 628</u>: Wings West six-two-eight, three-thousand-one-hundred, climbing. (1117:20)

<u>Center</u>: Wings West six-two-eight, cleared to San Francisco as filed, and climb and maintain seven-thousand, and Paso Robles altimeter [is] two-niner-niner-five. (1117:23)

<u>Wings West 628</u>: Wings West six-two-eight, understand [we are] cleared to San Francisco as filed, and climb and maintain seven thousand. (1117:32)

<u>Center</u>: [Unintelligible] six-two-eight, roger, you can expect [to be cleared to] eight-thousand [feet] shortly. (1117:38)

*[the C-99 airliner and the Rockwell **collide** at 1117:38; the impact totally destroys both aircraft]*

[1 minute later the controller notices the airliner secondary radar return has dropped off his scope]

<u>Center</u>: Wings West six-two-eight, I've lost your transponder, reset, squawk six-seven-two-one. (1118:40)

[there is no response]

The Collision: Complying with the SID, the C-99 airline pilots had climbed west on the localizer. At the same time the Rockwell pilots had descended east on the same localizer. The two aircraft flew head-on into each other:

Flight 628 and [the Rockwell] collided about 8 [miles] west-northwest of the San Luis Obispo airport . . . at 3,400 feet.

Beechcraft C-99 Commuter – note the bent propeller on the No. 1 engine on the left wing (photo courtesy of U.S. Federal Aviation Administration)

The Rockwell had rammed into the airliner cockpit. The small plane literally disintegrated. The larger airliner lost its cockpit and the top half of its fuselage. All aboard both planes died instantly:

> This was a non-survivable accident. The occupiable areas of the cockpits of both airplanes and the passenger cabin of the Beech C-99 were destroyed during the collision.

One witness who saw and heard the mid-air impact exclaimed: "It sounded like dynamite going off!" Wreckage from both planes rained down to the ground. Potential rescuers rushed to the rural crash site, but all aboard the two planes were beyond mortal aid. The next day the local *Telegram-Tribune* newspaper headline would read:

Mid-Air Collision Claims 17 Lives

One dazed firefighter was quoted in the newspaper:

> Scattered pieces of metal and broken bodies were strewn across about 500 acres of pasture and farmland. All the bodies were badly mangled, [and] I didn't see any body with its head on.

The Investigation: NTSB investigators concentrated on recorded radar data at Los Angeles Center. Following their flight plan after takeoff, the IFR airline pilots had tracked *outbound* on the localizer toward Crepe Intersection (intersection of the localizer and the 196 degree radial of Paso Robles VORTAC). At the same time the VFR Rockwell pilots had been flying *inbound* on the same localizer.

The controller had confirmed radar contact with the airliner only 22 seconds before the collision. He had seen the full data block from the airliner transponder, but he said he never saw a primary return from the Rockwell. However, recorded radar data confirmed:

> The Rockwell's return *had been displayed [emphasis added]*, and the reason the controller did not see it cannot be attributed to any failure of the [ATC] equipment.

The FAA Airman's Information Manual recommends that pilots monitor UNICOM or CTAF at uncontrolled airports. Pilots arriving or departing such airports should announce their intentions, and:

> Pilots wishing to make practice instrument approaches should notify the [ATC] facility indicated on the approach chart.

The Rockwell pilots never notified ATC. Consequently, neither the airline pilots nor the controller had known the Rockwell pilots were practicing an ILS approach. ATC had no inkling that any plane, other than the airliner, was anywhere near the airport.

Los Angeles Center had a computerized "conflict alert" program. When potentially conflicting traffic exceeded program parameters the system had been designed to alert the controller. Yet, in this instance:

> The Rockwell's track had not been inserted into the computer; it was an uncorrelated VFR target, and [consequently] the conflict alert system did not function.

Wings West policy required its airline pilots to tune one of their two communication radios to the Wings West company frequency at

all times. The pilots had done so. They had tuned their other radio
to UNICOM *before* takeoff, and then to Center *after* takeoff. Once
tuned to Center, the airline pilots no longer could hear any UNICOM
or CTAF transmissions from the inbound Rockwell.

Before the collision the planes had been closing on each other at
a rate of 544 feet per second. With no relative motion and only a
small frontal profile potentially visible, the NTSB opined:

> It may have been impossible for the pilots of each airplane to see
> and avoid the other.

Probable Cause: A year after the accident the NTSB adopted the
Aircraft Accident Report, NTSB-AAR-85-07, on 29 August 1985.
The airliner had been tracking outbound on the localizer while the
Rockwell had been flying inbound. ATC was not aware of the
Rockwell's presence. Pilots in both planes had been monitoring
different radio frequencies. They could not hear transmissions to or
from each other, and they were unaware of the presence of the other
plane. Approaching directly head-on, pilots in both planes had only
a slim chance of spotting the other aircraft. The board explained the
Probable Cause of this accident as follows:

> Failure of the pilots of both aircraft to follow the recommended
> communications and traffic advisory procedures for uncontrolled
> airports [in order] to alert each other to their presence and to
> enhance the controller's ability to provide timely traffic adviso-
> ries. Underlying the accident were the physiological limitations
> of human vision and reaction time. Also underlying the accident
> was the short time available [for] the controller to detect and ap-
> praise radar data and to issue a safety advisory.

The board explained a *Contributing Cause*:

> Contributing to the accident was the Wings West Airlines policy
> which required its pilots to tune one radio to the company
> frequency at all times.

Grand Canyon Scenic Tour
Mid-Air Collision over the Grand Canyon, Arizona
18 June 1986

<u>de Havilland Canada DHC-6 Twin Otter, Registration N76GC</u>
Grand Canyon Airlines sightseeing tour flight
20 aboard, 20 killed

<u>Bell 206 JetRanger, Registration N6TC</u>
Helitech Inc. sightseeing tour flight
5 aboard, 5 killed

<u>Synopsis</u>: A de Havilland Canada DHC-6 Twin Otter and a Bell 206 JetRanger flew aerial sightseers over the Grand Canyon in Arizona. The two aircraft collided in flight and disintegrated, killing all 25 people on board.

The de Havilland Canada DHC-6 Twin Otter Flight: Early in the morning a de Havilland Canada DHC-6 Twin Otter took off from Grand Canyon Airport, Arizona. Grand Canyon Airlines owned and operated the Twin Otter. The twin turboprop had been modified with larger windows for air tours. This flight was scheduled to be a 50 minute scenic air tour over the Grand Canyon National Park.

The Twin Otter captain planned to celebrate his 28th birthday in five days. He held an ATP certificate and had logged 5,970 hours of flight time. He had upgraded to captain three months before this final flight. His first officer, age 27, also held an ATP certificate and had logged 4,450 flight hours. He had been flying for Grand Canyon Airlines for six years. The 18 Twin Otter passengers included 12 people from The Netherlands, Switzerland, and South Africa.

The Bell 206 JetRanger Flight: Eighteen minutes after the Twin Otter began its flight, a Bell 206 JetRanger rotored skyward from a heliport five miles from the park. The helicopter was operated by Helitech Inc., a new air tour company. The rotary-wing aircraft carried a pilot and four sightseers for an aerial tour of the canyon. The flight had been planned to last 30 minutes:

Bell 206 JetRanger (photo courtesy of U.S. DOD Visual Information Center)

The Bell 206 JetRanger is the most popular light turbine helicopter in the world. Over 7,000 have been built. The first commercial model was delivered in 1967, and upgraded models have been developed ever since. Power to the rotor system comes from Allison turboshaft engines. Early models seated five persons, but later stretched variants will seat up to seven.

The JetRanger captain, age 40, held a Commercial Pilot certificate and had flown for various Grand Canyon air tour operators during the previous nine years. His logbook reflected 6,953 hours of flight time. He had been employed by Helitech for only five days. Helitech, a new air tour operator, had been in business for two weeks.

Trouble Ahead: Between 350,000 and 400,000 people had flown on scenic aerial tours over the Grand Canyon during the previous year. Forty-four aviation companies were in business to provide the scenic flights. During the flights all pilots would make voluntary position reports on 122.75 MHZ. The control tower at Grand Canyon Airport controlled arrivals and departures only at the airport. There was no air traffic control over the canyon for air tour flights.

The Twin Otter (call-sign, Canyon 6) and the JetRanger (call-sign, Tech 2) flew along their normal tour routes in uncontrolled airspace. The brief VHF radio transcript, below, is based upon statements of other pilots as documented in the subsequent NTSB report. The NTSB may have paraphrased part of the wording, so some verbiage may not be verbatim:

Canyon 6: ... Grand Canyon Airlines (Twin Otter)
Tech 2: ..... Helitech (JetRanger)

Canyon 6: Canyon six is passing the Holy Grail.

Canyon 6: This is Canyon six, passing Havasupai Point. (0928)

Tech 2: Tech two is at the Anasazi Ruins.

Canyon 6: Canyon six, we're at Scorpion.

Tech 2: Tech two is west of Mencius Temple at six-thousand-four-hundred, southbound. (apx. 0930)

Tech 2: Tech two is coming up on Crystal Rapids.

*[the Twin Otter and JetRanger **collide** at apx. 0933]*

The Collision: The two aircraft rammed into each other in straight and level flight high over the Colorado River near Crystal Rapids. The JetRanger main rotor chewed into the Twin Otter. The whirling rotor sawed off the nose gear and tore through fuel tanks on the twin turboprop. The JetRanger transmission ripped out of its mounts, the rotor blades shattered, and debris flew in all directions. The entire Twin Otter aft fuselage snapped off, and remains of both aircraft fell like stones into the chasm below:

Wreckage of the two aircraft [impacted] about 2,450 feet apart on the Tonto Plateau ... about 1½ miles north of the Crystal Rapids on the Colorado River.

The aerial impact had been explosive. The forward and rear fuselage sections of the Twin Otter fell 953 feet apart. Pieces of the JetRanger were discovered in the Twin Otter fuselage, and remnants of the Twin Otter were found in rubble of the JetRanger:

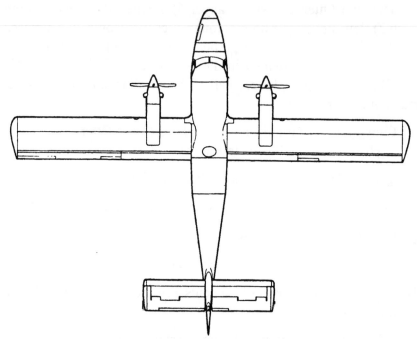

An illustration depicting an overhead view of a de Havilland Canada DHC-6 Twin Otter (illustration courtesy of U.S. NTSB)

Nearly 18 inches of the red main rotor blade spar cap of the Bell 206 was found embedded in the rear fuselage of the DHC-6 [and] the entire [JetRanger rotor] mast, which had separated from the transmission, was located near the tail section of the DHC-6.

No one aboard either aircraft had a chance to survive the mid-air collision, fall, and ground impact. Remains of both aircraft burned, and recovery of charred bodies would take several days. After post-mortem examinations the coroner would specify the cause of death for each of the 25 deceased persons as:

. . . multiple severe crushing and thermal injuries consistent with an airplane or helicopter crash.

The Investigation: Neither aircraft had a CVR or FDR, and there was no radar data. Consequently, an assessment of flight paths could not be made. However, the NTSB obtained a photograph of the post-collision vaporous cloud. It had been taken by a passenger on a raft near Boucher Rapids in the canyon. Using the picture and a technique known as photogrammetry, experts were able to pinpoint the location and altitude of the collision.

Helitech had no operations manual or training manual and had not obtained an operating certificate under 14 CFR Part 135. The new company employed one full-time pilot and two part-time pilots, and they operated two helicopters. Helitech flew sightseers "on-demand" whenever at least three paying customers showed up. There was no public address system on the JetRanger. The lone pilot had to take his eyes off the airspace ahead and turn his head to explain scenic wonders of the canyon to passengers in the rear seat:

> Other safety issues concern the lack of regulations to limit flight and duty times of pilots conducting scenic air tour flights, and the lack of a requirement for the pilots . . . to use intercoms or public address systems when narrating during the flights.

Grand Canyon Airlines had been flying tourists over the canyon for 60 years. Flights were scheduled each hour during daylight, and the company employed 10 pilots. On a typical day these pilots would be on duty for 12 hours and would accrue eight to nine hours of flight time. On each flight the non-flying pilot would use the Twin Otter public address system to explain scenic locations below.

The FAA Flight Standards District Office in Las Vegas, Nevada, was responsible for oversight of air tour companies at the canyon. Flights in the park were conducted in accordance with 14 CFR Part 135, and airspace usage rules were extremely liberal:

> Airspace in the Grand Canyon was unrestricted, and with the exception of the [Grand Canyon] airport traffic area [all airspace] was uncontrolled. As a result most flights in the Grand Canyon were required only to maintain 1 mile visibility, clearance from clouds, and a minimum safe altitude above the surface.

Probable Cause: Thirteen months after the accident the NTSB adopted the Aircraft Accident Report, NTSB-AAR-87-03. The board found that all three pilots were properly certificated and qualified for their flights. There were no known medical or behavioral factors that might have affected their ability to conduct the flights safely. They each had "considerable experience" in Grand Canyon sightseeing operations. Investigators had been unable to determine why the pilots had not seen each other in time to avoid the collision:

> The pilots of both aircraft should have been able to see and avoid each other.

> An absence of flight time and duty time limitations during the summer busy season was considered an adverse safety factor. Also, absence of an ICS or public address system on the JetRanger was deemed a safety breach.

The National Park Service had no authority over airspace usage at the Grand Canyon. However, the park service had "created the impression" that it would seek airspace restrictions to reduce noise in the park. This park service *jawboning* influenced air tour operators to move their air tour routes closer together. As a result the fixed-wing and the rotary-wing aircraft, with substantially different flight characteristics, shared the same airspace:

> The similarity of routes and the limited number of scenic points overflown by scenic air tours operators increased the risk of a mid-air collision.

The board found that the ***Probable Cause*** of this accident was:

> Failure of the flight crews of both aircraft to "see and avoid" each other for undetermined reasons.

The board found two ***Contributing Causes***:

> [First], the failure of the FAA to exercise its oversight responsibility in the Grand Canyon airspace, and [jawboning by] the National Park Service to influence the selection of routes by Grand Canyon scenic tour operators.

> [Second], modification and configuration of the routes of rotary wing operators resulting in their intersecting with the routes of Grand Canyon Airlines near Crystal Rapids.

Postscript: Helitech had been in business at the Grand Canyon for only two weeks. Several days after the accident, Helitech's insurance carrier terminated the company's coverage. Immediately thereafter Helitech closed its doors and went out of business.

Oh, [expletive]! This Can't Be!
Mid-Air Collision over Cerritos, California
31 August 1986

Douglas DC-9, Registration XA-JED
Aeronaves de Mexico (AeroMexico) Flight 498
64 aboard, 64 killed (plus 15 killed on the ground)

Piper PA-28 Cherokee, Registration N4891F
General Aviation pleasure flight
3 aboard, 3 killed

Synopsis: A Douglas DC-9 descended to land as a Piper PA-28 Cherokee plodded through the Terminal Control Area without an ATC clearance. In clear weather the two aircraft collided. The impact and subsequent crashes killed all 64 people on the DC-9, the 3 Cherokee occupants, and 15 people on the ground.

The Douglas DC-9 Flight: A Douglas DC-9, Aeronaves de Mexico (AeroMexico) Flight 498, left Mexico City, Mexico, and headed for Los Angeles, California. The twin-jet airliner swooped down for a quick stop at Tijuana, Mexico. Then the aircraft took off again for the short trip to Los Angeles. High over the congested urban sprawl of southern California the 58 passengers and 6 crewmembers fell under the guiding hand of Los Angeles Approach Control.

The captain, age 46, held two ATP certificates. One had been issued by the government of Mexico, and the other had been issued by the FAA in the United States. The captain had logged 10,641 flying hours, almost half of which were in the DC-9. His young first officer, age 26, held a Commercial Pilot certificate issued by Mexico and another issued by the FAA. He had logged 1,463 hours.

The Piper PA-28 Cherokee Flight: A Piper PA-28 Cherokee pilot and two passengers took off from Torrence, California, on a flight to Big Bear, California. Although he never activated it, the Cherokee owner-pilot had filed a VFR flight plan with a local flight service station. At 9,500 feet the proposed route would take him to Long Beach, then to Paradise VORTAC, and then to Big Bear.

The pilot, age 53, held a Private Pilot certificate. He was properly qualified to make this flight and had logged 231 flying hours. His Cherokee was a single engine four-seat plane with fixed landing gear.

Trouble Ahead: The IFR DC-9 approached Los Angeles to land. Meanwhile, the Cherokee had drifted off course and had blundered into the Los Angeles Terminal Control Area (TCA). The pilot was not in radio contact with any ATC facility.

The portion of the DC-9 CVR and ATC transcript, below, begins as Flight 498 first contacts Los Angeles Approach Control. The first officer is flying the DC-9:

> Capt: . . . . Captain, AeroMexico Flight 498
> FO: . . . . First Officer, AeroMexico Flight 498
> Radio-: . . . . (prefix) Radio transmission, Flight 498
> AppControl: . . . . Los Angeles Approach Control
> American 333: . . American Airlines Flight 333

Radio-Capt: Los Angeles Approach [Control], good morning, this is AeroMexico four-ninety-eight . . . seven-thousand. (1147:38)

AppControl: AeroMexico four-ninety-eight, [this is] Los Angeles Approach, depart Seal Beach three-two-zero [for] vector [to] ILS [for Runway] two-five left, final approach. (1147:39)

Radio-Capt: Affirmative, two-five-left [is the] runway. (1147:46)

Capt: Course two-four-nine, flight director up. (1148:15)

FO: Flight director up. (1148:16)

[1 minute and 49 seconds pass]

AppControl: AeroMexico four-ninety-eight, reduce speed to two-one-zero [knots]. (1150:05)

Radio-Capt: [Reduce] to two-one-zero, four-ninety-eight. (1150:08)

[in the clear sunny sky each aircraft is now within the other's field of vision; both will remain visible until the collision]

AppControl: AeroMexico four-ninety-eight, reduce speed to one-niner-zero, then descend and maintain six-thousand. (1151:03)

Radio-Capt: [Reduce speed to] one-niner-zero and then descend and maintain six-thousand. (1151:09)

AppControl: AeroMexico four-ninety-eight, maintain your present speed. (1151:46)

Radio-Capt: Roger . . . uuuhhh, what speed do you want? We're reducing to one-niner-zero. (1151:48)

AppControl: OK, you can hold what you have, sir, and we have a change in plans here, stand by. (1151:57)

*[at 6,560 feet the DC-9 and the Cherokee **collide** at 1152:09]*

[the DC-9 dives toward the ground, out of control]

Capt: Oh, [expletive]! This can't be! *[screamed]* (1152:10)

[the controller does not know the two aircraft have collided; he makes a routine radio call to the DC-9]

AppControl: AeroMexico four-ninety-eight, expect the ILS two-four right approach, localizer frequency is . . . . (1152:12)

[there is no response]

AppControl: AeroMexico four-ninety-eight, turn left [to] heading two-eight-zero. (1152:58)

AppControl: AeroMexico four-ninety-eight, turn left [to] heading two-eight-zero. (1153:08)

AppControl: AeroMexico four-ninety-eight, [this is] Los Angeles Approach. *[a query]* (1153:24)

[there is no response]

[the controller makes 4 more unsuccessful attempts to contact the DC-9; he then asks another airliner for assistance]

AppControl: American three-thirty-three heavy, I want you to look around at eleven o'clock and about five miles, I just lost contact with a DC-9, let me know if you see anything. (1156:05)

American 333: What altitude? (1156:17)

AppControl: He was last assigned to six-[thousand feet, but] he's no longer on my radarscope. (1156:21)

American 333: OK, I see – uuuhhh – a very large smoke screen off on the left side of [my] aircraft, abeam the nose of [my] aircraft, right off [to] our left – it is a very large smoke – uh – column. (1156:26)

The Collision: The American Flight 333 pilot had spotted the fiery grave of the DC-9. In a cloudless sky with a reported visibility of 14 miles, the DC-9 and the Cherokee had collided. The aerial impact had torn the vertical stabilizer and horizontal stabilizer off the DC-9 airliner. Collision damage was confined to the DC-9 tail and its control surfaces, so the passengers and crew survived the collision. However, their aircraft nosed over, dove into a residential area of Cerritos, California, and exploded:

> The airplane cockpit and passenger cabin were destroyed by massive impact forces and [a] post-crash fire. Although the occupants of the DC-9 survived the mid-air impact, this was an unsurvivable [ground impact].

The crash killed all 64 people on the DC-9 and another 15 people on the ground. The impact and the fire on the ground destroyed five homes and damaged many others. The bodies of the pilots were too "fragmented" to allow an autopsy or toxicological testing. The Los Angeles County coroner ruled that all occupants died from "multiple blunt force trauma."

The Cherokee engine and fuselage had rammed through the DC-9 vertical stabilizer. The impact had peeled back the top of the Cherokee fuselage, decapitating the pilot and his two passengers. Remains of the Cherokee fell to earth in a schoolyard about 1,700 feet from the DC-9 wreckage.

The coroner would ascribe deaths of the three Cherokee occupants to "multiple trauma due to, or as a consequence of, blunt force." Human remains from the Cherokee occupants were found embedded in the DC-9 horizontal stabilizer almost a half-mile away.

The Investigation: Accident reconstruction experts from the NTSB poured over recorded radar data. The DC-9 had been descending to the northeast. The Cherokee had been flying southeast on a track

Douglas C-9 Nightingale, a military variant of the civil Douglas DC-9 airliner (photo courtesy of U.S. DOD Visual Information Center)

perpendicular to that of the DC-9:

> The airplanes collided at a 90 [degree] angle at an altitude of 6,560 feet, and in visual meteorological conditions.

The Cherokee had flown straight into the left side of the vertical stabilizer on the DC-9. Even though the Cherokee was a much smaller and lighter aircraft, its heavy engine and propeller unit had torn through the main support structure of the DC-9 T-tail. Both the vertical and horizontal stabilizers ripped off airliner, and further controlled flight was not possible:

> The evidence was conclusive that the collision occurred within the Los Angeles TCA; that the Piper pilot had entered the TCA without having been cleared to do so; that the controller did not advise Flight 498 of the position of the Piper; and that [none of the three] pilots tried to perform any type of evasive maneuver.

The radar track of the Cherokee revealed it had flown almost directly to the scene of the collision after taking off from Torrence. The plane had been equipped with a non-Mode C transponder, which had been set to Code 1200. The transponder displayed only position information, not altitude, on ATC radar. The controller testified he never saw the Cherokee radar return.

The NTSB conducted an exhaustive cockpit visibility study. It revealed that each aircraft had been within the other's field of vision for at least 1 minute and 13 seconds before the collision. The

Cherokee had been visible through the DC-9 center windshield. The DC-9 had been visible through the right windshield of the Cherokee. The visibility study concluded:

> The AeroMexico flight crew should have had an almost unobstructed view of the Piper PA-28. *[and also]* The Piper pilot had an 80 percent probability of seeing the DC-9.

Probable Cause: The NTSB adopted its Aircraft Accident Report, NTSB-AAR-87-07, on 7 July 1987. In the VFR conditions inside the Los Angeles TCA, FAA regulations required pilots in both aircraft to "see and avoid" each other. Yet, there was no evidence to indicate that any of the three pilots had seen the other aircraft. Neither aircraft tried to avoid the other, for both had been in straight and level flight when they collided.

The board found the Cherokee had not been cleared to enter the TCA, but that the entry was inadvertent. The presence of the plane in the TCA was a "casual factor" to the accident. The board opined that the radar return from the Cherokee "may not have been displayed, or may have been displayed weakly" on ATC radar.

The board found that responsibility for this accident fell upon the shoulders of the air traffic control system in the United States. The board explained the *Probable Cause* in the following manner:

> The Probable Cause of the accident was the limitations of the Air Traffic Control system to provide collision protection through both air traffic control procedures and automated redundancy.

The board highlighted two *Contributing Factors*:

> [First], the inadvertent and unauthorized entry of the [Cherokee] PA-28 into the Los Angeles TCA.

> [Second were the inherent age-old] limitations of the "see-and-avoid" concept to ensure traffic separation [during VMC flight].

I'm in Trouble Down Here!
Mid-Air Collision over Kearns, Utah
15 January 1987

<u>Fairchild SA-226 Metro II, Registration N163SW</u>
SkyWest Airlines Flight 1834
8 aboard, 8 killed

<u>Mooney M-20, Registration N6485U</u>
Pilot proficiency training flight
2 aboard, 2 killed

Synopsis: A Fairchild SA-226 Metro II commuter airliner, in radio and radar contact with ATC, descended to land. A small Mooney M-20 flew into the Airport Radar Service Area and rammed into the SA-226 airliner. The aerial impact demolished both planes and killed all 10 people on board.

The Fairchild SA-226 Metro II Flight: SkyWest Airlines Flight 1834, a sleek Fairchild SA-226 Metro II, took off from Pocatello, Idaho. The popular commuter airliner carried a slim load of only six passengers. After 40 minutes the two pilots began a descent toward Salt Lake City International Airport in Utah:

The Metro II sprang from a Swearingen design in the late 1960s. Fairchild took over Swearingen in 1971, and the new Metro II featured many improvements over the original aircraft. Garrett turbine engines drove three-blade propellers. The pressurized turboprop regional airliner could seat 19 passengers and could cruise at over 250 knots.

The captain, age 39, held an ATP certificate and had been flying with SkyWest since 1984. He had logged 3,885 flight hours. His first officer, age 34, also had earned an ATP certificate and had 4,555 hours of flight experience. Their Metro II had been manufactured in 1980 and was among 35 such aircraft in the SkyWest fleet.

The Mooney M-20 Flight: Accompanied by a flight instructor, an owner-pilot took off from Salt Lake City Municipal Airport, which is located 10 miles south of the much larger international airport. The

Mooney M-20 (photo courtesy of Adrian R. Pingstone)

pilot practiced touch-and-go landings at the municipal airport for a half-hour. Without contacting ATC he then flew north and entered the restricted Airport Radar Service Area (ARSA) surrounding the sprawling international airport:

> *The Mooney M-20 is a high performance light aircraft. Designed in the 1950s, it is best known for its speed. Retractable gear, an aerodynamic fuselage, and a Lycoming IO-360 engine allow the plane to reach 171 knots. Most M-20s were purchased as cross-country cruisers and equipped with the latest avionics and instrumentation. The Mooney factory built over 5,000 of these distinctive forward-swept-tail speedsters.*

The Mooney pilot, age 40, had logged 301 hours of flight time and held a Private Pilot certificate. He had failed his first Private Pilot check-ride in 1981, but later passed after getting additional flight instruction. He took the Instrument Rating written test four years later and got a failing grade. He took the written test again the next year and passed, but he never took the necessary check-ride. The flight instructor, age 55, held Commercial Pilot and Flight Instructor certificates and had 2,547 hours of flight experience. He was a retired U.S. Army major.

Trouble Ahead: Salt Lake City Approach Control was busy with numerous airline, military, and corporate arrivals and departures. The controller planned to vector the inbound Metro II pilots for a visual approach to Runway 34 Left. Meanwhile, without contacting ATC the Mooney droned northwest into the ARSA. Neither Approach Control nor the Metro II pilots were aware of its presence.

The portion of the ATC transcript, below, contains radio traffic between the Metro II, another SkyWest flight, and Approach Control. Transmissions to and from numerous other aircraft have been omitted in the interest of clarity. The transcript begins as the Metro II pilots contact Approach Control. The captain makes the radio transmissions and the first officer flies the aircraft:

SkyWest 1834: . . SkyWest Airlines Flight 1834
SkyWest 575: . . . SkyWest Airlines Flight 575
AppControl: . . . . Salt Lake City Approach Control
Tower: . . . . Salt Lake City Intl. Airport Control Tower
Phone-: . . . . (prefix) ATC interphone message

SkyWest 1834: Approach, SkyWest eight-thirty-four is with you, descending to nine-thousand, we have the airport [in sight]. (1247:28)

AppControl: SkyWest eight-thirty-four, descend and maintain seven-thousand. (1249:19)

SkyWest 1834: [We are] out of nine-[thousand] for seven-[thousand], SkyWest eight-thirty-four. (1249:22)

AppControl: SkyWest eight-thirty-four, traffic, ten to nine o'clock, four miles, six-thousand [feet], Boeing seven-thirty-seven. (1250:28)

SkyWest 1834: SkyWest eight-thirty-four has the traffic. (1250:33)

AppControl: SkyWest eight-thirty-four, plan to follow that traffic, there's traffic south of him, eleven o'clock, six miles, northbound, [it is another Boeing] seven-thirty-seven, out of seven-thousand-five-hundred for the right [runway]. (1250:35)

SkyWest 1834: OK, we'll follow the first one, SkyWest eight-thirty-four. (1250:42)

[the controller tells the pilots to report the <u>second</u> Boeing 737]

Fairchild RC-26, a military reconnaissance variant of the Fairchild SA-226
Metro II airliner (photo courtesy of U.S. DOD Visual Information Center)

AppControl: SkyWest eight-thirty-four – wait a minute – report the
second one in sight. (1250:44)

SkyWest 1834: [This is] eight-thirty-four, OK, we're looking for
him. (1250:48)

AppControl: SkyWest eight-thirty-four, [the traffic is at] ten o'clock,
four miles, seven-thousand-four-hundred. (1251:02)

SkyWest 1834: Eight-thirty-four, we're looking [for it]. (1251:07)

AppControl: SkyWest eight-thirty-four, turn left [to] heading zero-
seven-zero. (1251:15)

[the pilots acknowledge the heading change]

AppControl: SkyWest eight-thirty-four, traffic's [at] ten to eleven
o'clock, three miles. (1251:32)

AppControl: SkyWest eight-thirty-four, turn left [to] heading zero-
five-zero. (1251:43)

[for the _third time_ *the Metro II Captain tells the controller he
does not see the second Boeing 737]*

SkyWest 1834: Left to zero-five-zero, eight-thirty-four, [we] still
have no [visual] contact on that traffic. (1251:46)

AppControl: SkyWest eight-thirty-four, roger, turn farther left [to] heading three-six-zero. (1251:50)

SkyWest 1834: Left to three-six-zero. (1251:53)

[turning left and looking for the Boeing 737 on their left, the pilots do not see the Mooney approaching from their right]

[the Metro II Captain sees the Mooney the instant before impact and shouts into his microphone]

SkyWest 1834: Oh, [expletive]! (1251:58)

*[the Metro II and the Mooney **collide** at 1251:58]*

[the controller does not know the Metro II has hit another plane]

AppControl: SkyWest eight-thirty-four is cleared, visual approach, Runway three-four left. (1252:01)

[there is no response]

AppControl: SkyWest eight-thirty-four, cleared, visual approach, Runway three-four left. (1252:09)

AppControl: SkyWest eight-thirty-four, Salt Lake. (1252:18)

AppControl: SkyWest eight-thirty-four, Salt Lake. (1252:32)

[there is no response, and the Metro II radar return has dropped off the controller's scope]

[the controller radios SkyWest Flight 575]

AppControl: SkyWest five-seventy-five, there should be traffic at eleven o'clock, three miles . . . [he] was at seven-thousand, [but] I've lost him on the radar. (1252:36)

SkyWest 575: We're looking, no contact. (1252:43)

[Approach Control makes a phone call to the control tower]

Phone-AppControl: [Do] you see SkyWest eight-thirty-four out there south of the marker anyplace? (1252:49)

Phone-Tower: No – no – I don't. (1252:50)

Phone-AppControl: What happened to him? (1252:51)

<u>Phone-Tower</u>: I don't know . . . I just saw [the radar target] go into coast, and I just asked what happened to him. (1252:52)

<u>Phone-AppControl</u>: I don't know what happened to him! I'm in trouble down here! I lost SkyWest eight-thirty-four! I don't know where he is! (1253:10)

[the controller makes a final and futile radio call]

<u>AppControl</u>: SkyWest eight-thirty-four, Salt Lake. (1253:19)

[there is no response]

The Collision: The Metro II had been turning left toward a heading of 360 degrees. The Mooney had been flying northwest. Recorded radar data later would reveal the Mooney's track:

> A gradual left turn was noted until the [Mooney] target merged with the target of SkyWest 1834 at 1251:58 [Hours].

At a closure rate of 272 knots, 461 feet per second, the Mooney had torn into the forward right fuselage of the Metro II at 7,000 feet. This high speed impact had ripped both planes apart. Debris fell into a residential area of Kearns, Utah:

> A school, several homes, automobiles, and public utilities were damaged by [falling] debris from the airplanes.

Pieces of wreckage and parts of bodies fell into the parking lot at St. Francis Xavier Catholic Church. The Mooney's heavy engine rammed into the earth in front of Western Hills Elementary School, which was packed with children. Lighter pieces of debris showered down along a path two miles long. Later *The Salt Lake Tribune* newspaper would quote a witness on the ground:

> It was like somebody was throwing stuff down at you. The downpour of wreckage – including suitcases, clothing, and body parts – lasted several minutes.

Another witness told a newspaper reporter:

> I saw a body hit right there [in a parking lot]. I ran over to see if there was anything I could do. When I got there I saw that the only thing holding her together was her clothes.

Meanwhile, at the airport terminal several relatives of passengers waited for Flight 1834 to land. Four men at the gate, boarding passes in hand, waited to board the plane for the next leg of its flight. The Metro II was a half-hour late when a visibly nervous ticket agent announced that Flight 1834 had been "cancelled" due to "mechanical problems." One ticket-holder idly passed the time waiting for the next plane by watching *The Newlywed Game* on television. The game show got interrupted by a news bulletin about the tragic airplane accident, complete with video footage of carnage and chaos on the ground. The awed ticket-holder exclaimed to a bystander:

That was our flight! My God! No wonder they told us it was cancelled. It's been cancelled permanently!

By 1900 Hours all identifiable human remains had been bagged and taken to a temporary morgue in the Catholic church. The Utah State Medical Examiner later would perform autopsies on remains of the pilots and passengers. The cause of death for all 10 persons would be attributed to "multiple severe impact trauma." The day after the accident *The Salt Lake Tribune* headline read:

Midair Collision Kills 10 In Disaster Over Kearns

The Investigation: The Mooney pilots had not filed a flight plan. Conversely, the Metro II had been on an IFR flight plan and in radio and radar contact with ATC.

The small municipal airport had no control tower. Ten miles to the north, the huge international airport had full ATC service. It also had an ARSA with an excluded "keyhole" in the southern quadrant to accommodate traffic at the municipal airport:

Pilots operating within an ARSA must [maintain] two-way radio communications with the controlling facility, and they may not enter the ARSA without [ATC permission].

Investigators entered stored Approach Control radar data into the FAA retrack program computer. This program recreated the alphanumeric data displayed on the controller's scope. It depicted the Mooney as a VFR non-Mode C target. The program documented the Mooney's entry into the ARSA and its collision with the Metro II:

The [Mooney] VFR target became visible at 1251:10 [Hours and] it remained so for about 48 seconds, until it merged with the target of SkyWest 1834 and disappeared from the scope.

Notwithstanding the IFR flight plan of the Metro II, both aircraft had been flying in VFR conditions with unrestricted visibility. Under these conditions all pilots were required to maintain visual separation from other aircraft. The controller had issued traffic advisories to the Metro II pilots, telling them about two Boeing 737s on their left. The Metro II pilots had seen the first one. Four times the controller had given heading and distance advisories about the second plane. Three times the pilots had replied that they did not see it. They had been visually scanning to their left. They never saw the Mooney approaching from their right until the instant before impact:

They were concentrating their scan toward the area of the Boeing 737 because they needed to see it and advise [ATC] so they could continue their approach.

A visibility study indicated the Mooney pilots possibly could have seen the Metro II. However, the NTSB concluded:

The ability of the pilots of either airplane to have seen the other airplane in time to avoid the collision was marginal.

Probable Cause: Fourteen months after the accident the NTSB adopted the Aircraft Accident Report, NTSB-AAR-88-03, on 15 March 1988. The controller's first priority had been to separate the Metro II from other IFR traffic. He had provided traffic advisories for non-IFR aircraft based upon available information.

The controller had told the Metro II pilots about two Boeing 737s on their left. These planes were (1) in radio contact with ATC and were (2) ARSA participants with (3) full data block radar returns, including (4) altitude information. Conversely, the Mooney pilot had not been in contact with the controller and lacked Mode C (altitude encoding) capability. The controller never saw it on his scope. Given the Mooney's absence of Mode C capability, even if the controller had seen the radar target:

> . . . the Mooney was in a location that easily could have been interpreted by the controller as an aircraft operating in the traffic pattern at SLC 2 *[the municipal airport]*.

The "bottom line" was that the Mooney had flown into the ARSA without the knowledge of, and without clearance from, ATC. The board found the following ***Probable Cause*** of this accident:

> Lack of navigational vigilance by the Mooney instructor pilot, which led to the unauthorized intrusion into the Salt Lake City Airport Radar Service Area.

The board pointed out two ***Contributing Causes***:

> [First], the absence of a Mode C transponder on the Mooney.

> [Second], the limitations of the ATC system to provide collision protection under the circumstances of this accident.

No Conflicting Traffic?
Mid-Air Collision near Independence, Missouri
20 January 1987

Beechcraft U-21 Ute, Serial No. 67-18061
U.S. Army administrative flight
3 aboard, 3 killed

Piper PA-31 Navajo Chieftain, Registration N60SE
Sachs Electric Company business flight
3 aboard, 3 killed

Synopsis: In clear skies a military Beechcraft U-21 Ute and a civil Piper PA-31 Navajo Chieftain flew head-on toward each other. ATC failed to see the conflict, and none of the pilots saw the other aircraft flying toward them. The planes collided and disintegrated, killing the six people on board.

The Beechcraft U-21 Ute Flight: At Anniston, Alabama, a U.S. Army general, commandant of the Military Police School, climbed into a military executive transport. He was the only passenger in the cabin of the Beechcraft U-21 Ute. The pilots fired up their turbines, took off, and flew toward Fort Leavenworth in Kansas. Although they had filed an IFR flight plan the weather was clear with visibility estimated at 20 miles:

The U-21 Ute is a military variant of the famed Beechcraft A-90 King Air twin turboprop, which first flew in 1963. It has been in continuous production ever since, and over 6,000 King Air and U-21 planes have been built. The aircraft carries one or two pilots plus eight passengers and cruises at about 250 knots. Armed forces of countries around the globe use the U-21 for liaison transport and VIP flights.

The pilot was a civilian, age 40, who was employed by the Army. He held an ATP certificate, was qualified in both helicopters and fixed-wing aircraft, and had 5,983 hours of flight experience. His copilot was an Army major, age 35, who had earned a civil ATP certificate and had logged 6,266 flight hours.

Beechcraft U-21 Ute (photo courtesy of U.S. DOD Visual Info. Center)

The Piper PA-31 Navajo Chieftain Flight: The chief pilot for Sachs Electric Company, one of the largest electrical contractors in the United States, flew two company executives to Kansas City, Missouri. After completing their business they returned to Kansas City Downtown Airport for a trip back to their home in St. Louis, Missouri. Their company's Piper PA-31 Navajo Chieftain took off from Runway 19 and climbed to the east:

> *The Navajo entered the commercial aviation market in 1967. A subsequent stretched and upgraded variant, the Navajo Chieftain, was powered by two 350 horsepower turbocharged engines that drove counter-rotating propellers. The new Chieftain proved to be a financial success. Small scale cargo and commuter operators found that ease of maintenance and a low seat-cost-per-mile made the speedy new aircraft an economic winner.*

The Chieftain pilot, age 42, had been employed by Sachs Electric Company for over 10 years. He held an ATP certificate and had logged 7,418 hours of flight time. No copilot was required in the twin engine Chieftain.

Trouble Ahead: The IFR Army U-21 neared its destination under the control of Kansas City Center. As the military turboprop cruised west through clear skies, Center told the pilots to expect a visual approach to Runway 33 at Sherman Army Airfield.

Meanwhile, the civil Chieftain climbed to the east. Although the pilot had flown IFR to Kansas City, he had not filed any type of flight plan for this return trip to St. Louis.

The portion of the ATC transcript, below, contains radio traffic between the Army U-21 and air traffic control facilities. Some of the transmissions have been *reconstructed* from the subsequent NTSB report, so some verbiage may not be verbatim. The transcript begins as ATC clears the U-21 pilots for an en route descent:

U-21: . . . . Beechcraft U-21 Ute (call-sign, Army 061)
Center: . . . . . . Kansas City Center (ARTCC)
AppControl: . . Kansas City Approach Control

Center: . . . descend to and maintain seven-thousand. (1218)

U-21: Army zero-six-one, descending to seven-thousand. (1218)

Center: . . . contact Kansas City Approach Control on . . . . (1221)

[the pilots make radio contact with Approach Control]

AppControl: Roger, Army zero-six-one, expect visual approach to Runway three-three at Sherman, sky clear, . . . wind two-six-zero at seven [knots], altimeter three-zero-two-six. (1221)

AppControl: Army zero-six-one, cleared direct to Kansas City VOR, depart the VOR on a heading of three-one-zero degrees. (1222)

U-21: Wilco. (1222)

[the Chieftain takes off from Kansas City Downtown Airport and climbs east at 1222]

AppControl: Zero-six-one, traffic, a twin Cessna, twelve o'clock at five miles, eight-thousand, heading southwest IFR. (1225:09)

U-21: Approach, zero-six-one, traffic in sight. (1225:23)

[the Army U-21 passes well clear of the twin Cessna]

*[the Army U-21 and the Chieftain **collide** at 1227:58]*

The Collision: The Army U-21 had cruised west at 7,000 feet. The Chieftain had flown east and was climbing through 7,000 feet. The Chieftain fuselage rammed into the Army U-21 at windshield height. The Chieftain disintegrated. The top half of the Army U-21 cockpit and cabin, as well as the entire empennage, ripped away. Wings, engines, propellers, and debris showered down from the sky:

> Army 18061 and [the Chieftain] collided in visual meteorological conditions over the Lake City Army Ammunition Plant, Independence, Missouri, at 7,000 feet msl. [Wreckage of] both planes fell to the ground within the confines of the ammunition plant.

The planes had torn into each other almost directly head-on at a closure rate of over 300 knots, instantly killing the six occupants of both aircraft. Debris fell along a path two miles long. The next day *The Topeka Capital-Journal* newspaper carried a headline:

General, Five Others Die in Mid-Air Collision

The newspaper provided details about the collision and crash:

> Brig. Gen. David H. Stem, commandant of the military police school and deputy commandant-general of Fort McClellan, Ala., was presumed dead . . . . *[and also]* Wreckage from the [Army] plane was scattered over an area of about one mile on the 3,900 acre ammunition plant. The main part of the U-21 crashed and burned within 20 feet of a building where workers at the plant change clothes . . . .

Later the Jackson County Medical Examiner and the Armed Forces Institute of Pathology conducted autopsies on the mutilated human remains. The cause of death for all victims was "multiple severe traumatic injuries."

The Investigation: The Army U-21 had been on an IFR flight plan under the control of ATC. The Chieftain pilot had not filed a flight plan, and none was required. However, he had made radio contact with Kansas City Tower. After takeoff at 1222 Hours (six minutes before the collision) the tower controller had told the Chieftain pilot to turn left and depart to the east. The pilot acknowledged those instructions, and that was his last known radio transmission.

Kansas City Approach Control operated ASR-8 radar. The radar had tracked the civil Chieftain as it climbed east while squawking Code 1200, the customary VFR code. The plane's Mode C transponder allowed ATC radar to display the Chieftain's altitude as well as its position. At 1227:58 Hours the secondary radar return suddenly vanished over the ammunition plant at 7,000 feet, and:

> Kansas City [Center] radar began to display multiple primary radar returns [from the falling pieces of wreckage] in the area where the secondary targets of [the Chieftain] and Army 18061 had been previously presented.

Two FAA controllers had been watching the Army U-21 on the same radarscope in Approach Control. They *claimed* they never saw a primary or secondary radar return from the Chieftain. However, NTSB investigators viewed the recorded radar data on the FAA retrack program computer. The transponder returns from the Army U-21 and the civil Chieftain were clearly visible. They slowly drew closer and closer and closer – and then merged:

> Army 18061 appeared on the scope at 1221:40 [Hours] and remained on the scope until the collision. N60SE *[the Chieftain]* appeared on the scope at 1222:45 [Hours] and remained on the scope until the collision. *[and also]* The U-21 was displayed as a computer tracked target on the controller's scope. The PA-31 [Chieftain] was displayed as a Code 1200 with Mode C altitude information on the same controller's scope.

The NTSB conducted a cockpit visibility study. In the 60 seconds before the collision both planes had maintained a constant flight path. The Army U-21 had been tracking 296 degrees at 7,000 feet and 190 knots. The Chieftain had been tracking 93 degrees at 140 knots and climbing. To the Army pilot-in-command, the Chieftain would have appeared stationary on his windshield, 13 degrees to his left. To the Chieftain pilot, the Army U-21 would have appeared stationary, 18 degrees to his right. Neither the sun nor its glare would have impaired the view of either pilot:

> Weather conditions . . . were characterized by high scattered clouds and excellent visibility.

The accident took place in VMC weather, so pilots in both aircraft should have been scanning for traffic. Yet, the planes had been flying toward each other at a closure rate in excess of 300 knots. Because they were closing almost directly head-on each aircraft would have presented a minimal visual profile to pilots in the other plane. In addition, because of their constant tracks no relative motion would have been discernable. Investigators noted:

> The "see and avoid" concept provided marginal opportunity to the pilots of both airplanes to avert the collision.

Probable Cause: Slightly over 12 months after the accident the NTSB adopted an Aircraft Accident Report, NTSB-AAR-88-01, on 3 February 1988. The board pointed out that the Chieftain pilot could have requested VFR flight-following service. If the ATC workload had permitted it, such a request would have activated the Approach Control automated "conflict alert" subprogram. However, the pilot did not request flight-following, and there was no requirement for such a request.

The board found the actions of the pilots had been proper. Also, neither weather, electronic aids, nor aircraft anomalies had contributed to the collision. The board attributed the ***Probable Cause*** of this accident solely to failures and deficiencies in the ATC system in the United States as follows:

> [First], failure of the [Kansas City Approach Control] controllers to detect the conflict and to issue traffic advisories or a safety alert to the flightcrew of the [Army] U-21.

> [Second], deficiencies in the see-and-avoid concept as a primary means of collision avoidance.

> [Third], the lack of automated redundancy in the air traffic control system [which would have been necessary] to provide conflict detection between participating and non-participating aircraft.

Cleared to Land? Is that Correct?
Runway Collision at Atlanta, Georgia
18 January 1990

Boeing 727, Registration N8867E
Eastern Airlines Flight 111
157 aboard, 0 killed

Beechcraft A-100 King Air, Registration N44UE
Epps Air Service charter flight
2 aboard, 1 killed

Synopsis: A tower controller did not maintain proper separation between arriving aircraft at night. While landing, a Boeing 727 struck a Beechcraft A-100 King Air that was taxiing on the active runway. No one aboard the airliner was killed, but the accident decapitated the King Air pilot.

The Boeing 727 Flight: Eastern Airlines Flight 111 accelerated and climbed skyward from La Guardia Airport in New York and cruised south toward Atlanta, Georgia. The tri-jet Boeing 727 carried 149 passengers, 5 flight attendants, 2 pilots, and a flight engineer. The en route phase of the IFR night flight proved to be routine:

> Upon arrival in the Atlanta area [Eastern Flight] 111 was vectored for . . . the ILS Runway 26 Right approach . . . .

The captain, age 49, held an ATP certificate and was qualified to fly both helicopters and fixed-wing aircraft. He had accumulated 13,320 hours of flight time and had been flying with Eastern since 1967. The first officer and flight engineer had earned ATP certificates and had logged a total of 12,818 hours in the air. Repairs on an inoperative anti-skid system had been deferred, as allowed, and the Boeing airliner had no other discrepancies.

The Beechcraft A-100 King Air Flight: Epps Air Service operated a fleet of on-demand charter aircraft. After dark one of its Beechcraft A-100 King Air planes took off from DeKalb/Peachtree Airport in Atlanta on an IFR flight plan. The two pilots planned to land at nearby Hartsfield International Airport, pick up passengers, and fly

Beechcraft T-44, a military variant similar to the civil Beechcraft A-100 King Air (photo courtesy of U.S. DOD Visual Information Center)

them to Albany, Georgia. The controller cleared the King Air pilots for an ILS approach to Runway 26 Right – the same runway to be used by the Boeing 727 airliner.

The young King Air pilot, age 30, held an ATP certificate and had logged 1,653 flight hours. His experienced copilot normally flew Piper PA-31 Navajo planes for Epps Air Service. This would be a mere familiarization flight for him, because the King Air is certified for single pilot operation.

Trouble Ahead: When passing the final approach fix the King Air pilot reduced speed to 94 knots for landing. After a normal touchdown he quickly slowed to taxi speed. He intended to turn off onto Taxiway D, only 3,000 feet down the runway.

Behind the King Air, in the dark, the Boeing 727 had slipped down the ILS glideslope. The pilots had flown their approach at 140 to 145 knots and had closed on the slower King Air ahead. The airliner had made a normal touchdown 1,200 feet down the runway, and the pilots had deployed their spoilers.

The portion of the ATC transcript, below, includes radio messages between the tower and the Boeing 727, plus CVR verbiage. Radio messages to and from other aircraft are excluded in the interest of clarity. The transcript begins as the tower controller clears the Boeing 727 to land. The captain is flying the airliner:

Capt: . . . . . . Captain, Eastern Airlines Flight 111
FO: . . . . First Officer, Eastern Airlines Flight 111

FltEngr: ... Flight Engineer, Eastern Airlines Flight 111
Radio-: .... (prefix) Radio transmission, Eastern Flight 111
PA-: ...... (prefix) Public address system, Eastern Flight 111
Tower: .... Hartsfield International Airport Control Tower
Maint: ..... Eastern Airlines Maintenance Office

Tower: Eastern one-eleven, you are in sight, cleared to land [on Runway] two-six right. (1901:57)

[the First Officer acknowledges the landing clearance, which failed to mention the slower King Air ahead]

Capt: [We are] coming up on the [outer] marker, gear down, final check, please. (1902:14)

FO: Gear down. (1902:17)

[sound of landing gear extension]

FltEngr: ... landing final checklist – no smoking [sign]. (1902:32)

FO: It's on. (1902:36)

FO: Got three green [lights], pressure and quantity check. (1902:41)

FltEngr: Fuel panels set. (1902:48)

Capt: Flaps thirty, please. (1903:05)

FO: Flaps thirty. (1903:05)

FO: Well, it's a little bit scuddy down there. (1903:08)

Capt: [One]-thousand feet. (1903:13)

[the Boeing 727 touches down on the runway at 1904:27]

[7 seconds later the First Officer sees the King Air ahead]

FO: Oh, [expletive]! Ooohhh! (1904:34)

*[**sound of impact**; the airliner's right wing slices through the much smaller King Air at 1904:37]*

FO: God bless! (1904:37)

Capt: ... what the hell was that? (1904:38)

FO: That was, ah, some kind of – aaahhh – two-oh-two – (1904:41)

Capt: [Expletive]! Tell 'em we had an accident. (1904:46)

Radio-FO: Tower – aaahhh – Eastern one-eleven – aaahhh – we just hit an aircraft on the runway. (1904:48)

Tower: Say again? (1904:51)

Radio-FO: There was an aircraft on the runway . . . . (1904:52)

[the tower controller does not respond]

Capt: [Expletive]! (1904:57)

Capt: Tell [the tower] we hit somebody on the runway. (1905:09)

FO: I just did. (1905:11)

Capt: Tell 'em again! (1905:12)

Radio-FO: Tower, Eastern one-eleven, aaahhh, you better keep the traffic off of two-six right, there's an airplane there. (1905:14)

Tower: Thank you. (1905:25)

[the Boeing 727 turns off the runway onto a taxiway and stops]

Radio-FO: OK, Tower, Eastern one-eleven, we just hit an airplane, [it] looked like [a] four-oh-two or somethin'. (1905:30)

Tower: One-eleven, roger. (1905:32)

FO: We were cleared to land! (1905:47)

Capt: I know! (1905:48)

FO: We were cleared to land! (1905:50)

PA-Capt: Folks, this is captain Orlando, I'm sorry to say there was an airplane on the runway – a little plane – and our right wing did catch that airplane, so – ah – we're holding here for –ah – further instructions and passenger assistance. (1905:54)

Capt: . . . let's see what's going on, why don't we have an "A" [hydraulic] system? (1906:07)

FltEngr: I don't know. (1906:17)

Capt: Call the company. [Expletive]! We don't have any steering, tell 'em we need a tug. (1906:18)

Radio-FltEngr: Atlanta ramp, Eastern one-eleven. (1906:25)

Maint: One-eleven, [go ahead]. (1906:29)

Radio-FltEngr: [This is] one-eleven, ah, there's been an accident, ah, we hit an aircraft on the runway and we don't have any steering, we need a tug, we're off to the side of Runway two-six right. (1906:31)

Capt: God Almighty! He had no lights! (1906:34)

FO: He had no lights. (1906:36)

Capt: I didn't see him 'till he was right there! (1906:39)

FO: He was – yeah – there were no lights. (1906:40)

Radio-Capt: Tower, Eastern one-eleven . . . we don't have any hydraulics and we're calling for a tug . . . . (1906:54)

Tower: Eastern one-eleven, just, ah, hold in position then. (1907:03)

Capt: We were on [the tower] frequency, and we were cleared to land? Is that correct? (1907:15)

FO: That's correct. (1907:18)

FltEngr: That's correct. (1907:19)

The Collision: After touchdown the Boeing landing lights suddenly had revealed the outline of a plane on the runway ahead. At over 100 knots the captain had swerved to the left, but it was too late. The Boeing 727 right wing sliced through the King Air:

> The collision occurred on Runway 26 Right about 3,000 feet from the approach end. *[and also]* The empennage and left wing of the King Air were severed . . . . The cabin roof was severed from the point of impact forward . . . . The left pilot's seat of the King Air was crushed by impact forces that were not survivable.

After stopping on a taxiway the Boeing captain had noticed loss of an hydraulic system. Moments later he saw fluid streaming down from his right wing. The pilots shut down the aircraft and called for

Boeing 727 (photo courtesy of U.S. NASA)

busses to carry their passengers to the terminal.

Airport rescue and firefighting crews had rushed to the accident site. The King Air pilot had suffered severe "craniocerebral injuries" that killed him instantly. An ambulance raced the comatose copilot to South Fulton Medical Center. He would survive, but:

> [He had] sustained serious injuries and recalled very little about the flight.

The next day *The Atlanta Constitution* newspaper headline read:

Eastern Jet Hits Plane, Kills 1

An accompanying photograph showed the King Air nose-down on the runway. The cockpit and the top of the fuselage – and the pilot's head – had been ripped off by the Boeing wing.

The Investigation: Neither weather nor mechanical problems had played a role in this accident. NTSB investigators faced three main questions. <u>First</u>, why had the controller not enforced proper longitudinal separation between the planes during final approach? <u>Second</u>, as the Boeing pilots approached to land, why had the controller not warned them that the King Air was still on the runway? <u>Third</u>, why had the Boeing pilots not seen the King Air lights in time to abort the landing and avoid a collision?

> Neither the [Boeing 727] flight crew nor tower personnel made visual contact with the King Air airplane on Runway 26 Right during the critical period preceding the collision. *[and also]* The controller was relieved of duty following the accident.

The controller had become preoccupied with a DC-9 that had a communications problem. He had never noticed that proper landing interval was slipping away. Also, investigators learned that in the night VFR weather the King Air had been *virtually invisible* (1) from behind and (2) from the tower:

> The King Air was not in compliance with airworthiness requirements because of deficiencies in the anticollision lighting system.

The (1) red anticollision light atop the vertical stabilizer and the (2) bright flashing strobe light in the tailcone had been inoperative. Further, examination of light bulb filaments showed the pilot had turned off the entire anticollision lighting system to prevent glare during the landing and rollout, an action allowed by FARs. From the rear the only visible King Air light had been the small white low-intensity position light:

> Only limited [King Air] conspicuity would [have been] afforded in a field of view that included a variety of runway, taxiway, and other lights.

With no flashing lights the King Air had not been seen by the tower controller or the Boeing pilots. Hartsfield International Airport had Airport Surface Detection Equipment (ASDE) radar to monitor the location of all planes on the airport during inclement weather. However, visibility had been estimated at five miles, so the ASDE had been turned off.

Probable Cause: Sixteen months after the accident the NTSB adopted the Aircraft Accident Report, NTSB-AAR-91-03, on 29 May 1991. The board found the absence of lighting on the King Air had not been a root cause of the accident. Both planes had been on IFR flight plans. Even in zero-zero visibility conditions, controllers have the ability to maintain separation between landing aircraft.

The controller had been distracted by communications problems with another aircraft. When he cleared the Boeing 727 to land he did not mention preceding traffic information, as required. This deprived the Boeing pilots of knowledge of the slower King Air ahead. Also, neither the final controller nor the radar monitor controller had given proper airspeed reduction instructions to the Boeing pilots:

> [This] led to a speed differential that resulted in a loss of separation between [the Boeing 727] and [the King Air].

The board found the *Probable Cause* of this mishap to be twofold:

> [First, the] failure of the Federal Aviation Administration to provide air traffic control procedures that adequately take into consideration human performance factors such as those which resulted in the failure of the north local controller to detect the developing [longitudinal separation] conflict between [the King Air] and [the Boeing 727].

> [Second, the] failure of the north local controller to ensure the separation [between] arriving aircraft which were using the same runway.

The board outlined two *Contributing Causes*:

> [First], failure of the local controller to follow the prescribed procedure of issuing appropriate traffic information [about the King Air] to [the Boeing 727 pilots].

> [Second], failure of the local controller and the radar monitor controller to issue timely speed reduction [instructions] to maintain adequate separation between [the two] aircraft on final approach.

Lost in Blinding Fog
Runway Collision at Detroit, Michigan
3 December 1990

Boeing 727, Registration N278US
Northwest Airlines Flight 299
154 aboard, 0 killed

Douglas DC-9, Registration N3313L
Northwest Airlines Flight 1482
44 aboard, 8 killed

Synopsis: Douglas DC-9 pilots got lost while taxiing in dense fog and blundered onto the active runway. A Boeing 727 accelerated down the runway to take off, and its wing tore into the Douglas plane. No one in the Boeing aircraft was injured, but the DC-9 burst into flame, killing 8 of the 44 people on board.

The Boeing 727 Flight: Fog shrouded Detroit Metropolitan/Wayne County Airport as Northwest Airlines Flight 299 pushed back from the gate. Headed for Memphis, Tennessee, the Boeing 727 carried eight crewmembers and 146 passengers as it lumbered out toward Runway 3 Center. The takeoff visibility minimum for that runway was one-quarter mile, which coincided with the visibility broadcast by ATIS. Tower controllers could not see the airliner through the fog as it slowly followed the "yellow line" toward the runway.

The captain, age 42, had logged 10,400 hours and had been flying with Northwest for seven years. His first officer and flight engineer collectively had logged 8,700 hours of flight time. Each member of the flight deck crew held an ATP certificate.

The Douglas DC-9 Flight: Four minutes after the Boeing airliner left the gate a Douglas DC-9, Northwest Airlines Flight 1482, began taxiing toward the same runway. Bound for Pittsburgh, Pennsylvania, the Douglas airliner carried two pilots, two flight attendants, and 40 passengers as it slowly crept through dense blinding fog:

The captain stated that visibility was deteriorating as they began taxiing, but he was able to [see] the [taxiway centerline].

The DC-9 captain, age 52, had been hired by Pacific Airlines in 1966. Several mergers later he wound up flying for Northwest. A kidney stone problem had caused him to be removed from flight status in early 1984, and he had not returned to the cockpit until two months before this final flight. He had logged about 23,000 flight hours. The first officer, age 43, had been a B-52 bomber pilot. After leaving the Air Force he had been hired by Northwest seven months before this flight. His logbook reflected 4,685 hours of flight time.

Trouble Ahead: The Boeing pilots knew the weather had deteriorated as they neared Runway 3 Center and stopped at the hold line. Detroit Tower cleared the flight for takeoff at 1344:15 Hours, and the tri-jet began accelerating down the runway.

Meanwhile, the DC-9 pilots had become lost in the fog. Not knowing where they were, they had blundered onto the active runway and stopped. The captain had radioed Ground Control:

> We're out here, we're stuck – we can't see anything out here, aaahhh . . . it's so foggy out here, we're completely stuck here.

Post-accident ATC and CVR recordings would be available from both aircraft. In the interest of clarity, only verbiage pertaining to the DC-9 is included in the transcript, below. This lengthy part of the transcript begins as Ground Control issues the first taxi instructions to the pilots. The transcript illustrates how the pilots became hopelessly lost. The captain is taxiing the aircraft, and the first officer initially makes all radio transmissions:

> Capt: . . . . Captain, Northwest Airlines Flight 1482
> FO: . . . . First Officer, Northwest Airlines Flight 1482
> Radio-: . . . . (prefix) Radio transmission, Flight 1482
> GndControl: . . . Detroit Ground Control

GndControl: Fourteen-eighty-two, right turn out of parking, taxi [to] Runway three center, exit ramp at Oscar six . . . . (1335:31)

Capt: I'm gonna just kinda wind around here, and Oscar six is gonna be just right around the corner here – OK – Jim, you just watch and make sure I go the right way. (1335:51)

FO: Just kinda stay on the ramp here. (1336:25)

Capt: Until [we get to] the yellow line, I guess, huh? That fog is pretty bad here. (1336:32)

FO: Hey, it looks like it's going zero-zero out there. (1337:02)

[the Captain intercepts the taxiway centerline]

GndControl: Fourteen-eighty-two, what's your position? (1337:05)

Radio-FO: [We are] right by the fire station. (1337:09)

GndControl: Roger, Northwest fourteen-eighty-two, taxi inner Oscar, six Fox, report making the, aaahhh, right turn on X-ray. (1337:11)

[the Captain begins to rely heavily on his First Officer as the two pilots try to follow taxi instructions]

FO: [I] guess we turn left here. (1337:37)

Capt: Left turn? Or right turn? (1337:39)

FO: . . . this is the inner here, we're still going for Oscar. (1337:46)

Capt: So, a left turn? (1337:46)

FO: [Yes, as] near as I can tell, I can't see [expletive] out here – go that way – OK – there's Oscar six right here. (1337:48)

Capt: OK, so what do we do here? (1338:12)

FO: Go – we take Oscar six to Foxtrot. (1338:13)

Capt: Right turn? Right here? (1338:16)

FO: Yeah, [go] right out there. (1338:18)

Capt: What runway [are] we going to? (1338:50)

FO: [Runway] three center. (1338:52)

Capt: [The visibility has] gotta be below minimums! (1338:56)

FO: I think they'll tell us – six hundred feet now? We can see six hundred feet? [Do you] think so? (1339:02)

Capt: Naw, I don't think we got six hundred feet – well anyway, flaps twenty and takeoff check when you get time. (1339:22)

[the First Officer completes six items on the takeoff checklist]

GndControl: Fourteen-eighty-two, what's your position? (1339:37)

Radio-FO: We are . . . on Oscar six. (1339:40)

GndControl: You're approaching Oscar six and Runway niner-two-seven? (1339:48)

Radio-FO: Ah, we're headed eastbound on Oscar six here. (1339:52)

GndControl: Northwest fourteen-eight-two, report crossing Runway niner-two-seven on Fox. (1340:01)

[the First Officer makes his first admission of a mistake]

Radio-FO: OK, I think we might have missed Oscar six. [I] see a sign here that says – aaahhh – the arrows to Oscar five, [I] think we're on Foxtrot now. (1340:06)

GndControl: Northwest fourteen-eighty-two, ah, you just approach Oscar five and you – are you on the outer? (1340:16)

Radio-FO: Yeah, that's right. (1340:23)

GndControl: Northwest fourteen-eighty-two, at Oscar four make the right turn on X-ray and then report crossing [Runway] nine-two-seven. (1341:05)

[the pilots abandon references to the Jeppesen airport diagram and rely totally on airfield signs]

Capt: This is [Runway] nine-two-seven, huh? (1341:17)

FO: That says X-ray right there. (1341:19)

Capt: So – what's [the controller] want us to do here? (1341:24)

FO: You can make the right turn, he said, and report crossing [Runway] two-seven, and then I'll ask him – there's Oscar four, this is X-ray. (1341:25)

[the Captain taxis toward the Oscar 4 intersection]

Capt: This is a right turn here, Jim? (1342:00)

FO: That's the runway. (1342:01)

Capt: OK, we're goin' right over here then? (1342:02)

[the First Officer becomes more and more confused]

FO: Yeah, that way – well – wait a minute. Oh [expletive]! This, uuuhhh – aaahhh – I think we're on X-ray here now . . . this is [Runway] nine. We're facing one-six-zero – yeah. [We're] cleared to cross it. (1342:03)

Capt: We're cleared to cross [the runway]? (1342:42)

FO: Yeah, we're cleared to cross. (1342:44)

Capt: When I cross this – which way do I [turn], right? (1342:50)

FO: Yeah. (1342:51)

[the Captain voices grave doubts about his position]

Capt: This is the active runway here, isn't it? (1342:56)

FO: This should be [Runway] nine-two-seven – it is – yeah – this is [Runway] nine-two-seven. (1342:59)

Capt: Follow this – [expletive]! We're cleared to cross this thing? [Are] you sure? (1343:14)

FO: That's what [the controller] said, yeah – but this taxi light takes us – is there a taxiway over there? (1343:18)

[the Captain stops and sets the parking brake]

Capt: Naw, I don't see one. Give [the controller] a call and tell him that – aaahhh – we can't see nothin' out here. (1343:24)

[the First Officer fails to comply, but the Captain releases the parking brake and creeps forward]

GndControl: Northwest fourteen-eighty-two, Ground [Control], say your position? (1343:45)

Radio-FO: [We] believe we're at the intersection of – aaahhh – X-ray and [Runway] nine-two-seven. (1343:48)

GndControl: . . . OK, are you southbound? (1343:58)

Radio-FO: Yeah, we're holding short of [Runway] nine-two-seven here right now. (1344:01)

GndControl: Cross [Runway] nine-two-seven, Northwest fourteen-eighty-two. (1344:05)

[the Captain blunders onto Runway 3 Center]

Capt: Now, what runway is this? This is a runway? (1344:23)

FO: Yeah, turn left over there – naw, that's a runway too! (1344:32)

[realizing he is totally lost, the Captain stops on the runway]

Capt: Tell [the controller] we're out here, we're stuck. (1344:35)

[once again the First Officer __fails to comply__ with his Captain's order; the Captain makes all subsequent radio transmissions]

Radio-Capt: Ground, fourteen-eighty-two, we're out here, we're stuck, we can't see anything out here. (1344:47)

GndControl: Northwest fourteen-eighty-two, just to verify, you are proceeding southbound on X-ray now, and you are approaching [Runway] nine-two-seven? (1344:58)

Radio-Capt: Aaahhh, we're not sure, it's so foggy out here, we're completely stuck here. (1345:02)

GndControl: OK – ah – are you on a taxiway or runway? (1345:05)

[hidden in the fog, the Boeing 727 begins its takeoff roll down Runway 3 Center at 1345:05]

Radio-Capt: We're on a runway, we're right by zero-four. (1345:07)

GndControl: Northwest fourteen-eighty-two, roger, are you clear of Runway three center? (1345:12)

Radio-Capt: It looks like we're on two-one center. (1345:17)

Capt or FO: [Expletive]! (1345:20)

GndControl: Northwest fourteen-eight-two, [do] y'say you are on [Runway] two-one center? (1345:27)

Radio-Capt: I believe we are, we're not sure. (1345:29)

[Ground Control tells the pilots to get off the runway]

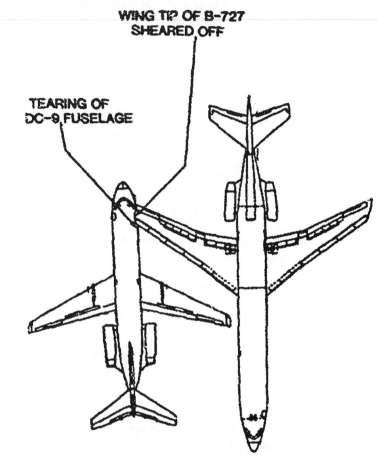

An illustration of the collision (illustration courtesy of U.S. NTSB)

GndControl: Northwest fourteen-eighty-two, roger, if you are on two-one center, exit that runway immediately, sir! (1345:33)

[the DC-9 pilots hear the roar of the approaching Boeing 727]

Capt or FO: Oh, [expletive]! Oh, [expletive]! (1345:38)

*[2 seconds later at 1345:40 the accelerating Boeing 727 **collides** with the DC-9 parked on the runway]*

The Collision: As the lost DC-9 blocked the runway, the onrushing Boeing 727 had materialized out of the thick mist. There was no time for the pilots of either aircraft to take evasive action:

The right wingtip of the Boeing 727 struck just below the first officer's middle window on the DC-9. .... The Boeing 727 was nearly on the centerline and the DC-9 was near the right edge of Runway 3 [Center].

Accelerating through the fog at 211 feet per second, the Boeing wing tore through the DC-9 cockpit and forward fuselage. The wing began to shear off as it raked along the fuselage and ripped the No. 2 engine off the DC-9. The Boeing pilots hit their brakes and managed to stop at the end of the runway. There was no fire on the Boeing plane. No one had been injured. Damage was confined to the right wing, which had been demolished.

The DC-9 and its occupants were not as lucky. Their fuselage had been slashed open from the cockpit to the aft bulkhead. Fire broke out under the aircraft and quickly spread up into the cabin:

The cabin of the DC-9 was consumed by fire. .... Fuel lines that were feeding fuel to the DC-9 right engine . . . were ruptured when the engine separated from the fuselage.

Dense fog prevented tower controllers from seeing the collision and fire. However, as the Boeing plane had braked to a stop its captain had radioed the tower and reported the accident. The airport fire department dispatched five fire trucks and two ambulances. Driving through fog with visibility "as low as 50 to 100 feet," firefighters finally found the Boeing 727 at the end of the runway:

[Firefighters] saw fuel leaking from the damaged right wing, and they blanketed the wing and . . . the ground with foam.

By the time firefighters found the DC-9 its cabin was a raging inferno. They tried to squirt foam into the fuselage through broken windows and emergency exits, but they were unable to make headway against the flames:

A short time later fire breached the roof of the DC-9, and overhead turrets applied foam into the fuselage [through the gaping hole in the top of the plane].

Most DC-9 occupants had been able to escape through emergency exits, but seven passengers and one flight attendant had died in the cabin. Three had been killed by "massive blunt force trauma." Five more succumbed to "asphyxia secondary to smoke and soot inhalation." Thirty-six others suffered severe trauma and burns.

The Investigation: NTSB investigators determined that no aircraft component or design feature contributed to the accident:

> The DC-9 was positioned on active Runway 3C/21C, and the Boeing 727 was on its takeoff roll.

Investigators learned the DC-9 captain had been on a medical furlough for seven years. This mishap flight would have been his first unsupervised trip as a captain since returning to flight status. He was in an unfamiliar environment with new manuals, new checklists, and new procedures. To complicate the matter, before starting to taxi the first officer had made exaggerated statements about his military flying accomplishments. As a result of these circumstances the captain relied too heavily upon his first officer:

> The first officer began to dominate decision-making in the cockpit. *[and also]* A nearly complete and unintentional reversal of command roles took place in the cockpit of the DC-9 shortly after taxiing began. . . . [The captain's] over-reliance on the first officer without effectively using other available resources . . . amounted to a relinquishment of his command responsibilities.

CVR and radio transcripts conclusively showed the DC-9 pilots had been lost in the fog. As early as 1341 Hours they had abandoned attempts to use the Jeppesen airport diagram and had relied totally on airfield signs. During the investigation the NTSB examined signs in the X-ray and Oscar 4 area. Even during daytime in VFR conditions, investigators found the signs to be confusing and could not agree on their meaning. Also the taxi lines were faded and "nearly invisible." Investigators concluded:

> The captain . . . believed that the first officer knew where he was, and the first officer apparently could not bring himself to admit, or was not aware, that his assertive directions had placed the airplane in this predicament.

Tower controllers had been unable to see either aircraft. They said visibility had been "the lowest in which they had ever controlled

Boeing C-22, a military variant of the civil Boeing 727 airliner (photo courtesy of U.S. DOD Visual Information Center)

traffic." Visibility had dropped to only 100 feet at many places on the airport. Controllers had not used the visibility reference chart to determine the prevailing visibility.

By radio, Ground Control had learned the DC-9 was on the active runway during the Boeing 727 takeoff roll. The ground controller had told the DC-9 pilots to get off the runway. Yet, he failed to warn the tower controller and the Boeing pilots. The tower supervisor initially told the NTSB that she had been standing and "observing the overall operation." Yet, she later admitted that she had been seated at a desk and doing paperwork.

Probable Cause: The NTSB adopted the Aircraft Accident Report, NTSB-AAR-91-05, five months after the accident on 25 June 1991. A host of multiple interrelated circumstances had caused this runway collision. The board explained the *Probable Cause* as follows:

> Lack of proper crew coordination, including a virtual reversal of roles by the DC-9 pilots, which led to their failure to stop taxiing their airplane and alert the ground controller of their positional uncertainty *[meaning, that they were lost in the fog]* in a timely manner before and after intruding onto the active runway.

The board found the following ***Contributing Causes***:

[First], deficiencies in the air traffic control services provided by the Detroit Tower, including:

A: Failure of the ground controller to take timely action to alert the local controller of the possible runway incursion.

B: Inadequate visibility observations.

C: Failure to use progressive taxi instructions in low visibility conditions.

D: Issuance of inappropriate and confusing taxi instructions compounded by inadequate backup supervision . . . .

[Second], deficiencies in the surface markings, signage, and lighting at the airport, and the failure of the Federal Aviation Administration to detect or correct any of these deficiencies.

[Third], failure of Northwest Airlines to provide adequate cockpit resource management training to their line aircrews.

He Went Up in Flames
Runway Collision at Los Angeles, California
1 February 1991

Boeing 737, Registration N388US
USAir Flight 1493
95 aboard, 22 killed

Fairchild SA-227 Metroliner, Registration N683AV
SkyWest Airlines Flight 5569
12 aboard, 12 killed

Synopsis: At night a controller forgot that a Fairchild SA-227 Metroliner was on the runway, and she cleared a Boeing 737 to land. In the dark the Boeing 737 rammed into the Metroliner. The explosion and fire killed all 12 people on the Metroliner and 22 of the 95 people on the Boeing airliner.

The Boeing 737 Flight: USAir Flight 1493 left Columbus, Ohio, and winged its way west toward Los Angeles, California. The Boeing 737 carried 89 passengers, four flight attendants, and two pilots. The aircraft cruised leisurely at flight level 350 and approached the Los Angeles area shortly after twilight had faded into total darkness:

The Boeing 737 twin-jet is the most popular airliner in the world. The cabin is based upon the earlier Boeing 727 design, and it can seat six persons abreast. The plane first flew in 1967, and it has been in continuous production ever since. There have been nine major passenger, freight, and military variants, and they involve a variety of engines, cockpit layouts, and fuselage stretches. The landmark 5,000th Boeing 737 was delivered in February 2006.

The captain, age 48, had begun his airline career with Mohawk Airlines, a forerunner of USAir, and had logged 16,300 flying hours. The first officer had been hired by USAir in 1988 and had 4,316 hours of flight experience. By coincidence each of the four flight attendants had been flying with USAir since 1989.

The Fairchild SA-227 Metroliner Flight: As the big Boeing 737 neared Los Angeles International Airport to land, a Fairchild SA-227

Fairchild C-26, a military variant of the civil Fairchild SA-227 Metroliner (photo courtesy of U.S. DOD Visual Information Center)

Metroliner taxied toward the runway to take off. The twin turboprop airliner, SkyWest Airlines Flight 5569, carried two pilots and only 10 passengers, most of whom were evening rush-hour commuters. They were headed for home in Palmdale, California, a short flight north from the urban sprawl of the Los Angeles basin.

The captain had been employed by SkyWest for five years and had 8,806 flight hours in his logbook. The first officer had logged over 8,000 hours. Although SkyWest Airlines operated under Part 135 rules, both pilots had earned ATP certificates.

Trouble Ahead: It had become totally dark as the Metroliner taxied toward Runway 24 Left. Meanwhile, the Boeing 737 approached the airport and slipped down the ILS glideslope toward the same runway.

At Los Angeles, a Level V (highest traffic level) ATC facility, 13 controllers had been scheduled for the evening shift. The airport averaged over 100 IFR operations per hour:

The airport has dual parallel runways between 9,000 and 12,000 feet long. Runways 25 Left and [25] Right comprise the south runway complex, and Runways 24 Left and [24] Right are referred to as the north runway complex.

The controller handling the Metroliner and the Boeing 737 was also juggling five other aircraft. She cleared the Metroliner into "position and hold" on Runway 24 Left at the Taxiway 45 intersection. Then, busy with other traffic and a misfiled flight progress strip, she simply *forgot* about the Metroliner.

The ATC transcript, below, includes radio messages to and from the Boeing 737, the Metroliner, and ATC, plus CVR verbiage from the Boeing airliner. In the interest of brevity the transcript excludes transmissions to and from most other aircraft:

USAir 1493: . . . . USAir Flight 1493 (Boeing 737)
CVR-: . . . . (prefix) CVR, USAir Flight 1493
SkyWest 5569: . . SkyWest Airlines Flight 5569 (Metroliner)
Tower: . . . . Los Angeles Airport Control Tower
AppControl: . . . . Los Angeles Approach Control
GndControl: . . . . Los Angeles Ground Control
Unk: . . . . Transmission from unknown source
PD 80: . . . . Police department helicopter 80

GndControl: [SkyWest fifty-five-sixty-nine], turn right on Tango, and then at forty-five transition to Uniform, taxi to Runway two-four left. (1802:43)

AppControl: USAir fourteen-ninety-three . . . contact Los Angeles Tower, one-three-three-point-niner at Romen, good night. (1803:05)

CVR-USAir 1493: Gear down . . . flaps one . . . five. (1803:20)

SkyWest 5569: [Tower], Skywest five-sixty-nine at forty-five, we'd like to go *[meaning, take off]* from here if we can. (1803:38)

Tower: SkyWest five-sixty-nine, taxi up to and hold short of [Runway] two-four left. (1803:40)

SkyWest 5569: OK, [we'll] hold short. (1803:44)

USAir 1493: [Tower], USAir fourteen-ninety-three [is] inside of Romen *[meaning, the outer marker]*. (1804:33)

Tower: SkyWest five-sixty-nine, taxi into position and hold, Runway two-four left, traffic will cross down-field. (1804:44)

SkyWest 5569: 'Kay, [Runway] two-four left, position and hold, SkyWest five-sixty-nine. (1804:49)

[the Metroliner is now in takeoff position on the runway]

CVR-USAir 1493: Gear, flaps, landing clearance remains. (1805:11)

USAir 1493: USAir fourteen-ninety-three for the left side, [Runway] two-four left. (1805:29)

CVR-USAir 1493: [We are] out of a thousand feet. (1805:41)

[the busy controller handles several other aircraft and looks for a missing flight progress strip; she forgets the Metroliner is still on the runway and clears the Boeing 737 to land]

Tower: USAir fourteen-ninety-three, cleared to land, Runway two-four left. (1805:51)

USAir 1493: [We are] cleared to land, [Runway] two-four left, fourteen-ninety-three. (1805:55)

CVR-USAir 1493: Looks real good . . . aaahhh, you're coming out of five-hundred feet, bug [speed] plus twelve, sink is seven. (1806:07)

CVR-USAir 1493: Lights on. (1806:19)

[as the Boeing 737 touches down on the runway the First Officer sees the Metroliner dead ahead]

CVR-USAir 1493: [An unintelligible shout]. (1806:57)

*[the Boeing 737 **collides** with the Metroliner at 1807:00]*

[the Boeing 737 explodes into flames, drags the crushed Metroliner down the runway, veers left into the grass, and crashes into an abandoned building]

[the controller and many pilots see the explosion and fire]

Unk: What the hell? (1807:04)

PD 80: [This is] PD eighty, you need any help over there? (1807:30)

Tower: Right now we don't know. (1807:32)

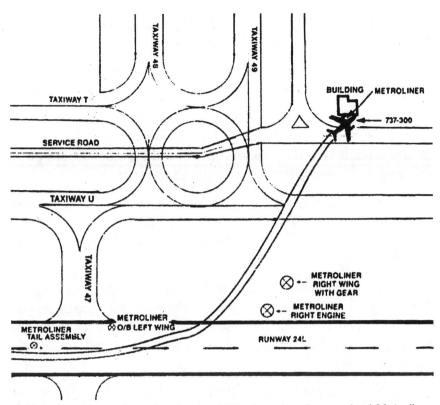

Diagram depicting how the Boeing 737, dragging the crushed Metroliner underneath its fuselage, veered left off the runway and crashed into a building (diagram courtesy of U.S. NTSB)

Tower: ... we had an aircraft go off the runway in flames. (1807:39)

[the controller knows the Boeing 737 has crashed, but she does not realize the Metroliner is involved]

Tower: ... hold short of Runway two-four left until we find out what happened. (1808:09)

Tower: OK, we just had a [Boeing] seven-thirty-seven land and blow up – he went up in flames – he's off the runway now, [Runway] two-four left is closed. (1808:31)

The Collision: In the dark the Metroliner had waited on the runway for takeoff clearance. Unable to see the Metroliner, the Boeing 737 pilots had swooped down to land. The Boeing airliner crossed the threshold at 130 knots, flared, and softly touched down on the runway

1,500 feet beyond the approach end. At that instant the first officer had seen the Metroliner straight ahead of him:

> The collision occurred [when the Boeing] nosewheel contacted the runway. . . . There was an explosion and fire upon impact.

The speeding Boeing airliner crushed the Metroliner and drug it down the runway. With the smaller plane jammed beneath it, the big Boeing aircraft veered off the runway. It skidded 1,200 feet through the grass and slammed into an abandoned fire station:

> The major portion of the Metroliner had been crushed beneath the Boeing 737. . . . The [Metroliner] was totally destroyed by the initial impact, the subsequent dragging along the ground . . . and by ground fire.

Airport firefighters rushed to the burning wreckage. Upon arrival they saw 40 to 50 people already out of the Boeing airliner, many of them horribly burned. Firemen used foam in an effort to quench fires burning outside the Boeing fuselage, but flames raged unchecked inside the cabin:

> The entire aft section of the [Boeing 737] was lying [flat] on the ground and had rotated counterclockwise.

In the dark neither the controller nor the firefighters realized another aircraft was involved. However, one fireman soon noticed a propeller blade protruding from the Boeing right engine intake. His supervisor radioed the tower and asked if the Boeing airliner had hit another plane before slamming into the vacant building. Initially the tower did not know. However, controllers eventually realized the Metroliner was not responding to radio calls:

> About 1814 [Hours, seven minutes after the crash,] the tower indicated that a Metroliner might be involved. The Incident Commander initiated a search of the runway [and] found five fatally injured persons and debris scattered along the path of the Boeing 737. As firefighters extinguished the fire under the Boeing 737, the fuselage of the Metroliner was found crushed under [the Boeing airliner].

The T-43, a military variant of the Boeing 737 (official U.S. Air Force photo)

Although 134 firefighters had battled the flames, the fire inside the Boeing cabin had been hard to extinguish. Twenty-two people, including the captain and a flight attendant, had been unable to escape from the burning plane. Post-mortem examination later would show that the captain died from "multiple traumatic injuries." All other fatalities on the Boeing airliner resulted from smoke inhalation and the fire. In the crushed Metroliner all 12 occupants had been killed. A SkyWest spokesman would lament:

> The USAir airplane is sitting right on top of our airplane, and [our airplane is] smashed flat. There are no survivors.

The Investigation: Investigators found no evidence of improper action by the crews of either plane. The Metroliner had been cleared onto Runway 24 Left, and the Boeing plane had been cleared to land on the same runway. Investigation quickly focused on ATC.

Los Angeles had Airport Surface Detection Equipment (ASDE) radar designed to detect airplanes and vehicles on the surface of the airport. This system was intended to augment visual observations, but it was inoperative on the night of the accident.

The tower controller had been talking with the Metroliner pilots. Her recorded statements confirmed she had known the Metroliner had been holding in takeoff position on the runway. However, she had been searching for the missing flight progress strip while directing several other aircraft:

She subsequently "forgot" that [the Metroliner] was on the runway and misidentified [another plane]. . . . She developed a mental picture and a reasonable expectation that the runway was clear and issued the landing clearance to [the Boeing 737].

Six weeks before the accident an ATC supervisor had appraised the job performance of the controller. The supervisor's report identified several deficiencies, listed below. Two of these same deficiencies had surfaced again on the night of the accident:

1. A loss of awareness of aircraft separation.

2. Misidentification of an aircraft . . . .

3. Failure to complete two required coordinations . . . .

4. Failure to issue a required advisory to an aircraft.

Official sunset had occurred at 1723 Hours, 44 minutes before the accident. The Metroliner had (1) landing lights, (2) strobe lights, (3) navigation lights, (4) a taxi light, (5) ice detection lights, and (6) a red rotating anticollision light on the vertical stabilizer. Yet, only the navigation lights and anticollision light were turned "on." SkyWest policy required that all strobe lights, landing lights, and taxi lights be turned "off" until after receipt of a takeoff clearance.

During the investigation an identical Metroliner was positioned on the runway at night with its lights set to mimic the condition at the time of the collision. Investigators made landing approaches to simulate the Boeing 737 approach. They discovered the Metroliner navigation lights blended into the runway lighting. The flashing anticollision light, located on the forward edge of the vertical stabilizer, was not visible from directly behind the Metroliner:

The [Metroliner] light sources [were] virtually indistinguishable when viewed from directly above and behind.

Investigators asked the surviving Boeing first officer why he had not seen the Metroliner on the runway in front of him. The first officer testified: "It wasn't there, it was invisible!"

Probable Cause: The board adopted the Aircraft Accident Report, NTSB-AAR-91-08, on 22 October 1991 almost 10 months after the accident. ATC procedures in the tower had not conformed to FAA standards, which required that flight progress strips be processed through the Ground Control position. Also, FAA evaluations had not detected that "essential redundancy was absent" in the tower:

> [The Metroliner was in] takeoff position on Runway 24 Left . . . [and] this situation went undetected. *[and also]* She *[the controller]* misidentified an airplane and issued a landing clearance that led to the runway collision.

Yet, the board did not crucify the errant controller. Conversely, board members explained how the *system* had failed:

> The expectation that controllers can perform for any length of time without error is unwarranted. In addition, the FAA's expectation of flawless human performance is unrealistic in rapidly changing and dynamic environments that exist at airports such as [Los Angeles].

The board found that the ***Probable Cause*** of this accident was twofold, as follows:

> [First], failure of the Los Angeles Air Traffic Facility Management to implement procedures that provided redundancy comparable to the requirements contained in the National Operational Positions Standards.

> [Second], failure of the FAA Air Traffic Service to provide adequate policy direction and oversight to its air traffic control facility managers.

The board explained the result of these failures:

> [They] created an environment in the Los Angeles Air Traffic Control tower that ultimately led to the failure of the local controller to maintain an awareness of the traffic situation, culminating in the inappropriate clearances and subsequent collision of the USAir [aircraft] and SkyWest aircraft.

The board pointed out a ***Contributing Cause***:

> Failure of the FAA to provide effective quality assurance of the ATC system.

Military Low-Level Training Flight
Mid-Air Collision near Kendal, Cumbria, United Kingdom
23 June 1993

Panavia GR-1 Tornado, Serial No. ZG-754
Military low-level navigation training flight
2 aboard, 0 killed

Bell 206 JetRanger, Registration G-BHYW
Industrial pipeline inspection flight
2 aboard, 2 killed

Synopsis: A civil Bell 206 JetRanger helicopter pilot conducted a low-level pipeline inspection. A Panavia GR-1 Tornado fighter flew a low-level navigation training mission. In VFR conditions the two aircraft collided less than 400 feet above the ground. The damaged Tornado landed safely, but the JetRanger crashed and killed the two people on board.

The Panavia GR-1 Tornado Flight: Two supersonic Panavia GR-1 Tornados left Bruggen, Germany, and headed west. After flying over the English Channel the Royal Air Force (RAF) warplanes practiced bombing profile maneuvers at the Cowden air-to-ground weapons range on the United Kingdom (U.K.) east coast. Then they began a low-level navigation training flight toward the RAF base at Leuchars, Scotland, where they planned to refuel:

The Tornado had been built by Panavia, a multinational consortium backed by the U.K., Germany, and Italy. The two-seat twin engine plane has variable sweep wings, giving it good low speed flight characteristics while also allowing Mach 2 dash speed. It was designed for ultra-low-level penetration of Warsaw Pact airspace. The Tornado is powered by two afterburning Turbo Union jet engines. The plane sports a huge Mauser 27mm cannon. It is capable of carrying a wide range of conventional stores including the JP233 anti-airfield weapon, the ALARM anti-radar missile, and laser guided munitions. An enhanced model, the GR-4, is equipped with FLIR and HUD instrumentation and upgraded weapons systems.

Two Panavia GR-4 Tornado fighters, similar to the earlier Panavia GR-1 Tornado (official U.S. Air Force photo)

The RAF pilot, age 25, had logged 718 flight hours, 421 of which were in the Tornado. His navigator, seated in a rear tandem seat, had over 1,000 hours of flight experience. For this lengthy mission their fighter had been fitted with three 1,500 liter external fuel tanks.

The Bell 206 JetRanger Flight: Thirty-five minutes after the two Tornados left Germany, a Bell 206 JetRanger pilot pulled in collective pitch, eased his cyclic control forward, and rotored skyward from Blackpool Airport in the U.K. The helicopter pilot and his passenger, a pipeline superintendent, began inspecting the Ethylene pipeline between Stanlow and Grangemouth:

> The inspections, which were carried out fortnightly *[meaning, each 14 days]*, were conducted at heights between 300 feet and 700 feet agl and at a speed of approximately 60 knots.

Lakeside Helicopters Ltd., owner of the JetRanger, had contracts with three major U.K. chemical companies to conduct their pipeline aerial inspections. The pilot, age 37, held an ATP certificate and had logged 4,868 flight hours. He had been hired on a free-lance basis for the pipeline inspection flights and had flown such missions on seven previous occasions.

Trouble Ahead: At 440 knots the Tornado wingman veered away from his leader to wind his way through a valley near Farleton Knott.

In the rear seat the navigator peered to the right toward Kendal, where his charts showed an active hang-gliding site.

Meanwhile, the JetRanger pilot had followed his scheduled route to a location where contractors were working on the pipeline. He circled to the left to give the superintendent a good view of the work activity. Down below the contractors stopped working to watch the helicopter. They could see the superintendent's face as he waved at them. Flying less than 400 feet above the ground, the JetRanger pilot slowly circled around the construction site at 40 knots.

The brief VHF radio transcript, below, includes messages to and from the JetRanger and ATC. Military UHF radio transmissions were not documented:

> JetRanger: . . . Bell 206 JetRanger
> LARS: . . . . . . Lower Airspace Radar Service (ATC)

JetRanger: . . . just outbound from Blackpool, resuming the low level pipeline [inspection], three hundred [feet] at Inskip . . . we'll be resuming [a] northbound track back towards Edinburgh.

> *[ATC reports that radar following is available; the JetRanger pilot contacts Warton Approach Radar]*

JetRanger: . . . on the pipeline, working our way northbound. (1024)

LARS: Thank you, keep a good lookout. (1024)

JetRanger: [Unintelligible transmission]. (1041)

> *[the JetRanger and the Tornado **collide** at 1049]*

> *[out of control, the JetRanger plunges to the ground; the Tornado remains flyable and is able to land safely]*

The Collision: After circling the construction site the JetRanger pilot had prepared to continue his flight along the pipeline. However, at that moment the Tornado was screaming through the valley:

> The helicopter rolled out in a level attitude at about 300 to 400 feet [agl] and proceeded northbound. Moments later it was hit by the Tornado. . . . The Tornado [was] flying at a speed of 440 knots on a heading of 304 degrees, and the helicopter [was] flying at a speed estimated to be 60 knots on a heading of 36 degrees.

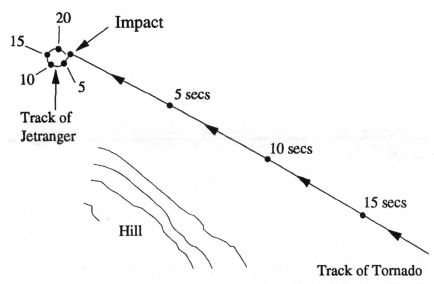

Diagram depicting the JetRanger circular flight track, upper left, and the Tornado 304-degree flight track, lower right to upper left (diagram courtesy of U.K. Department for Transport)

The Tornado pilot never saw the civil JetRanger helicopter. The Tornado radome severed the entire JetRanger tail boom. Debris from the collision caused the Tornado No. 2 engine to fail immediately:

> The [Tornado] pilot reported there was a loud bang. Presuming that his aircraft had suffered a bird strike, and noticing significant damage to the radome, the pilot . . . initiated a climb.

Flying on only the No. 1 engine, the Tornado pilot diverted to the British Aerospace (BAe) airfield at Warton and landed safely. He and his navigator climbed out to inspect their aircraft:

> They discovered metal fragments embedded in the nose section. It was only at this time that they realized they had been involved in a mid-air collision with another aircraft.

The JetRanger pilot and his passenger were not as lucky as the military crew. With their tail boom gone, with no anti-torque control, with the CG now far beyond the forward limit, the helicopter made three rapid spirals and dropped vertically to the ground. The impact instantly killed the pilot and pipeline superintendent:

Bell OH-58 Kiowa, a military variant of the civil Bell 206 JetRanger (photo courtesy of U.S. DOD Visual Information Center)

Lack of damage to the main rotor blades indicated that they had very little rotational energy at [ground] impact. *[and also]* Decelerative forces experienced by the pilot and observer [upon ground impact] . . . were well beyond human tolerance.

Missing parts of the JetRanger were found in a trail of debris 250 meters long. However, investigators were unable to find the heavy tail-rotor gearbox. The mystery would be solved when the Tornado was examined at the BAe factory. Most of the missing gearbox was found *inside* the Tornado No. 2 engine, which had disintegrated.

The Investigation: The Air Accidents Investigation Branch of the Department for Transport began an investigation on the afternoon of the collision. Weather, ATC, and mechanical problems were quickly eliminated as potential factors in this accident. Post-mortem and toxicological examination of the JetRanger pilot's body revealed no medical issues which might have played a role. The cause of the collision boiled down to an elementary dilemma. The pilots in each aircraft had not seen the other ship prior to impact.

The Tornado was equipped with a FDR which documented voice

and flight data. Prior to the mid-air impact there had been no verbal outcry or indication that the military pilot had seen the JetRanger ahead. The FDR confirmed that no evasive action had been taken.

A conspicuity study showed that under *optimum* conditions the Tornado pilot could not have seen the JetRanger until five seconds before impact. The JetRanger pilot could not have seen the Tornado overtaking him from behind.

Logic indicated the JetRanger pilot had been looking at the ground, the pipeline, and construction site. The Tornado pilot, racing through the valley at high speed and low altitude, had been concentrating on maintaining clearance from the ground flashing by below him, while scanning the high terrain on his left and right. Visually clearing airspace *ahead* had been a secondary consideration:

> While "see-and-avoid" undoubtedly prevents many collisions, the principle is far from reliable. . . . Cockpit workload and other factors reduce the time that pilots can spend in traffic scans.

The collision had taken place in uncontrolled airspace where the see-and-avoid principle applied. Military low-level flight training in this airspace is an essential element of U.K. national defense policy, as is the case in most developed countries. Civil aviation has an equal need to use the same airspace. Indeed, the instant civil flight was required to maintain the pipeline infrastructure necessary for U.K. society and industry to function. The only aerial separation in force had been the two pilots' ability to see-and-avoid each other:

> There was no technical system available to alert either crew to the proximity of the other aircraft. . . . The collision was clear evidence that, in this situation, the "see-and-avoid" principle had failed.

Probable Cause: Eleven months after the mid-air collision the Chief Inspector for Air Accidents submitted his Aircraft Accident Report, File No. 2-94, to the Department for Transport. U.K. aviation policy, of necessity, permitted use of the airspace in question by both military and civil aircraft. The collision had occurred at low altitude with a closure speed in excess of 400 knots. Both aircraft had been flown in a professional manner. The JetRanger had slowly flown in a circle, resulting in an ever-changing spatial relationship for both pilots. Under such conditions the right-of-way rules did not apply. With admirable candor the board stated:

> The collision was the tragic consequence of what must be considered a *risky operational environment [emphasis added]*.

The board found the following *Causes* of this accident:

> [First], neither pilot saw the other aircraft in time to avoid the collision.

> [Second], unsuitability of the "see-and-avoid" principle in these circumstances failed to ensure the necessary separation.

> [Third], there were no routine procedures . . . to inform either pilot about the presence of the other aircraft prior to impact.

Cleared To Land
Mid-Air Collision at Pope AFB, North Carolina
23 March 1994

<u>Lockheed C-130 Hercules, Serial No. 68-10942</u>
U.S. Air Force pilot training flight
5 aboard, 0 killed

<u>General Dynamics F-16 Fighting Falcon, Serial No. 88-0171</u>
U.S. Air Force pilot training flight
2 aboard, 0 killed (24 killed on the ground)

Synopsis: A Lockheed C-130 Hercules approached the runway. Simultaneously ATC cleared a General Dynamics F-16 Fighting Falcon to land, and the two aircraft collided on final approach. The two F-16 pilots ejected and survived. Their then-pilotless fighter crashed into a parked Lockheed C-141 Starlifter and a group of paratroopers, killing 24 and hospitalizing over 100 more. The damaged C-130 eventually landed safely.

The Lockheed C-130 Hercules Flight: At Pope AFB, North Carolina, a Lockheed C-130 Hercules crew practiced VFR maneuvers. The U.S. Air Force pilots made two full-stop landings in an "assault zone." Then they executed a touch-and-go on Runway 23:

The C-130 transport seems to be ageless. The first of the four engine turboprop planes was delivered in 1956, and the aircraft is still in production over 50 years later. The C-130 has an aft loading ramp that can accommodate wheeled vehicles, palletized cargo, and troops. It can airdrop loads weighing up to 42,000 pounds and can operate from primitive dirt airstrips. The most recent model features digital avionics, integrated defensive systems, composite-material "scimitar" propellers, plus glass cockpit LCD displays and HUD instrumentation.

The aircraft commander, a captain, was an instructor pilot. Flying with him were a copilot, flight engineer, loadmaster, and flight surgeon. The copilot was flying the aircraft. After the touch-and-go he climbed and turned downwind for another approach.

Lockheed C-130 Hercules (official U.S. Air Force photo)

The General Dynamics F-16 Fighting Falcon Flight: A U.S. Air Force fighter took off from Pope AFB on a practice bombing sortie. At a nearby air-to-ground range the General Dynamics F-16 Fighting Falcon pilots made attack runs on an aircraft-on-a-runway target. Then the two-seat F-16 headed for home. Pope Tower cleared the pilots for a simulated flame-out (SFO) approach to Runway 23:

> *The F-16, a multi-role fighter with all-weather capability, had first flown in 1976. With 27,000 pounds of thrust, the warplane has Mach 2 speed. It offers excellent pilot visibility, fly-by-wire flight controls, and side-stick control. Armament consists of a 20mm cannon, up to six air-to-air missiles, and a variety of conventional air-to-ground munitions.*

The pilot, a captain, had 1,201 hours of flight time. He carried an "observer" pilot behind him in the rear seat. As fate would have it, this was the observer pilot's *very first* flight in an F-16.

Trouble Ahead: As the big C-130 (call-sign, Hitman 31) began its approach the tower told the pilots to "keep it in tight." Moments later the tower instructed the C-130 to "fly through final" and then make a left 270 degree turn. Only seven seconds later another controller changed those instructions and told the C-130 to "continue straight in and make a low approach." The C-130 pilot would later testify:

> It sounded like they *[meaning, tower controllers]* were having some training going on because sometimes you could hear somebody overriding someone else.

Meanwhile, the F-16 (call-sign, Webad 3) was shooting a SFO to the same runway. The tower cleared the F-16 to land. Descending rapidly at about 250 knots, speed brakes extended, the fighter quickly caught up with the slower C-130. The tower, with a clear view of the converging planes, (1) did not transmit a warning to either aircraft and (2) did not order either of them to abort their approach.

The ATC radio transcript, below, begins as the tower instructs the C-130 pilots to "keep it in tight." The tower is controlling several other aircraft in the traffic pattern. In the interest of clarity most transmissions to and from these other aircraft are omitted:

F-16: . . . . General Dynamics F-16 (call-sign, Webad 3)
C-130: . . . . . . Lockheed C-130 (call-sign, Hitman 31)
Tower: . . . . . . Pope AFB Control Tower
AppControl: . . Fayetteville Approach Control
Phone-: . . . . . . (prefix) ATC interphone transmission
Unk: . . . . Radio transmission from unknown source

Tower: Hitman thirty-one . . . keep it in tight, [you'll] be turning inside [another] C-one-thirty [that is at the] four mile initial for the random steep [approach]. (1407:54)

C-130: Roger, [this is] Hitman thirty-one. (1408:02)

[the F-16 makes initial radio contact with Pope Tower]

F-16: Pope Tower, Webad three, [at] ten DME, [requesting a] straight-in SFO low approach. (1408:38)

Tower: WeBad one [sic] – aaahhh – SFO approved – aaahhh – report – aaahhh – short – correction, report five DME. (1409:02)

F-16: [This is] WeBad three, [I] copy. (1409:08)

[the controller tells the F-16 pilot about a C-130 at 4,500 feet that will land after him; the controller fails to mention the C-130 (Hitman 31) that is on downwind for landing ahead of him]

Tower: WeBad one [sic], traffic is [a] C-one-thirty holding at four-thousand-five-hundred over the field, making one turn for a random steep, he'll be following you. (1409:12)

F-16: Webad three, tally-ho *[meaning, I see him]*. (1409:18)

> *[The F-16 pilot acknowledged seeing the C-130 high overhead at 4,500 feet. He does not know about the other C-130 (Hitman 31) that is making an approach ahead and below him]*

Tower: Hitman thirty-one – aaahhh – fly through final . . . make a left two-seventy, report back – aaahhh – on final. (1409:34)

> *[6 seconds later the tower changes its instructions to the C-130]*

Tower: Hitman thirty-one, continue straight in, make a low approach, . . . traffic – aaahhh – F-sixteen, four miles. (1409:40)

C-130: Understand you want us to – aaahhh – make, uuuhhh, the approach straight through – aaahhh – low approach? (1409:48)

Tower: Affirm. (1409:53)

C-130: [Hitman] three-one, wilco. (1409:54)

F-16: Webad three [is at] five DME. (1409:56)

Tower: Roger, [Webad three], you got C-one-thirty traffic, short final on the go. (1409:57)

> *[the F-16 pilot can not see the C-130 below him]*

> *[despite the C-130 on _short final approach_ to the _same runway,_ the tower controller clears the F-16 to land]*

Tower: And, Webad, check wheels down, wind one-niner zero at one-five, cleared to land. (1410:01)

Tower: Hitman thirty-one – aaahhh – present position right closed, approved. (1410:13)

C-130: Present position, right closed for 'thirty-one. (1410:17)

> *[300 feet above the ground, one-half mile short of the runway, the F-16 and the C-130 **collide** at 1410:24]*

> *[the damaged C-130 climbs straight ahead]*

[the damaged F-16 pitches up, then rolls out of control]

Unk: Oh, [expletive]! (1410:30)

Unk: Holy [expletive]! (1410:30)

Unk: Tower, you [see] that mid-air? (1410:31)

Unk: Eject! Eject! Get out of there! Get out of there! (1410:36)

[both F-16 pilots eject]

[the pilotless F-16 crashes and skids into a parked C-141 and a group of several hundred U.S. Army paratroopers]

Unk: We got an ejection! (1410:39)

Unk: Got two chutes! (1410:43)

Phone-Tower: Fayetteville [Approach Control], Pope [Tower], the runway's closed – ooohhh – my God! (1410:48)

Unk: We got two good chutes, two good chutes! (1411:04)

Phone-Tower: [Approach Control], the runway is closed, do you copy? (1411:06)

Phone-AppControl: Yeah. (1411:08)

[pilots in the damaged C-130 radio the controller]

C-130: Pope Tower, this is Hitman thirty-one. (1411:08)

Tower: Hitman thirty-one, this is Pope Tower, go ahead. (1411:12)

C-130: Our systems are good right now, but we need to get this thing on the ground! (1411:14)

Tower: Hitman thirty-one, ah, you can, ah, turn around and make an approach into Runway five if you need to land now. (1411:25)

C-130: Tower, Hitman thirty-one [has an] emergency! (1412:26)

Tower: Hitman thirty-one, roger, continue inbound, report right base for landing. (1412:28)

Phone-AppControl: . . . what we got going on over there? (1413:25)

Phone-Tower: We had an accident, a C-one-thirty and an F-sixteen on a straight-in SFO . . . and the runway is closed. (1413:30)

Tower: Hitman thirty-one [on] base, your discretion? (1413:54)

C-130: Tower, 'thirty-one, emergency – we're just, ah, trying to figure out if we got enough controllability to [land], stand by . . . definite vibrations in the controls. (1413:56)

Tower: Hitman thirty-one, say again? (1414:03)

C-130: . . . have, aaahhh, the fire department standing by . . . definite vibrations in the controls. (1414:05)

Phone-AppControl: [Tower, are] you still talking to that C-one-thirty *[Hitman 31]* up there at twenty-five-hundred? (1414:06)

Phone-Tower: Affirmative, I think that's Hitman thirty-one instead of Hitman zero-five. (1414:11)

[the C-130 eventually lands safely]

The Collision: On final approach the F-16 radome had blocked line-of-sight vision below the aircraft, so the F-16 pilots had been unable to see the C-130 below them. Also, the C-130 crew had not been able to see the F-16 overtaking them from above and behind. The tower controller *could* see the two aircraft converging; however:

> There were no [tower] transmissions directing Webad 03 to de-conflict with Hitman 31. There were no radio calls made [for] sequencing the F-16 or the C-130 on final.

The F-16 radome had ripped through the C-130 right horizontal stabilizer and elevator. Then the C-130 pilots had seen the damaged F-16 zoom by the right side of their cockpit. They had aborted their approach and climbed straight ahead. As they accelerated through 150 knots the control column began to vibrate. They climbed to 2,500 feet, made a controllability check, and eventually landed safely. The F-16 pilot would later explain the mid-air impact:

> I didn't have time to react . . . . When I saw [the C-130, I] hit [it], and the next thing I know I'm staring out at the nose of my aircraft, which is no longer there.

General Dynamics F-16 Fighting Falcon (official U.S. Air Force photo)

The F-16 had pitched up, then down. Only 300 feet above the ground the two pilots had ejected. One landed in a parking lot and the other floated down into a tree. Both were shaken but unhurt.

The pilotless F-16 had tumbled down onto the airport. Engulfed in a searing fireball, the wreckage had ricocheted across the airfield. An independent technical report would explain the tragic result:

> The accident sent the fighter skidding across the pavement and into a parked [Lockheed] C-141 Starlifter, then into an area known as the Green Ramp, filled with [over 300] paratroopers.

The C-141 fuel tanks had ruptured, fuel ignited, and the transport exploded into flames. Many of the U.S. Army paratroopers, who hailed from the 82nd Airborne Division at nearby Fort Bragg, writhed on the ground. Some had been cut down by cartwheeling F-16 wreckage. Over 100 others were drenched in burning jet fuel.

Uninjured soldiers ran to assist their brothers-in-arms. They used their hands and bodies to smother the flames. The pungent stench of burned flesh quickly permeated the air. Firefighters, triage teams, and ambulances flocked to the scene. Eventually 24 young paratroopers would succumb to their injuries, and over 100 more would require extensive hospitalization. William "Bill" Clinton, President of the United States, would visit the burned survivors at Womack Army Medical Center at Fort Bragg.

The scandal-hungry news media monitored technical information about the tragedy. The Air Force soon released transcripts of the ATC tapes, and the *Associated Press* opined:

> Obviously, there was a failure to communicate [because] both pilots thought they were cleared to land.

The Investigation: On 15 June 1994 the Ninth Air Force approved the Accident Report that stemmed from an Initial Investigation. An independent technical study noted:

> There were multiple causes of the mid-air collision, the majority of which occurred in air traffic control.

An Inspector General review confirmed these findings. Yet, the review concluded the Accident Report failed to properly examine the role of the F-16 pilot. Further investigation was ordered.

The Re-Investigation: The CVR from the C-130 could no longer be found. The C-130 FDR was in hand, but the tape had jammed and had not recorded. The flight path of the C-130 had to be reconstructed based upon the recollection of the crew.

Analysis of the F-16 FDR showed the parameters of the SFO approach were within acceptable limits. Weather had not been a factor. With respect to ATC the investigators concluded:

> The last 4 radio transmissions made by the tower in the 26 seconds prior to the collision . . . did not build SA *[situational awareness]* for anyone in the pattern and only caused confusion.

Investigation produced "clear and convincing evidence" that the C-130 was in the line-of-sight of the F-16 pilot for only 12 seconds early during the approach. That 12 second window took place with the C-130 flying downwind while the F-16 was almost three miles away. The tower had not radioed the (1) existence, (2) identification, or (3) location of the C-130 to the F-16 pilot. Also, the C-130 was painted in a European green-gray camouflage pattern. Investigators found by "clear and convincing evidence" that:

> While Hitman 31 *[the C-130]* was technically within [the F-16 pilot's] field of view for a few seconds, due to numerous factors [the F-16 pilot] could not have seen the C-130 . . . .

In addition, some of the more significant investigatory conclusions, findings, and opinions are quoted below:

1. Two different tower personnel each mixed up Webad 3's call-sign, once by calling him "Webad 1" when he initially checked-in, and again at 1409:12.

2. Two different tower personnel each mixed up Hitman 31's call-sign, once by calling him "Felix 3" at 1409:22, and again at 1409:27. . . . This mix-up occurred while Hitman 31 was starting its turn on [the] base leg of its approach pattern.

3. None of the voices of tower controllers ever indicated urgency with any control instructions.

4. Webad 3 was told that the [only] C-130 he had in sight would be *following [emphasis added]* him.

5. [After the "short final on the go" transmission, Webad 3] did not query the controller to determine the other aircraft's exact position. *[however]* The controller instructed [Webad 3] that he was "cleared to land," leading him to believe that no traffic conflict existed . . . .

The board was of the *opinion* that the F-16 pilot, although he had been cleared to land, should have aborted his SFO approach after hearing the "short final on the go" transmission. The message was unspecific. Yet, the board *opined* that it should have alerted the pilot that there was another plane – somewhere – that he could not see.

Probable Cause: The board found *scores* of causative factors and circumstances, some of which have been detailed above. Obviously, both ATC and plane-to-plane communication had been a nightmare:

> Hitman 31 was broadcasting [and] monitoring only UHF, as was [Webad 3]. Yet, the other C-130 in the local pattern, Hitman 05, was broadcasting only on VHF.

In total, these circumstances snared the F-16 and C-130 pilots in a web of misfortune and led to the mid-air collision. Three years after the accident, investigators re-interviewed the F-16 pilot in 1997. The gist of the pilot's testimony was that (1) he was cleared to land, (2) he could not see the C-130 below him, and (3) the tower failed to tell him about it. The board sought the pilot's *opinion* and asked: "To what do you attribute this mishap?" The pilot replied:

> . . . a bad use of call-signs, a bad use of radios, a bad use of, basically, traffic management – that allowed a C-one-thirty to get in front of me in a position where I couldn't see him. The tower didn't point him out to me . . . and the C-one-thirty didn't know that I was there to do anything on its own. The only people with the big picture [were in the] tower, and they didn't communicate that big picture to either me, the [C]-one-thirty, or anybody else.

Photograph depicting damage to the C-130 right horizontal stabilizer and elevator, most of which are ripped away (official U.S. Air Force photo)

There's An Airplane!
Runway Collision at St. Louis, Missouri
22 November 1994

Douglas DC-9, Registration N954U
Trans World Airlines Flight 427
140 aboard, 0 killed

Cessna 441 Conquest, Registration N441KM
Superior Aviation Inc. charter flight
2 aboard, 2 killed

Synopsis: *Because of confusing airport layout and lighting, a Cessna 441 Conquest taxied onto the wrong runway at night. A Douglas DC-9 accelerated down the runway and tore through the Conquest at 100 knots. The collision injured eight people on the DC-9 and killed both people aboard the Conquest.*

The Douglas DC-9 Flight: At night a heavily loaded Douglas DC-9 began taxiing toward Runway 30 Right at Lambert/St. Louis International Airport, Missouri. Trans World Airlines (TWA) Flight 427 carried 132 passengers, 5 flight attendants, and 3 pilots. The third pilot was not part of the crew. He was tagging along in the jump seat on the flight to Denver, Colorado.

The captain, age 57, had been flying with TWA for 29 years and had logged 18,651 hours of flight time. His first officer, age 38, had logged an impressive 10,353 hours. Every seat in the cabin was filled as the twin-jet DC-9 lumbered toward the active runway.

The Cessna 441 Conquest Flight: Superior Aviation Inc., based at Iron Mountain, Michigan, operated an on-demand charter service. After dark one of the company's Cessna 441 Conquest aircraft flew a business executive from Iron Mountain to St. Louis. The Conquest landed on Runway 30 Right, taxied to the general aviation terminal, and dropped off the executive. Then the pilot taxied toward the runway to take off for the night flight back to Iron Mountain:

The Conquest was the first turboprop aircraft built by Cessna, and the factory delivered the first new airplane in 1977. The pressurized craft could carry 8 to 10 passengers at speeds approaching 300 knots. Two Garrett turbines drove huge four-blade McCauley propellers and gave the plane a service ceiling of 35,000 feet. The Conquest filled the gap between Cessna's piston engine twins and its twin-jets.

The Conquest pilot, age 56, held a Commercial Pilot certificate and had logged 7,940 hours in the air. The Conquest did not require a copilot. Nonetheless, a non-company pilot rode along in the right seat on an unofficial "familiarization flight." He was an accountant by profession, he held a Private Pilot certificate, and his wife worked as a receptionist at Superior Aviation.

Trouble Ahead: Ground Control cleared the Conquest pilot to taxi to *Runway 31*. However, the pilot erroneously taxied onto adjacent parallel *Runway 30 Right* at a taxiway intersection. He lined up on the runway centerline and got ready to take off.

Meanwhile, the DC-9 was holding at the approach end of the same runway. In the dark the DC-9 pilots could not see the Conquest on the runway a half-mile ahead. At 2201:23 Hours the tower cleared the airliner for takeoff.

The portion of the ATC transcript, below, includes radio messages between the tower, ground control, and the two aircraft. It also includes content from the DC-9 CVR. Radio messages to and from other aircraft are excluded in the interest of clarity. The transcript begins as Ground Control separately clears the planes to taxi:

<u>Capt</u>: . . . . Captain, TWA Flight 427 (DC-9)
<u>FO</u>: . . . . First Officer, TWA Flight 427
<u>OffDuty</u>: . . . . Off duty pilot, TWA Flight 427
<u>FltAtt</u>: . . . . . . Flight Attendant, TWA Flight 427
<u>Radio-</u>: . . . . . (prefix) Radio transmission, TWA Flight 427
<u>PA-</u>: . . . . (prefix) Public Address system, TWA Flight 427
<u>Conquest</u>: . . . Cessna 441 Conquest (call-sign, 441 KM)
<u>Tower</u>: . . . . . St. Louis Airport Control Tower
<u>GndControl</u>: . St. Louis Ground Control
<u>FireTruck</u>: . . St. Louis Airport fire truck

<u>GndControl</u>: One Kilo Mike, roger, back-taxi into position and hold, Runway three-one, let me know on this frequency when you're ready

for departure. (2156:18)

[the Conquest pilot acknowledges the taxi clearance, but he does not specify the runway to which he will taxi]

GndControl: TWA four-twenty-seven, cross Runway three-zero left, taxi [to] Runway three-zero right. (2157:34)

[the DC-9 pilots acknowledge the taxi clearance]

[at a taxiway intersection, the Conquest pilot <u>erroneously taxis onto Runway 30 Right</u> and prepares for takeoff]

[the DC-9 pilots taxi onto the <u>same runway</u> at the approach end and prepare for takeoff; in the dark they can not see the Conquest holding 2,500 feet down the runway ahead of them]

PA-FltAtt: Ladies and gentlemen, we're ready for departure, cabin attendants, take your seats for takeoff. (2200:16)

Capt: We're ready [for takeoff], aren't we? (2201:18)

Radio-FO: TWA four-twenty-seven's ready [for takeoff]. (2201:21)

Tower: TWA four-twenty-seven, winds two-seven-zero at seven, Runway three-zero right, turn right heading three-three-five [after takeoff], cleared for takeoff. (2201:23)

[the DC-9 First Officer acknowledges the clearance]

Capt: Before takeoff checklist. (2201:54)

[the DC-9 pilots complete the before takeoff checklist]

[sound of the DC-9 accelerating down the runway at 2202:27]

[simultaneously the Conquest pilot radios the tower]

Conquest: Kilo Mike's ready to go. (2202:28)

[the controller <u>assumes</u> the Conquest is on Runway 31]

Tower: Roger, I can't roll you simultaneously with the, uh, traffic departing the right [runway], just continue holding in position, I'll have [takeoff clearance] for you in just a second. (2202:30)

Conquest: Kilo Mike. *[acknowledgment of the delay]* (2202:37)

Capt: Eighty knots. (2202:40)

OffDuty: There's an airplane! *[a shouted warning]* (2202:44)

*[**sound of impact** as the DC-9 hits the Conquest at 2202:47]*

[sound of reverse thrust at 2202:53]

Radio-Capt: [Tower], roll the emergency equipment! TWA four-twenty-seven hit the other airplane on the – aaahhh – runway, roll the emergency equipment! (2202:59)

[the DC-9 brakes to a stop on the runway at 2203:10]

GndControl: Everybody stand by, we've had an aircraft accident on the runway. (2203:17)

FireTruck: St. Louis Ground, this is [fire] truck forty-two, do you have a location on this accident? (2204:06)

GndControl: It's at the intersection of Runway three-zero right and taxiway November, we need you to move there now . . . . (2204:10)

Radio-Capt: Tower, TWA four-twenty-seven.

Tower: TWA four-twenty-seven, the equipment's rolling right now.

Radio-Capt: Do you see any fire or smoke around [my] aircraft?

Tower: No, sir, I don't. (2204:13)

Tower: [That other plane] was supposed to be on Runway three-one, I did not see the aircraft – (2204:16)

[the Captain interrupts the tower controller]

Radio-Capt: OK, we'll handle that later, [right now] I just want to make sure everything's safe here. (2204:19)

Tower: Roger. (2204:22)

Capt: *[speaking to the off duty pilot]* Go back and see if there's any damage back there. (2204:23)

PA-FltAtt: Ladies and gentlemen, everyone seated, please, everyone seated, please take your seats – everyone please stay in your seats, we will inform you as soon as possible. (2204:38)

Capt: I'm shuttin' [down] the right engine. (2204:39)

[the off duty pilot returns to the cockpit with bad news]

OffDuty: Gasoline [is] running under the airplane. (2204:51)

FO: We better get off this thing! (2204:53)

[the crew and passengers safely evacuate the DC-9]

The Collision: The DC-9 had begun its takeoff roll. One-half mile ahead – on the same runway – the Conquest had waited for takeoff clearance. Accelerating down the runway, the DC-9 pilots saw the silhouette of the Conquest materialize out of the darkness:

> [The DC-9 pilots] first saw [the Conquest] when it was illumi-nated by their [landing] lights. *[and also]* Superior Aviation procedures required that illumination of the strobe, taxi, and landing lights take place after receipt of a takeoff clearance.

The captain and first officer had hit their brakes and applied left rudder, but they were going far too fast to stop in time. At 100 knots the DC-9 right wing had sliced through the Conquest and cut off the upper half of the fuselage. The two Conquest pilots literally lost their heads as well as their lives; they got decapitated:

> The pilot and pilot-rated passenger on board the [Conquest] died of severe craniocerebral injuries.

Both planes had been on the runway heading, facing northwest. There was no way the Conquest pilots could have seen the airliner bearing down on them from behind.

The DC-9 captain had managed to stop his craft 1,450 feet beyond the point of impact. The airliner sustained heavy damage to the right wing leading edge devices, flaps, upper and lower airfoil surfaces, and spar. Part of the severed Conquest left wing was wrapped around the DC-9 right main landing gear strut. Eight DC-9 passengers suffered injuries, but no one aboard had been killed.

The Investigation: The NTSB quickly found out exactly *what* had taken place. The night weather had been clear. The DC-9 had been cleared for takeoff on Runway 30 Right, and the Conquest had been

Douglas C-9 Nightingale, a military variant of the civil Douglas DC-9 airliner (photo courtesy of U.S. DOD Visual Information Center)

cleared into "position and hold" on Runway 31. Yet, the Conquest pilot had taxied onto Runway 30 Right a half-mile from the approach end. He was out of sight of the DC-9 pilots, who were 2,500 feet behind him. Investigators concentrated on *why* the Conquest had taxied onto the wrong runway:

> The pilot of the [Conquest] unintentionally deviated from the taxi clearance he received from Ground Control and taxied onto the active runway being used by the [DC-9].

The airport has five runways, and three of them run parallel to each other. Runway 30 Right (on which the DC-9 was cleared for takeoff) is a primary runway 9,003 feet long and 150 feet wide. Runway 31 (on which ATC told the Conquest to hold) was built as a parallel taxiway. It is only 75 feet wide. In 1988 the airport had begun using this taxiway as an occasional departure-only runway for general aviation aircraft. The airport called this taxiway "Runway 31," but it still looked like a taxiway.

The Conquest pilot had landed on Runway 30 Right (a primary runway) only 18 minutes before the accident. While he was dropping off his passenger at the terminal, all takeoffs and landings had taken place on that runway. Also, ATIS was announcing Runway 30 Right as the active runway for all arrivals and departures. ATIS made no mention of any other runway – and it got worse:

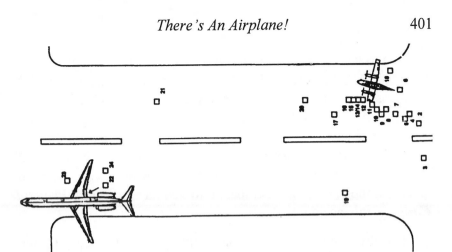

Diagram depicting the Douglas DC-9, lower left, and the Cessna 441 Conquest, upper right, and debris on Runway 30 Right after the collision (diagram courtesy of U.S. NTSB)

> St. Louis controllers did not treat Runway 31 *[the former taxiway]* as an active runway; for example, when the [Conquest] pilot cleared Runway 30 Right *[on which he had landed]* on his inbound flight, his taxi clearance . . . *did not [emphasis added]* include a clearance to "cross Runway thirty-one."

Runway 30 Right had bright runway lights. The lights along Runway 31 (the converted taxiway) were dim taxiway-type lights. The ground controller had twice said "Runway thirty-one" in his instructions to the Conquest pilot. Investigators concluded the pilot did not notice the change, or *assumed* that Runway 31 was the same as Runway 30 Right, on which he had landed minutes before. He had scant opportunity to read-back the entire clearance because:

> The [lone] controller was monitoring seven different frequencies [and there were] several instances of simultaneous transmissions.

Probable Cause: Nine months after the runway collision the NTSB adopted the Aircraft Accident Report, NTSB-AAR-95-05, on 30 August 1995. While waiting for his takeoff clearance the Conquest pilot had kept his anticollision lights and strobe lights turned off. Thus, against the background of runway and taxiway lights, the dim position lights on the Conquest had been invisible from the tower. For the same reason the DC-9 pilots a half-mile away on the same runway had been unable to see the Conquest.

The tower controller had *assumed* the Conquest was holding on Runway 31 (the former taxiway). However, the board concluded the Conquest pilot most likely did not know Runway 31 existed. He probably *assumed* he was cleared onto Runway 30 Right. Airport Surface Detection Equipment (ASDE) radar normally would have alerted ATC to the Conquest's position. Unfortunately the ASDE system had been out-of-service due to a inoperative computer hard drive. In the end, board members outlined an elementary *Probable Cause* of this accident:

> The Probable Cause of this accident was the [Conquest] pilot's mistaken belief that his assigned departure runway was Runway 30 Right, which resulted in his undetected entrance onto Runway 30 Right, which was being used by the [DC-9] for its departure.

The board then explained a *Contributing Cause*:

> Lack of ATIS and other ATC information regarding the occasional use of Runway 31 *[the former taxiway]* for departures.

Give It All You Got!
Bird Strike at Elmendorf AFB, Alaska
22 September 1995

<u>Boeing E-3 Sentry AWACS, Serial No. 77-0354</u>
U.S. Air Force surveillance training mission
24 aboard, 24 killed

<u>Flock of Canadian Geese *(Branta canadensis)*</u>
Sudden flight from a migratory roosting area near the runway
N.A. / N.A.

<u>Synopsis</u>: At liftoff a Boeing E-3 Sentry AWACS flew into a flock of Canadian Geese. Both engines on the left wing failed after ingesting several large geese. The four engine aircraft climbed to 200 feet above the ground, slowly lost airspeed, stalled, and crashed. Ground impact and fire killed all 24 men aboard.

The Boeing E-3 Sentry AWACS Flight: Shortly after daylight a Boeing E-3 Sentry AWACS prepared to take off from Elmendorf AFB in Alaska. This U.S. Air Force flight promised to be a routine six hour airborne surveillance training mission. The 24 U.S. Air Force men aboard the E-3 ranged in rank from airman to lieutenant colonel. The plane's radio call-sign was Yukla 27 (Yukla means "Eagle" in a native Alaskan Indian dialect):

The E-3 is based upon a modified Boeing 707 airframe. These four engine Airborne Warning and Control System (AWACS) planes are distinguished by the "flying saucer" appearance of the large radar dome atop the rear fuselage. The powerful radar system has a range of over 250 miles and permits surveillance from the ground to the stratosphere. E-3 planes are designed for surveillance, identification, and weapons control. They are the premier aerial command-and-control platforms in the world.

The pilot, a captain, had logged 1,922 flight hours since completion of basic pilot training. He had been certified as a mission ready E-3 pilot for over a year. His copilot, also a captain, had logged 1,258 hours since his basic training. The plane and the crew were assigned to the 962nd Airborne Surveillance Squadron:

This incident [would be] the first loss of an E-3, so local and national media attention was high.

The Canadian Geese Flight: Two minutes before the E-3 began its takeoff roll, a Lockheed C-130 Hercules had accelerated down the same runway. The Hercules had flushed a flock of Canadian Geese *(Branta canadensis)* that had been roosting in the infield grass. The tower controller observed this flock but did not notify the E-3 pilots or Airfield Management:

> *Canadian Geese are native to North America and breed primarily in Canada and the high Arctic tundra. There are seven subspecies, the largest of which weighs up to 24 pounds and has a six foot wingspan. By nature these geese are migratory, and they usually fly south to winter in the United States. They have adapted to urban environments with large lawns and ponds, such as golf courses, city parks, and airports. These waterfowl are known for their V-formation flights and loud "honking" calls. They have become a nuisance in many urban areas.*

Trouble Ahead: At 0745 Hours the big E-3 began rumbling down Runway 5 with the copilot at the controls. During the takeoff roll another flock of geese rose from a roosting area in the tall grass and began flying across the runway in front of the aircraft.

The portion of the CVR and ATC transcript, below, begins as the E-3 pilots receive their takeoff clearance. The transcript has been *reconstructed* from multiple technical sources, so some verbiage may not be verbatim:

Pilot: . . . . Pilot, Boeing E-3 Sentry (call-sign, Yukla 27)
Copilot: . . . . Copilot
FltEngr: . . . . Flight Engineer
UnidCrew: . . Unidentified crewmember
Radio-: . . . . . (prefix) Radio transmission, Boeing E-3 Sentry
Tower: . . . . . Elmendorf AFB Control Tower

Tower: Yukla two-seven heavy, wind three-one-zero [degrees] at one-one [knots], cleared for takeoff, Runway five, traffic is a C-one-thirty, three miles north of Elmendorf, northwest-bound, climbing out of two-thousand [feet]. (0745:29)

Pilot: [Traffic] in sight. (0745:39)

Radio-Copilot: Yukla two-seven heavy [is] cleared for takeoff, traffic in sight. (0745:41)

Copilot: Cleared for takeoff – crew? (0745:44)

FltEngr: Check complete. (0745:45)

Pilot: Engineer, set takeoff power. (0745:47)

[sound of engines spooling up; the takeoff roll begins]

Copilot: Eighty knots, copilot's aircraft. (0746:09)

Pilot: Your airplane. (0746:11)

Pilot: V-one. (0746:20)

Pilot: Rotate. (0746:28)

[a huge flock of geese flies in front of the accelerating E-3]

Copilot: [Look at] all the birds! (0746:28)

FltEngr: Lotta birds here. (0746:31)

[as the E-3 lifts off, several large geese are sucked into the two engines on the left wing; both engines fail immediately]

*[the E-3 has **lost all thrust** from the No. 1 and No. 2 engines; it yaws left but continues a shallow climb]*

Pilot: [Expletive], we took one. (0746:33)

FltEngr: We took two of 'em. (0746:36)

Copilot: What do I got? (0746:36)

Pilot: We got two motors [inoperative]. (0746:37)

FltEngr: [Let's do a] flight start. (0746:37)

Copilot: Roger that. (0746:38)

[as the aircraft climbs, slowly losing airspeed, the pilot takes over the flight controls]

Pilot: Take me to override. (0746:40)

Radio-Copilot: Elmendorf Tower, Yukla two-seven heavy has an emergency, [we] lost two engines, we've taken some birds. (0746:43)

FltEngr: You're in override . . . . (0746:44)

Pilot: Thank you. (0746:48)

FltEngr: Start [to] dump fuel? (0746:49)

Pilot: Start dumping. (0746:51)

Tower: Yukla two-seven heavy, roger, say intentions. (0746:52)

[the stick-shaker activates at 0746:55]

Radio-Copilot: Yukla two-seven heavy is coming back around for an emergency return. (0746:56)

Tower: 'Two-seven heavy, roger. (0746:58)

Copilot: Lower the nose, lower the nose, lower the nose! (0746:58)

[the aircraft stalls and mushes down toward the ground]

Pilot: Goin' down! (0747:00)

UnidCrew: Oh, my God! (0747:02)

Pilot: Oh [expletive]! (0747:04)

Copilot: Give it all you got! Give it all you got! (0747:04)

Radio-Copilot: 'Two-seven heavy, emergency! (0747:06)

Pilot: We're going in – we're going down! (0747:11)

[sound of impact with the ground at 0747:12]

The Collision: Witnesses later would report that over 100 geese had swarmed into the path of the E-3. As the pilot had called for rotation, the copilot had remarked: "[Look at] all the birds!" Three seconds later the flight engineer had chimed in: "Lotta birds here."

> The aircraft struck a flock of Canadian Geese, several of which were ingested by the No. 1 and No. 2 engines. The result was an immediate, unconfined, catastrophic failure of the No. 2 engine as well as compressor stalls in the No. 1 engine.

Boeing E-3 Sentry AWACS (official U.S. Air Force photo)

The E-3 had lost all thrust from both engines on the left wing. Flying low and slow, gear and flaps down, the pilots struggled to gain altitude and adequate flying speed – a losing fight.

The aircraft had climbed to an altitude of only 200 feet above the trees. Airspeed slowly decayed. The stick-shaker signaled imminent peril. The pilots were powerless to counter (1) the loss of two engines, (2) the asymmetric thrust condition, and (3) the decaying airspeed. The E-3 stalled and began a horrifying arc down toward the ground. One independent report would later explain:

> The $270 million aircraft and its 24 crewmembers disappeared in the dense woods beyond the runway. A large fireball erupted skyward as the aircraft, loaded with 125,000 pounds of jet fuel, slammed into the earth.

Air Force firefighters made their way to the crash site on foot, for there were no roads to the wooded location one mile beyond the end of the runway. The fire burned for several hours. Later that day a recovery team probed through charred debris and stuck scores of tiny flags into the ground. Blue flags marked airplane parts. Red flags marked human remains. Medical examinations later would reveal that "the crewmembers died instantaneously and did not suffer."

The Investigation: Air Force investigators soon eliminated contaminated fuel and other such potential causes. Analysis clearly showed that two engines had failed due to ingestion of several huge geese. Flight simulator checks confirmed that loss of thrust from both engines on the left wing, at rotation, would render an E-3 incapable of sustained flight. Based upon the CVR transcript and the aircraft response, investigators determined:

> The flight deck aircrew performed their emergency procedures flawlessly in an attempt to fly this aircraft out of an unflyable scenario.

Investigators delved into the presence of Canadian Geese on the airfield. The Air Force had a Bird Aircraft Strike Hazard (BASH) program aimed at reducing the possibility of bird-aircraft collisions. At Elmendorf AFB it was common knowledge that migratory geese arrived every September and October. Airfield Management had reacted to scare away the geese when flocks were seen, but this response was found to be reactive, not proactive. Air Force regulations had mandated:

> Birds on runways, taxiways, or infields create an immediate safety hazard and must be dispersed before flying operations can safely continue.

The military Chief of Conservation and Environmental Planning estimated that about 900 geese were on the base on the day of the accident. Most had been roosting in tall grass near Runway 5. On two recent occasions the Air Force had resorted to shooting the geese with shotguns to force flocks to disperse. One witness stated:

> I've really never been at a location, except Elmendorf, that's had so many birds. I mean, not only on the flight line, of course, but all over the place. I mean, they're everywhere!

Probable Cause: The Air Force approved the Aircraft Accident Report on 8 January 1996. All flight crewmembers had been fully qualified, and neither material nor maintenance factors had played a role in the accident. The board found that the presence of geese on the airfield resulted from "systemic failure."

Static deterrents for birds, such as sound cannons, had not been used. Also, the BASH group had not taken any action to prepare for the annual migratory season. The board found "the worst possible combination" of factors had existed:

> There were infrequent patrols of the airfield, nearly no checking of infields, and no placement of static deterrents. This combination of factors resulted in geese roosting in infields without being detected.

A tower controller had seen the C-130 flush a flock of geese two minutes before the E-3 had begun its takeoff roll. Yet, the controller had not warned the E-3 pilots that geese were present in infield grass near the runway. With respect to the *Probable Cause* of this accident the board opined:

> The accident was directly caused by ingestion of Canadian Geese into the [No.] 1 and [No.] 2 engines. The resultant loss of thrust rendered the aircraft incapable of controlled flight.

In the 17 days prior to the accident, Airfield Management had dispersed roosting geese in the infield on seven occasions and had shot some geese that refused to fly away. However, the board found these efforts were inadequate and "lacked perseverance." The board explained two *Contributing Factors* as follows:

> [First], perseverance is the key to [bird] dispersal. [The] lack of perseverance, especially in the infield areas, was a contributing factor to this accident.

> [Second, the tower controller] had a duty to warn the [E-3 pilots], and his failure to do so was a contributing factor to this accident.

When Giants Collide
Mid-Air Collision over Charki Dadri, Bhiwani District, India
12 November 1996

<u>Boeing 747, Registration HZ-AIH</u>
Saudi Arabian Airlines Flight 763
312 aboard, 312 killed

<u>Ilyushin Il-76, Registration UN-76435</u>
Kazakhstan Airlines Flight 1907
37 aboard, 37 killed

<u>Synopsis</u>: Under the control of ATC, a Boeing 747 took off and climbed toward its cruising altitude. On the same airway an Ilyushin Il-76 approached to land. The two giant transports collided head-on in the clouds, killing all 349 people aboard.

The Boeing 747 Flight: At the Indira Gandhi International Airport in Delhi, India, 23 crewmembers boarded a Boeing 747. Some of their 289 passengers (215 Indians, 53 Nepalis, 13 Saudis, 4 Pakistanis, 3 Americans, and 1 Briton) were oilfield workers returning to jobs in the Middle East. Yet, the majority were Muslims making a pilgrimage to Mecca. Their aircraft, Saudi Arabian Airlines Flight 763, was bound for Dhahran, Saudi Arabia, four flying hours away. The big jumbo-jet lumbered down the runway, rumbled low over the city, and began climbing toward its cruising altitude.

The experienced pilot, age 45, held an ATP certificate and had logged 9,837 flight hours. He had flown the Delhi-to-Dhahran route many times and anticipated no problems. He flew the Parvi SID, and then Approach Control guided him and his first officer as they climbed through the night sky.

The Ilyushin Il-76 Flight: Kazakhstan Airlines Flight 1907, a big Ilyushin Il-76, left Chimkent Airport in Kazakhstan with a crew of 10 and an extremely slim load of 27 passengers. The big transport could have carried ten times as many people. The non-scheduled flight had been chartered by a clothing conglomerate in Kazakhstan. The passengers were businessmen traveling to Delhi to purchase wool goods for resale at home:

Ilyushin Il-76 (photo courtesy of U.S. DOD Visual Information Center)

The Russian-built Ilyushin Il-76 (NATO code-name, Candid) was conceived as a heavy freighter able to cope with severe weather in Siberia. The design is similar to the American-built Lockheed C-141 Starlifter. Four Soloviev turbofan engines pump out 106,000 pounds of thrust. The wings span 166 feet, and the plane has a maximum takeoff weight of 346,000 pounds.

The Il-76 served as a freighter, passenger airliner, and military troop transport with operators around the world. Variants have flown as waterbombers, cruise missile carriers, tankers, and AWACS platforms. In August 2005 when levees in Louisiana succumbed to Hurricane Katrina, two Il-76 freighters from the Russian Federation flew relief supplies to the United States.

The Kazakh pilot, age 44, held a "First Class Pilot" certificate and had logged 9,229 hours of flight time. The flight to India progressed routinely. The crew radioed Delhi Approach Control and began a long en route descent toward the airport.

Trouble Ahead: Flying *outbound* on Airway G-452, the Boeing pilots were cleared to flight level 140 (14,000 feet). The controller planned to hold them at that altitude until the *inbound* Il-76, descending to flight level 150 (15,000 feet) on the same airway, passed over them. Then the controller intended to clear the Boeing pilots to climb to their cruising altitude.

The portion of the ATC transcript, below, contains radio messages between Approach Control and the two aircraft. Some transmissions have been *reconstructed* from multiple technical sources, so some verbiage may not be verbatim. The transcript begins as the inbound Il-76 descends through flight level 230:

> KazAir 1907: . Kazakhstan Airlines Flight 1907 (Il-76)
> Saudi 763: . . . . Saudi Arabian Airlines Flight 763 (Boeing 747)
> AppControl: . . Delhi Approach Control

KazAir 1907: Good evening, [this is KazAir] one-nine-zero-seven, passing through [flight level] two-three-zero for [flight level] one-eight-zero, seventy-four miles from [the airport].

AppControl: Roger, KazAir nineteen-zero-seven, descend to and report [flight level] one-five-zero.

> *[the Boeing 747 takes off and climbs toward the Il-76; the Boeing pilots radio Approach Control as they near 10,000 feet]*

Saudi 763: Delhi Approach, Saudi seven-six-three is approaching [flight level] one-hundred. *[10,000 feet]*

AppControl: Saudi seven-sixty-three, roger, climb to [flight level] one-four-zero and maintain [flight level] one-four-zero, stand by for [a] higher [altitude].

AppControl: KazAir one-nine-zero-seven, Delhi Approach, say your distance from Delhi now.

KazAir 1907: Delhi Approach Control, one-nine-zero-seven is at [flight level] one-five-zero, forty-six miles from the airport.

AppControl: Roger, maintain one-five-zero, identified traffic [is at] twelve o'clock, [on a] reciprocal [heading, a] Saudi Boeing seven-forty-seven, fourteen miles, report [when it is] in sight.

> *[although Approach Control told the Il-76 pilots to maintain flight level 150, the Il-76 pilots descend to flight level 140]*

> *[the Boeing and Ilyushin transports are now (1) flying toward each other on the (2) same airway at the (3) same altitude]*

KazAir 1907: Report? – How many miles?

AppControl: Fourteen miles now, one-nine-zero-seven.

AppControl: Traffic is thirteen miles, level at one-four-zero.

<u>KazAir 1907</u>: One-nine-zero-seven. *[an acknowledgment]*

*[the 2 aircraft **collide head-on** in the clouds at flight level 140]*

The Collision: The two jumbo-jets had raced toward each other through a heavy cloud deck. Above the overcast the pilot of a U.S. Air Force C-130 transport had looked down at solid clouds below. He later would explain his view of the collision:

> A cloud [deck] at our two o'clock position lit up. The light was orange in color, and its intensity continued to increase and involve the entire cloud. I remarked [to my copilot] that it must be a rocket launch. Then a plume of fire *[the burning fuselage of a plane]* came out of the cloud . . . .

The American pilot had witnessed the fiery death of the two giant jetliners. Their wings had torn into each other at a closure speed of about 700 knots. The wings disintegrated, and the fuselages ripped apart. Several hundred-thousand pounds of jet fuel ignited and set the whole sky ablaze. Remains of the two transports spun to earth in spectacular twin fireballs.

Villagers in the rural hamlet of Charki Dadri, Bhiwani District, Haryana, India, heard the deafening noise of the aerial collision three miles overhead. One resident said the night sky turned red "like the morning sun." Shredded pieces of wreckage, bodies, and baggage began raining down into the village. The next day many newspapers in the faraway United States would carry the same headline:

Mid-Air Collision Leaves 349 Dead in New Delhi

The aerial fire had incinerated most passengers before they hit the ground. Wreckage and baggage were strewn across six miles. The Dadri Government Hospital had geared up to treat survivors, but there were none. Everyone aboard both planes had perished. Charred mortal remains littered unplanted chick-pea and mustard fields. The stench of hundreds of burned corpses mingled with the pungent odor of jet fuel. One witness wrote:

> Wearing handkerchiefs and mufflers around their noses, searchers walked shoulder-to-shoulder across the fields, collecting severed limbs and bits of flesh and placing them on stretchers . . . .

Indian officials began stacking human remains atop blocks of ice in makeshift morgues. Most remains were burned too badly to be identified. Over 100 bodies, or pieces of bodies, were buried in a mass grave in a Muslim cemetery. Another 94 were cremated or buried after Hindu and Christian ceremonies. A final tally of the dead proved to be impossible. Most remains were merely unidentifiable bits and pieces of burned flesh and bone.

The Investigation: The Prime Minister of India ordered a Court of Inquiry to determine what caused the accident. He appointed Justice Ramesh Lahoti, of the Delhi High Court, to preside over the investigation. The Prime Minister promised his countrymen:

> We will not spare any culprit, if any individual was to blame.

At the time, India and nearby Pakistan were on the brink of war. Therefore, because of military security concerns the Indian Army had mandated a single air corridor for all inbound and outbound flights at Delhi. Operating with antiquated equipment, Indian ATC had to vertically separate two-way traffic on this single airway.

The outdated radar system at Delhi could not display an aircraft's altitude. Primary target "blips" on the radarscope gave a controller only distance and bearing information. For vertical separation, controllers asked each pilot to report his altitude by radio. Controllers then would issue clearances based upon the pilot's radio report. An Indian ATC guild report stated:

> Some foreign pilots . . . refer to airspace here as "the black hole" and bemoan the low quality of information they receive.

Boeing 747 (photo courtesy of U.S. DOD Visual Information Center)

Commercial airliners in the Western World are equipped with automated collision avoidance systems. Independent of ATC, these "last chance" electronic marvels aurally and visually alert pilots when other aircraft pose a collision hazard. However, neither the Saudi nor Kazakh planes had been equipped with such devices.

The controller had told the Saudi pilots to fly at flight level 140. He told the Kazakh pilots to maintain flight level 150. In accordance with ICAO mandates, this would have given the planes 1,000 feet of vertical separation as they zipped past each other on the same airway. Yet, FDR data from both planes revealed that they collided at flight level 140. Why had the Kazakh flight crew descended below flight level 150? An independent technical report explained:

> Indian authorities fingered the Kazakh pilots . . . . The aviators from the former Soviet republic, they said, did not understand English instructions clearly, flew badly maintained aircraft, and had been involved in near collisions in the past.

The ICAO mandates English as the international ATC language. Use of a common language enables pilots and controllers to avoid the biblical "Tower of Babel" *[see Genesis 11:1-9]* problem. However, verbal nuances on the ATC tape made it clear to investigators that the Kazakh pilots had trouble understanding English. The pilots had not understood that they should have stopped their descent at flight level 150. They descended to flight level 140 and struck the Saudi airliner head-on in the clouds. An Indian ATC guild reported:

> The controller . . . could only watch as the two blips on his screen
> hurtled toward each other at a combined speed of 800 miles per
> hour *[about 700 knots]*, hoping they would pass each other safely.

Probable Cause: The Court of Inquiry submitted its report to the
Indian Central Government in July 1997. The government then
waited over a year and finally "accepted" it in October 1998. The
report was critical of ATC coordination, but it said the controller gave
proper clearances to both planes. The report found the Saudi pilots
acted properly, but the Kazakh pilots failed to understand instructions
from the controller. The report spelled out the *Root and Approxi-
mate Cause* of the accident:

> Unauthorized descending by the Kazakh aircraft to FL-140 and
> failure to maintain the assigned FL-150.

The report attributed several *Contributing Causes* to the crew of
the Kazakh aircraft:

> [First], inadequate knowledge of [the] English language.

> [Second], poor airmanship.

> [Third, the] casual attitude of the crew.

The Court of Inquiry pointed out the absence of modern radar at
Delhi and explained that more air corridors to the airport would
enhance safety. The report explained that Indian controllers had no
licensing system, and Indian military and civil controllers were not
held to the same proficiency standards. The report acknowledged that
traffic volume at Delhi had been high, and yet a single controller had
been handling all arriving and departing flights.

Max Reverse!
Runway Collision at Quincy, Illinois
19 November 1996

<u>Beechcraft 1900 Beechliner, Registration N87GL</u>
United Express Flight 5925
12 aboard, 12 killed

<u>Beechcraft A-90 King Air, Registration N1127D</u>
Pilot proficiency training flight
2 aboard, 2 killed

<u>Synopsis</u>: A Beechcraft 1900 Beechliner landed in clear weather at night. At the same time a Beechcraft A-90 King Air was taking off on an intersecting runway. The two aircraft collided at the runway intersection, and the post-crash fire killed all 14 people aboard both planes.

The Beechcraft 1900 Beechliner Flight: On a clear evening, United Express Flight 5925 took off from O'Hare Airport at Chicago, Illinois. The Beechcraft 1900 Beechliner (Beech 1900) regional airliner made a quick stop at Burlington, Iowa. After an exchange of passengers and baggage the plane flew toward Quincy, Illinois, with 2 pilots and 10 passengers aboard:

The Beech 1900 pressurized twin turboprop is based upon the highly successful Beechcraft King Air. An air-stair cabin door and a large T-tail give the aircraft a "big airplane" image. Two Pratt & Whitney PT6A engines each pump out 1279 horsepower to power the four-blade Hartzell composite propellers. The plane cruises at about 260 knots. It has become the most popular 19 passenger regional airliner in the world.

The captain, age 30, held an ATP certificate and had logged about 4,000 hours of flight time, 700 of which were as pilot-in-command in the Beech 1900. She had begun flying for United Express three years earlier and had upgraded to captain nine months before this final flight. Her young first officer, age 24, had earned a Commercial Pilot certificate and had been flying with the airline for 14 months. He had logged 1,950 hours in the air.

Beechcraft 1900 Beechliner flown by the United Nations (photo courtesy of U.S. DOD Visual Information Center)

The Beechcraft A-90 King Air Flight: Meanwhile, at the airport at Quincy two Beechcraft A-90 King Air pilots prepared to take off on Runway 4. The pilot-in-command, age 63, was a retired U.S. Air Force and retired Trans World Airlines (TWA) pilot. He had logged 25,647 flight hours. However, six months before this flight he had been involved in a gear-up landing mishap. In addition:

> TWA records [show that] the pilot was transferred from the status of captain to that of flight engineer . . . because of flying deficiencies . . . which resulted in a failed proficiency check and a failed special line check.

The young copilot, age 34, worked as a ground instructor for Flight Safety International, and she held a Commercial Pilot certificate. As the King Air taxied into position on Runway 4 she radioed in the blind on the common traffic advisory frequency (CTAF), stating the King Air was "holding short" on the runway.

Trouble Ahead: There was no control tower at Quincy. As the Beech 1900 neared the airport in the dark, the captain transmitted in the blind on the CTAF. She reported the airliner was 30 miles north of the airport and inbound for landing on Runway 13. At that time the King Air was holding near the threshold on Runway 4. These two runways intersected near the middle of the airport.

The portion of the Beech 1900 radio and CVR transcript, below, begins as the captain transmits on the CTAF. The first officer is flying the airliner. The flight operates under a code sharing agreement with United Airlines and uses the call-sign, Lakes Air 251:

Capt: . . . . . . Captain, United Express Flight 5925 (Beech 1900)
FO: . . . . First Officer
Radio-: . . . . (prefix) Radio t'mission, United Exp. Flight 5925
King Air: . . Beechcraft A-90 King Air
Cherokee: . . Piper PA-28 Cherokee

Radio-Capt: . . . two-fifty-one is a Beech airliner, just about thirty miles to the north of the field, inbound for landing, Runway one-three at Quincy, any traffic in the area, please advise. (1652:07)

[there is no response, 3 minutes and 12 seconds pass]

[the young female King Air copilot transmits in the blind]

King Air: Quincy traffic, King Air one-one-two-seven delta's taxiing out [for] takeoff on Runway four [at] Quincy. (1655:19)

FO: She sounds like a little kid. (1655:26)

Capt: She's a little baby, an Elmer Fudd girl. (1655:29)

[a small Piper PA-28 Cherokee is taxiing far behind the King Air, and the Cherokee pilot transmits on the CTAF]

Cherokee: Quincy traffic, Cherokee seven-six-four-six juliet, back taxi – uuuhhh – taxiing to Runway four, Quincy. (1655:40)

Capt: They're both using Runway four. (1655:48)

FO: [The airport is] in sight. (1656:13)

Capt: [We'll be landing on Runway] one-three, right? You're planning on [Runway] one-three still, right? (1656:42)

FO: Yeah, unless it doesn't look good, then we'll just do a downwind for [Runway] four, but right now plan on one-three. (1656:46)

[the Beech 1900 Captain again transmits on the CTAF]

Radio-Capt: Quincy traffic, Lakes Air two-fifty-one is a Beech airliner currently ten miles to the north of the field, we'll be inbound to enter a left base for Runway one-three at Quincy, any other traffic, please advise. (1656:56)

[there is no response; 2 minutes and 7 seconds pass]

King Air: Quincy traffic, King Air one-one-two-seven delta, holding short of Runway four, [we'll] be takin' the runway for departure and [then] heading southeast [from] Quincy. (1659:03)

Capt: . . . she's takin' Runway four right now? (1659:14)

FO: Yeah. (1659:22)

Radio-Capt: Quincy area traffic, Lakes Air two-fifty-one is a Beech airliner, currently just about to turn about a six mile final for Runway one-three . . . at Quincy. (1659:29)

Capt: Landing gear. (1659:52)

FO: Down, three green, flaps approach. (1659:52)

Capt: Full [flaps] indicated. (1659:59)

Radio-Capt: Quincy traffic, Lakes Air two-fifty-one's on short final for Runway one-three. The aircraft . . . [are you] gonna hold on Runway four, or [are] you guys gonna take off? (1700:16)

> *[instead of a response from the King Air holding on Runway 4, a response comes from the <u>Cherokee</u>, which is <u>not on the runway</u> and <u>not ready to take off</u>]*

Cherokee: Seven-six-four-six juliet – uuuhhh – holding – uuuhhh – for departure on Runway four . . . . (1700:26)

> *[the Beech 1900 Captain <u>assumes</u> she is talking to the King Air that is holding on Runway 4]*

Radio-Capt: OK, we'll get through your intersection in just a second, sir, we appreciate that. (1700:37)

Capt: Landing gear's down, three green, flaps are at landing, your yaw damp is off, finals are complete. (1700:42)

> *[the King Air pilots do not see the Beech 1900 approaching; without radioing their intentions <u>they begin their takeoff roll</u>]*

> *[the Beech 1900 touches down on Runway 13 at 1700:59; the Captain sees the approaching King Air about 2 seconds later]*

Capt: Max reverse! Oh, [expletive]! (1701:01)

FO: What? – Oh, [expletive]! (1701:03)

Capt: Oh, [expletive] me! (1701:07)

*[the two aircraft **collide** at the intersection of Runway 4 and Runway 13 at 1701:08]*

The Collision: Both planes were going far too fast to stop in time. They slammed into each other at the runway intersection. Both aircraft skidded to the edge of Runway 13. Both fuselages remained upright on their landing gear, but fuel from ruptured tanks leaked down onto the ground and ignited.

Three pilots at the airport terminal saw the collision and ran out to the accident site. When they arrived the King Air was totally ablaze, but only a small fire burned on the right side of the Beech 1900. The potential rescuers ran to the front of the regional airliner. The captain had her head and one arm out of the cockpit window. She screamed to them in desperation: "Get the door open!"

Witnesses who ran to the scene . . . stated that they heard sounds of life from within the cabin of the Beech 1900 and that the captain talked to them from the cockpit.

The rescuers ran to the forward air-stair door, the main entrance and exit on the Beech 1900. The exterior door handle was already in the unlocked position. People inside were frantically pounding on the door. The rescuers could hear them shouting and screaming as they pushed and kicked the jammed door, trying to open it. Standing on the ground outside, a rescuer twisted the handle with all his strength. The door refused to budge. Another rescuer gave it a try, but the door still would not open.

Slowly the fire grew more intense. The Beech 1900 cabin filled with jet-black smoke, then orange flames. Gradually the pitiful pleas and screams of those trapped inside the burning airliner fell silent:

The Quincy Fire Department was 10 miles away, and it took 14 minutes for firefighting units to arrive.

The fire department dispatched two fire trucks and a firefighting contingent. By the time they arrived the conflagration had consumed

Photograph depicting the burned rubble of the Beechcraft A-90 King Air the morning after the fire (photo courtesy of U.S. NTSB)

both aircraft. Firefighters used foam to smother the flames, but they were too late to save the pilots and passengers. Everyone inside both aircraft was already dead.

The Adams County Coroner performed an autopsy on each of the 14 victims. He found that the collision had not seriously injured any of them. They all had died in the post-crash fire:

> The cause of death for 10 occupants was carbon monoxide intoxication from inhalation of smoke and soot. The cause of death for [the remaining four] occupants was inhalation of products of combustion.

The Investigation: Investigators examined physical evidence at the collision site. Pilots on both planes desperately had tried to stop. On Runway 13 the tires on the Beech 1900 had left skid marks for 475 feet leading up to the point of collision. On Runway 4 the King Air had left skid marks for 260 feet.

The Beech 1900 CVR had documented a standard pre-recorded passenger briefing prior to takeoff from Burlington:

> There are three clearly marked over-wing exits, two over the right wing and one over the left wing. To open, pull down on the handle and pull the exit inward. The main cabin door through which you entered is also an emergency exit. To open, push the

button next to the handle and then rotate the handle counter-clockwise and push the door open.

According to the coroner the Beech 1900 occupants were "not incapacitated" by the collision. Yet, they had been unable to open the air-stair cabin door. Also, the potential rescuers outside the plane had been unable to pull the door open. Investigators determined the door frame had been slightly deformed by the collision, jamming the door shut. There was no indication that occupants had tried to get out via any of the other three exits.

The NTSB conducted visibility studies to determine line-of-sight conspicuity of the Beech 1900 from the King Air cockpit. There were no physical limitations to visibility. The study concluded the pilots in the King Air should have been able to see the Beech airliner. Yet, investigators reasoned:

> The occupants of the King Air must have been unaware . . . that [the Beech 1900] airplane was about to land.

The Beech 1900 captain had made proper radio calls during the approach. She and her first officer mistook the Cherokee pilot's *erroneous message* (that he was holding for departure on Runway 4) as a reply from the King Air. Therefore, the Beech 1900 pilots had logically assumed the King Air was going to hold on Runway 4.

The Cherokee pilot had obtained a Private Pilot certificate nine months before the accident. He had logged 80 hours of flight time. The board stressed that his aircraft had *not* been on the runway and had *not* been in a position to take off. The board pointed out that his reply to the Beech 1900 pilots was "unnecessary and inappropriate" and "reflected his inexperience."

Probable Cause: Seven months after the accident the NTSB adopted an Aircraft Accident Report, NTSB-AAR-97-04, on 1 July 1997. The board found that failure of the King Air pilots to announce their intention to take off created the potential for the accident:

> It is clear that neither occupant of the King Air properly scanned for traffic.

The board concluded that the King Air pilots were "inattentive" and "distracted" from their duty to see and avoid other aircraft. The board concluded that they never heard radio transmissions from the Beech 1900 airliner.

Board members reasoned that, because of the Cherokee pilot's lack of experience, he had not realized a collision was imminent and had not broadcast a warning. Nonetheless, the board found that the *Probable Cause* of this accident was:

> Failure of the pilots in the King Air A-90 to effectively monitor the common traffic advisory frequency [and] to properly scan for traffic, resulting in their commencing a takeoff roll when the Beech 1900 . . . was landing on an intersecting runway.

The board found two *Contributing Factors*:

> [First], the Cherokee pilot's [erroneous] radio transmission.

> [Second], contributing to the severity of the accident and the loss of life were the lack of adequate aircraft rescue and firefighting services, and the failure of the air-stair door on the Beech 1900 to open.

Beechcraft C-12 Huron, an advanced military variant of the civil Beechcraft A-90 King Air (photo courtesy of U.S. DOD Visual Information Center)

No Traffic Reported
Mid-Air Collision over the South Atlantic Ocean
13 September 1997

<u>Lockheed C-141 Starlifter, Serial No. 65-9405</u>
U.S. Air Force humanitarian aid flight
9 aboard, 9 killed

<u>Tupolev TU-154, Serial No. 1102</u>
Luftwaffe international public relations flight
24 aboard, 24 killed

<u>*Synopsis*</u>: *A Lockheed C-141 Starlifter and a Tupolev TU-154 cruised over the South Atlantic Ocean. While transiting airspace controlled by various African ATC agencies, the two aircraft collided head-on at flight level 350. The collision killed all 33 people aboard the two military jet transports.*

The Lockheed C-141 Starlifter Flight: A huge Lockheed C-141 Starlifter took off from Windhoek, the capital city of Namibia. The U.S. Air Force transport (call-sign, Reach 4201) and its crew were based at McGuire AFB in New Jersey. On this humanitarian mission they had flown to war-torn Namibia to deliver a U.S. Army demining team and mine detection equipment. On the return flight to the United States the pilots flew northwest over the South Atlantic Ocean toward a planned refueling stop on Ascension Island:

The C-141 was a workhorse in the Air Mobility Command of the U.S. Air Force. Four Pratt & Whitney turbofans and a large fuel capacity gave the transport intercontinental range. With a maximum takeoff weight of 323,100 pounds it could deliver troops or cargo to virtually any spot on Earth.

The aircraft commander, an Air Force captain, was an experienced instructor pilot who had logged over 2,400 flight hours. His crew included two copilots, three flight engineers, two loadmasters, and a crew chief.

The Tupolev TU-154 Flight: As the American military plane flew northwest over the ocean, a Tupolev TU-154 (call-sign, GAF 074)

Tupolev TU-154 (photo courtesy of U.S. DOD Visual Information Center)

was flying southeast. The Luftwaffe (German Air Force) transport had taken off from Cologne, Germany, and had refueled at Diori Hamani Airport at Niamey, Niger, which one sage had called:

The place at the beginning of the end of the world.

Now the TU-154 winged its way toward another refueling stop at Windhoek. The final destination was to be Capetown, South Africa. The mission called for transporting 12 German Marines, and two of their spouses, to Capetown for a regatta celebrating the 75th anniversary of the South African Navy. The plane carried a crew of 10 in addition to its 14 passengers:

The Russian-built TU-154 looks similar to the American-built Boeing 727. Three Soloviev turbofans power the jetliner. Two are mounted on the rear fuselage, and a third is embedded in the base of the vertical stabilizer. The plane had begun commercial service with Aeroflot in 1972. It can seat up to 180 passengers and has a maximum takeoff weight of 220,460 pounds. Military variants are in use throughout Europe and Asia.

Trouble Ahead: The C-141 droned northwest over the ocean at flight level 350 (35,000 feet). The pilots remained in radio contact with ATC Agency-Windhoek in Namibia.

Meanwhile, the TU-154 flew southeast over the ocean toward Windhoek. The German plane cruised at flight level 350, the same altitude flown by the American plane. The TU-154 pilots had been in radio contact with ATC Agency-Luanda, located in Angola. They tried repeatedly to make radio contact with ATC Agency-Windhoek, but they never got a response.

The portion of the ATC transcript, below, includes explanatory text and post-accident activity:

C-141: .... Lockheed C-141 Starlifter (call-sign, Reach 4201)
TU-154: ... Tupolev TU-154 (call-sign, GAF 074)
Luanda: .... ATC Agency-Luanda
Windhoek: . ATC Agency-Windhoek

[the TU-154 takes off from Niamey, Niger, at 1035]

[the C-141 takes off from Windhoek, Namibia, at 1411]

Windhoek: *[transmitted to the C-141]* At position ILDIR, contact Luanda on frequency eight-eight-eight-eight.

Luanda: *[transmitted to the TU-154]* ... no traffic reported.

[the TU-154 pilots make numerous radio attempts to contact ATC Facility-Windhoek, but there is no response]

C-141: Reach four-two-zero-one checking-in at [flight] level three-five-zero.

TU-154: [Verbal exclamation upon seeing the C-141 the instant before impact].

*[the two aircraft **collide** at 1510]*

[between 1600 and 1700 the U.S. Tanker Airlift Control Center (TACC) repeatedly tries to call Windhoek to confirm departure of the C-141, but no one ever answers the telephone]

[between 1700 on 13 September and 0820 on 14 September, the TACC makes apx. 50 calls to numerous ATC agencies in an effort to locate the C-141]

[at 1000 on 14 September, TACC finally gets confirmation that the C-141 departed from Windhoek the previous day at 1411]

[at 1100 on 14 September, TACC declares the C-141 missing]

[at 1110 on 14 September the Air Force Operations Center begins SAR operations which include French, German, South African, and Namibian military forces]

[on 26 September the Air Force terminates rescue efforts and declares the missing crewmen to be deceased]

[on 27 September various military agencies begin Search and Recovery operations]

[Salvage operations on the ocean floor begin on 4 December and continue through 21 December]

The Collision: The C-141 had been flying northwest on a heading of 315 degrees. The TU-154 had been flying southeast on a heading of 137 degrees. The two aircraft had been racing head-on toward each other in clear daylight at a closure rate of almost 1,000 knots. They collided at flight level 350. The impact destroyed the Luftwaffe TU-154. Evidence later would indicate the Air Force C-141 fuselage had remained intact for about 13 seconds. Then it disintegrated:

A U.S. reconnaissance satellite reported a *bright flash [emphasis added]* at position 18.8 South, 11.3 East, at 1510 GMT. *[and also]* The C-141 and the TU-154 collided in mid-air . . . approximately 65 nautical miles west of the African coast near Namibia.

A French Air Force pilot had heard a desperate "mayday" distress call. Total silence had followed. Remnants of the military transports had rained down into the sea, and no one survived.

A Namibian patrol boat found debris, and partial human remains from one Luftwaffe flight attendant, floating on the water. Three months later more human remains, the flight recorders, and pieces of the two planes were lifted to the surface and recovered.

The Investigation: The U.S. Air Force and the German Luftwaffe conducted parallel investigations. The TU-154 pilots had activated their flight plan when they took off after refueling at Niamey. It called for an initial cruise at flight level 350, followed by a subsequent climb to flight level 390:

Both aircraft were on the flight route for which they had filed and to which they were assigned.

As the TU-154 neared Windhoek the heading had changed to southeast. According to ICAO standards an easterly heading, off airways, requires flight level 290, 330, 370, etc. Yet, the pilots had maintained flight level 350. They repeatedly tried to radio ATC Agency-Windhoek but never got a response. No ATC agency requested an altitude change. A technical report would relate:

> The South African Airline Pilots Association had labeled the Angolan airspace as "critically deficient." . . . Angola, in particular, has been unable to upgrade and maintain its aviation infrastructure to keep pace with increasing volumes of air traffic flying across Africa.

Investigation revealed the AFTN had been inoperative. Niamey had sent a departure message after the TU-154 took off, but it was not received by any other ATC agency. During the flight the German pilots had contacted ATC Agency-Luanda, which was (1) confused about the TU-154 route and (2) had no information about the C-141:

> During the final transmission between ATC Agency-Luanda and GAF 074, Luanda advised: ". . . no traffic reported."

The recovered C-141 CVR later showed the U.S. Air Force pilots *possibly* had heard some of the radio calls from the TU-154 as it tried to contact Windhoek. Nonetheless:

> . . . because the [radio] transmissions did not include GAF 074's location, Reach 4201 could not have known that GAF 074 was traveling toward them at the same altitude.

The TU-154 CVR revealed the German pilots *may* have heard the C-141 pilots when they checked-in with Windhoek at flight level 350. Yet, even if they heard the radio transmissions, the German pilots would have had no way to know the location of the C-141.

Probable Cause: The U.S. Air Force board finalized its Accident Investigation Report on 20 March 1998, six months after the mid-air collision. Neither aircraft had been equipped with TCAS, which could have prevented the mid-air collision. Yet, many planes flying

Lockheed C-141 Starlifter (official U.S. Air Force photo)

the world's airways do not have TCAS. It was not required, so its absence was not considered a factor in the accident.

ATC services had been deplorable. Luanda had not relayed the departure of the TU-154 to Windhoek, as required. The TU-154 had been unable to get a radio response from Windhoek. No ATC agency warned either plane that they were converging at the same altitude. One independent report noted the 17 fatal crashes in Africa during the previous year, plus an astounding "77 near-mid-air collisions":

> The regional ground communications links between the different African ATC agencies are unreliable. The forwarding of flight plans through AFTN . . . is complicated and sporadic.

The planes, flying in opposite directions, had been racing toward each other at the *same altitude* in violation of ICAO standards:

> When flying off airways the appropriate altitude is determined by the magnetic heading. . . . When the [TU-154] began flying in an easterly direction (0 degrees to 179 degrees) an altitude change was required in accordance with ICAO procedures. The correct altitude, according to ICAO procedures, was either 29,000 feet (FL 290), 33,000 feet (FL 330), 37,000 feet (FL 370), 41,000 feet (FL 410), etc.

The C-141 had been flying at a proper altitude for westbound traffic, flight level 350. The TU-154 had *not* been at a proper altitude for eastbound traffic. Had the TU-154 been at a proper altitude, the two aircraft would have been vertically separated by a minimum of 2,000 feet – they could not have collided. Therefore, the U.S. Air Force board found the following **Primary Cause** of the accident:

The Tupolev [TU]-154 [was flying] at the wrong cruise altitude of 35,000 feet *[flight level 350]*. This altitude was contrary to ICAO rules and was also contrary to the flight planned altitude. . . . Had the crew flown either the ICAO mandated altitude or the flight planned altitude, the accident would not have occurred.

The board found two **Contributing Factors**:

[First], ATC Agency-Luanda's poor management of air traffic through its airspace . . . . Luanda flight following procedures were weak, and they did not comply with procedures prescribed by ICAO directives . . . .

[Second], the complicated and sporadic operation of the AFTN. Routing of messages to affected air traffic control agencies [in parts of Africa] is not direct and is convoluted, creating unnecessary delays and unfortunate misrouting.

Woooppps!
Mid-Air Collision near Cavalese, Trento Province, Italy
3 February 1998

<u>Grumman EA-6 Prowler, Bureau No. 163045</u>
U.S. Marine Corps low-level navigation training mission
4 aboard, 0 killed

<u>Mount Cermis ski-lift cable car system</u>
Winter resort ski-lift cable car system
20 aboard, 20 killed

<u>Synopsis</u>: A military Grumman EA-6 Prowler sped along a low-level training route through the mountains. The plane struck and severed steel cables supporting a ski-lift cable car system. The passenger gondola plummeted 370 feet to the ground, killing all 20 occupants. The damaged Prowler flew back to its base and landed safely.

The Grumman EA-6 Prowler Flight: Scores of NATO military aircraft staged at Aviano Air Base in Italy to fly Operation Deliberate Guard missions over war-torn Bosnia. In August 1997 a U.S. Marine Corps squadron, Marine Tactical Electronic Warfare Squadron 2 (VMAQ-2), went to Aviano to join the NATO forces. Six months later the squadron scheduled a low-level training mission for one of its Grumman EA-6 Prowler aircraft. This would be the crew's first low-level training flight since arrival in Italy the previous year:

The Prowler is a new variant of the Grumman A-6 Intruder all-weather attack plane which first flew in the 1960s. The Prowler features an integrated warfare system with advanced electronic countermeasures, including radar and radio jamming. It is a command and control platform for strike missions.

The Prowler is crewed by a pilot plus three electronic counter-measures officers (ECMOs). The pilot and one ECMO, who doubles as the navigator, sit in the front cockpit. The other two ECMOs sit in a cockpit behind them. Two non-afterburning Pratt & Whitney J-52 jet engines power the warplane, which has a maximum takeoff weight of 61,500 pounds.

Grumman EA-6 Prowler from VMAQ-2 (photo courtesy of U.S. DOD Visual Information Center)

The Prowler crew (call-sign, Easy 01) had logged a total of 2,834 hours of flight time. After takeoff the pilot, a Marine Corps captain, checked in with Padova ATC and descended toward the first leg of the AV-047 low-level route. The mission called for practicing "threat avoidance and terrain masking" on a winding route through the northern Italian Alps. The crew calculated that after 1.5 hours they would be back on the runway at Aviano.

Mount Cermis Ski-Lift Cable Car System: Near the village of Cavalese in Trento Province, Italy, a ski-lift runs up and down the slopes of Mount Cermis. The cable car system is anchored by gigantic concrete stanchions at the top and bottom of the lift. A huge 2.25 inch diameter load-bearing transit cable supports a gondola designed to carry 40 people. A 7/8 inch diameter cable, powered by electric motors, drives the system. A third back-up cable is available in the event the primary drive system fails.

Trouble Ahead: The Prowler swooped down onto the first leg of the low-level route at 1440 Hours. The pilot whipped his aircraft along the zig-zag route while relying on his maps for navigation. Racing down a long snow-covered valley between two high mountains, he suddenly spotted cables suspended in the air ahead of him:

His immediate reaction was to push the nose down in an attempt to survive and in hopes of avoiding the cables altogether.

The Collision: The cockpit accelerometer snapped to negative 2.3 G's, but it was too late. The crew felt their craft shudder, and they heard a loud thud. The Prowler sliced through the 2.25 inch load-bearing cable and the primary drive cable, severing both:

> The cables of the Cermis gondola system were struck by . . . Easy 01 at an altitude no higher than 113 meters *[meaning, no higher than 370 feet above the ground]*.

The passenger gondola plummeted straight down to the valley floor. Impact with the ground ripped it apart. Then the massive cable hook assembly, weighing over a ton, tumbled down atop the wreckage. All 20 passengers (seven Germans, two Poles, five Belgians, three Italians, two Austrians, and one Dutch national) perished in the rubble. *CNN World News* would later describe the accident site:

> The remains of the cable car lay on the snowy floor of a valley, a mangle of flattened yellow metal. Broken skis, a pair of red gloves, ski boots . . . lay strewn about. Blood stained the snow.

After slicing through the cables the sturdy Prowler had remained flyable. The pilot and navigator knew they had clipped some type of wires, but they had no idea a passenger gondola was involved. They climbed and squawked their transponder emergency code. The pilot radioed Padova ATC, declared an emergency, and headed for home. At 1535 Hours he made an arrested landing on Runway 23 at Aviano Air Base. His Prowler had sustained severe damage to the right wing, vertical stabilizer, and one jamming pod, but none of the four crewmen had been injured.

The Investigation: The United States asserted its right to investigatory jurisdiction under the NATO Status of Forces Agreement. Members of a Command Investigation Board arrived at Aviano two days after the accident:

> The board observed a highly professional atmosphere within the squadron while conducting interviews and examining squadron procedures.

The cables had been severed only 370 feet above the ground, so the investigation centered on reasons why the pilot had been flying so low. Also, regulations called for a maximum speed of 450 knots on the training route. Yet, the mission data recorder showed the jet had flown at speeds up to 555 knots.

The AV-047 low-level route had been designed and approved by the Italian Air Force. Maps used by Italian military pilots showed the ski-lift cables crossing the training route. However, maps used by American crews did not depict the ski-lift or its cables.

Italian regulations previously had allowed pilots to fly as low as 500 feet agl on Leg 1 and Leg 2 of the route. However, there had been many public complaints about noise from low-flying jets. In August 1997 the Italian government had raised the minimum altitude to 2,000 feet agl. VMAQ-2 had arrived at Aviano that same month. No mention of the higher altitude limit had been made during the VMAQ-2 in-brief at Aviano. Further, VMAQ-2 did not have a read-and-initial program for such unclassified Italian regulations:

> There was no evidence to suggest that anyone in VMAQ-2 was aware of the Italian [2,000 foot minimum altitude].

The VMAQ-2 commanding officer described the mishap crew as "professional, extremely talented, and possessing excellent aircrew coordination skills." Until recently they had been authorized to fly many low-level routes at altitudes down to 500 feet agl:

> Prior to 19 September 1996, VMAQ-2 and all other EA-6 squadrons were allowed to conduct low-level flight training . . . as low as 500 feet agl on selected training flights.

After September 1996, Marine Corps flight procedures required a higher altitude, 1,000 feet agl. Accordingly, before the accident the Easy 01 pilot had briefed his crew for a minimum altitude of 1,000 feet above ground. Also the low-level procedures kneeboard card carried in the cockpit by the pilot and navigator stated in part:

> Minimum altitude over snow-covered mountainous terrain is 1,000 feet agl.

The squadron commanding officer had approved the flight, and the four crewmen separately told investigators they had *intended* to fly at 1,000 feet. The pilot said he set his radar altimeter warning pointer at 800 feet, a standard procedure. This would give the crew an aural warning tone if the jet descended to that height; yet:

All four mishap aircrew [members] stated that they did not hear the low altitude warning tone prior to impact [with the cables]. *[and also]* A fully functioning radar altimeter system is required for low-level navigation flight.

The only cockpit instrument that displays height above ground is the radar altimeter. The pilot and navigator told investigators the altitude needle had jammed at 2,000 feet early in the flight.

VMAQ-2 maintenance records showed the radar altimeter had become inoperative – and supposedly was repaired – 13 days before the accident. During the investigation, under supervision of investigators, technicians removed it from the aircraft and tested it. On the very first test the altitude needle jammed.

Probable Cause: A Command Investigation Board submitted Report No. 5800 dated 10 March 1998. In an endorsement of the findings the commander of Marine Corps forces in the Atlantic Theater, based in Norfolk, Virginia, stated:

The cause of this tragedy was that the Marine aircrew flew much lower than they were authorized to fly, putting themselves and others at risk.

In part, the board had reached the following ***Conclusion***:

The aircrew aggressively maneuvered their aircraft, exceeded the maximum airspeed, and flew well below 1,000 feet agl on the second and sixth legs.

Obviously the crew had flown too low. Yet, whether or not they *intended* to fly below 1,000 feet agl was subject to debate. When zipping along low over the ground most of a pilot's terrain clearance awareness must come from visual cues outside the cockpit. Flying over snow-covered terrain vastly complicates this task.

Many experienced pilots have inadvertently collided with terrain covered by snow. No non-flying layman can fully appreciate the optical illusion involved. A blanket of snow only 100 feet below an aircraft looks exactly like a blanket of snow 1,000 feet below. The *absence of visual contrast* robs a pilot of depth perception. A tragic example of this cruel "white-out" phenomenon involved a Douglas DC-10 jumbo-jet airliner, Air New Zealand Flight 901:

> *On an aerial sightseeing flight over Antarctica, the airline pilots flew straight and level 1,500 feet above the snow below. They cruised under a white overcast. The white clouds above and the white snow below visually blended together. The pilots looked into a featureless, dimensionless, monocolored white void – no horizon, no contrast, no shadows, no depth perception. Bright white glare seemed to come from all directions.*
>
> *In this white-out condition the pilots flew toward Mt. Erebus. Although forward visibility was estimated at 40 miles, the pilots could not see the snow-covered mountain ahead. It blended into the sea of white. It was invisible. The DC-10 FDR later revealed the pilots flew straight into the side of the mountain. The head-on impact killed all 257 people on board. The Aircraft Accident Report later would explain:*
>
>> *"The pilots probably assumed that they would be able to see any and all obstacles clearly with a 2,000 foot cloud base and 40 miles' visibility below the clouds. It is not likely that the white-out hazard was appreciated by the crew."*

In the Cavalese disaster the pilot had faced a similar situation. He had been flying over the floor of a snow-covered valley, with snow-covered mountains towering high overhead on both sides – a sea of white. With no visual contrast, he lost depth perception. He had relied on his radar altimeter to determine his height above the valley floor. Unfortunately the instrument malfunctioned.

Postscript: Despite the circumstances, the pilot and navigator faced 20 counts of negligent homicide and 20 counts of involuntary manslaughter. During a court martial at Camp Lejeune, North Carolina, the military prosecutor argued that the pilot had recklessly ignored altitude restrictions. Conversely, the defense attorney pointed out (1) the white-out phenomenon, (2) the defective radar altimeter, (3) absence of knowledge of the Italian minimum altitude (4) absence of the ski-lift cables on maps used by the crew, and (5) lack of knowl-edge of the 450 knot speed restriction. One international news media headline explained:

Witnesses Back Pilot in Cable Car Deaths

In March 1999 a military jury found the pilot "not guilty" on all counts. Charges against the navigator were dropped. However, both men later would face obstruction-of-justice charges in connection with the disappearance of a video-tape record of the flight.

Low-level navigation proficiency is essential for pilots of military fighter and attack aircraft. During the previous eight years there had been over 30 instances wherein military jets, including Italian military jets, had hit and severed power lines in the mountains during low-level training. With regard to the Cavalese tragedy, one aviation professional pointed out the total absence of wilful negligence and noted: "This was merely a terrible and tragic accident."

Author's Note: For more information about the accident in Antarctica, referenced above, refer to one of the author's previous books, MAYDAY: Accident Reports and Voice Transcripts from Airline Crash Investigations, and the chapter, "Air New Zealand Flight 901."

Chinese Air Force (PLAAF) Intercept
Mid-Air Collision over the ocean near Hainan Island, China
1 April 2001

Lockheed EP-3E Aries II, Bureau No. 156511
U.S. Navy signals intelligence mission
24 aboard, 0 killed

Jian F-8 II Apollo (Finback-B), Serial No. unknown
Chinese Air Force (PLAAF) intercept mission
1 aboard, 1 killed

Synopsis: A military Lockheed EP-3E Aries II cruised over the ocean 65 miles off the coast of China. A supersonic Jian F-8 II Apollo (Finback-B) interceptor closed on the Aries and collided with it. The Finback broke apart and tumbled into the sea, killing the pilot. The damaged Aries made an emergency landing in China. None of the 24 crewmembers on board were injured.

The Lockheed EP-3E Aries II Flight: Early on April Fool's Day (perhaps an omen of the misfortune to come) a Lockheed EP-3E Aries II took off from Kadena AFB on Okinawa, Rukuyu Islands, before daylight at 0500 Hours. The U.S. Navy aircraft, with ultra-secret intelligence gathering capability, flew toward air and naval military installations on Hainan Island, China. The Aries crew paralleled the Chinese coast and cruised 60 to 70 miles offshore over international waters of the South China Sea:

The Lockheed EP-3E Aires II is an aerial electronic reconnaissance platform flown by the U.S. Navy. It has been called a "giant electronic vacuum cleaner in the sky." From a radome under the fuselage and from scores of pods on the sides, antennas detect transmissions from radar, radios, cell phones, e-mail, and even fax machines. Everything transmitted into the air within 250 miles is sucked into the Aries' powerful computers and recording devices. This "signals intelligence" (SIGINT) equipment is one of the United States' most closely guarded military secrets.

The Aries airframe is based upon the Lockheed P-3 Orion, which in turn is based upon the old Lockheed L-188 Electra turboprop

Lockheed EP-3E Aries II (photo courtesy of U.S. DOD Visual Information Center)

airliner. Four Allison T-56 turboprop engines power the plane, which has a maximum takeoff weight of 142,000 pounds. The "Star Wars" interior features rows of radar consoles, huge racks of computers, myriad recording devices, and 19 work stations for selecting and interpreting electronic data.

On this mission the Aries carried 14 American servicemen in a combat reconnaissance crew, plus a 10 member naval security group. The mission commander and pilot-in-command was a U.S. Navy lieutenant. He and his aircraft were assigned to Fleet Air Reconnaissance Squadron One (VQ-1), known as the "World Watchers." The squadron was home-ported at NAS Whidbey Island, Washington, but the Aries and its crewmembers were stationed at Kadena AFB.

The Jian F-8 II Apollo (Finback-B) Flight: Chinese military radar on Hainan Island painted the Aries as it cruised 65 miles offshore. The Peoples Liberation Army Air Force (PLAAF) – usually called the Chinese Air Force – scrambled warplanes to intercept and shadow the potential intruder. Two supersonic Jian F-8 II Apollo interceptors roared skyward from their PLAAF airbase on Hainan Island. Within minutes they located the Aries and stabilized a mile off its left wing in a loose trail position:

The Jian F-8 II Apollo (NATO code-name, Finback-B) interceptor is a recent variant of the first warplane indigenously designed and built in China. It is based upon the Russian MIG-21, and it first flew in June 1984. The Finback is built at the Shenyang Aircraft Plant, which has been incorporating upgrades since production began. Yet, military analysts call the latest Finback variant an "advanced obsolete aircraft" comparable – at best – to the 1960s era American-made McDonnell F-4 Phantom.

The Finback has Mach 2.2 speed thanks to twin Wopen-13A turbojet engines. Recent variants have all-weather avionics and Russian-made pulse Doppler fire control radar. The interceptor is equipped with a 23mm cannon. Seven hard-points under the wings and fuselage can be used for fuel drop tanks, bombs, rockets, or air-to-air missiles. Although China had hoped to export the aircraft, at present the only two operators are the PLAAF and the Peoples Liberation Army Navy (PLAN).

The Finback section leader broke away from his wingman and swooped toward the Aries. He stabilized 10 feet off its left wing and hand-saluted the Aries crew. Then he rolled left and arced away to rejoin his wingman. However, within a minute he roared in again. He and Aries crewmembers could clearly see each other's faces. The Finback pilot, oxygen mask removed, gestured to the Aries crew with hand signals. Then, rolling left, he sped away again.

Trouble Ahead: The Aries crew cruised ahead, flying straight and level. Intercepts by Chinese warplanes were nothing new. Aviator to aviator, American pilots respected the professionalism normally evidenced by their PLAAF contemporaries.

Within minutes the Finback section leader returned for the third time. He zoomed toward the American plane – too fast!

On the F-8's third and final pass the F-8 pilot overshot [due to an excessive closure rate] his attempt to join up with [the Aries]. This placed the F-8 below and forward of [the Aries] port wing.

The Finback pilot momentarily stabilized *under* the front of the Aries left wing. He could see the forward fuselage of the Aries, but he could not see its huge wing above and behind him.

Jian F-8 II Apollo (photo courtesy of U.S. DOD Visual Information Center)

The Collision: Apparently thinking he would clear the Aries wing, the Finback pilot pulled up. In so doing his aft fuselage, in front of the vertical stabilizer, slammed into the Aries No. 1 propeller. The huge swirling propeller sawed into the ill-fated Finback and slung debris in all directions.

The front of the Finback snapped to the right and ripped the radome off the Aries nose. The slipstream caught the radome and whipped it rearward down the right side of the fuselage and into the No. 3 propeller. An independent technical report would explain:

> With two of its four propellers out of commission the [Aries] began rolling over . . . the plane went into a 130 degree roll . . . another few seconds and it would have been uncontrollable.

The Aries pilot wrestled his craft back to level flight. Air rushing against the flattened nose violently buffeted the plane. The No. 1 and No. 3 propellers were damaged, out of balance, and vibrating wildly. With the pitot tubes torn away the pilot had no indication of his airspeed. Wind shrieking through the ruptured forward pressure bulk-head made conversation almost impossible. The pilot radioed a distress message several times but never heard a response.

For the moment the Navy lieutenant had his wounded Aries under control. He dared not attempt to fly all the way back to his base on Okinawa. He thought about trying to limp to the Phillippines. Yet, he could not feather his out-of-control No. 1 propeller, and he knew the vibration might tear the engine off the wing. Consequently, he headed for the nearest airport, Lingshui PLAAF airbase on Hainan Island. As he neared the Chinese military base his crew frantically began destroying secret equipment and code-books:

> Destruction of classified material in flight included jettisoning [it] out of the starboard over-wing hatch . . . smashing equipment with the onboard axe . . . and hand-shredding classified papers.

With only two good engines, no flaps, and no airspeed indication the pilot managed to land safely. None of the 24 crewmembers, including three women, had been injured.

The Finback pilot had not been as lucky as the Aries crew. An independent technical report noted his fate:

> Sliced by the prop, the Chinese jet – minus its tail – broke into two [pieces] and fell away.

The Finback interceptor was last seen in two pieces, both trailing smoke and flames as they fell toward the sea. The body of the lone Chinese pilot was never found.

China detained the American crewmembers. With regard to the Aries and its secret equipment and code-books, one Chinese official claimed the American crew had flown the plane to China as a *gift* to the Chinese people. A Chinese diplomat proclaimed:

> It fell [into our country] from the heavens, so it is ours.

After 11 days of intense international debate, China released the American crew on 12 April 2001. However, the Aries and its super-secret equipment would remain in the hands of Chinese authorities for almost three more months.

The Investigation: The Chinese Foreign Ministry initially claimed the Aries had veered left and had rammed the Finback. In fact the Aries *had* rolled left, but only *after* the collision. That left roll, 40 degrees past vertical, caused the aircraft to lose 8,500 feet of altitude before the pilot could recover.

Before the collision the Aries had been flying straight and level on a heading of 070 degrees with the autopilot engaged. Neither the Finback pilot nor the Aries crew had reason for concern, for Chinese warplane intercepts were routine on these surveillance flights. Both

Chinese and American crewmembers had photographed each other while Finbacks were flying in close formation with Aries aircraft. After previous flights the now-deceased Chinese pilot had posted his close-up photographs of Aries planes onto the internet. However, a United States diplomat in Beijing would report:

> Over the past year or so before the collision there had been a pattern of *increasingly aggressive [emphasis added]* intercepts of U.S. surveillance aircraft by Chinese fighters.

A U.S. Judge Advocate General (JAG) investigation determined the accident resulted from a judgement error by the Finback pilot. JAG investigators also concluded:

> [First], the second F-8 was not a factor in the collision.

> [Second, the Aries crew] did not commit any dangerous or hazardous maneuvers.

> [Third, the Aries crew] did not provoke the lead F-8 pilot.

> [Fourth, the Aries crew] did not cause the collision between [the Aries] and the F-8.

> [Fifth, the Aries crew] was not negligent, not responsible, and not at fault for the collision between the F-8 and [the Aries].

Probable Cause: The JAG report spelled out the mechanics of the accident. On his third rendezvous with the Aries the Chinese pilot had an excessive closure rate. He was going too fast to stabilize off the left wing of the American aircraft, and he ended up:

> . . . below and slightly forward of [the Aries] port wing. The F-8 pilot slightly raised the nose of his aircraft in what appeared to be an effort to slow his aircraft. . . . The F-8 pitched up into [the Aries] number-one propeller, striking the propeller at the point where the F-8's vertical stabilizer and fuselage meet. The F-8 was immediately ripped in half . . . .

Postscript: Chinese and American diplomats worked out a plan for return of the Aries to the United States. A Lockheed recovery team flew to the Chinese military base on 15 June. They removed all the

The disassembled Aries fuselage at Lingshui airfield (photo courtesy of U.S. DOD Visual Information Center)

Aries' engines, wings, landing gear, vertical and horizontal stabilizers, and antennas. They loaded the disassembled Aries into a leased Russian-built Antonov An-124 Condor transport. On 3 July, three months after the accident, the Condor took off from Lingshui airbase before daylight. After refueling stops in the Phillippines and Hawaii the giant transport landed at Dobbins AFB, Georgia, on 5 July 2001. There the recovery team delivered what was left of the Aries to Lockheed Martin Aeronautics.

Author's Note: On 8 February 2006 in accordance with provisions of the U.S. Freedom of Information Act, the author contacted the Naval Safety Center and requested a copy of the Mishap Report dealing with this accident. By return letter dated 23 February 2006 the Department of the Navy acknowledged the request. For 16 months thereafter the author followed-up with six more letters plus numerous e-mails and telephone calls. Officially, the ultra-sensitive Mishap Report is "not in file." Consequently, this chapter is based upon (1) multiple independent technical sources privy to the investigation, (2) public domain data from the Navy Office of Information, and (3) statements from knowledgeable parties.

TCAS vs Air Traffic Control
Mid-Air Collision near Ueberlingen, Germany
1 July 2002

<u>Tupolev TU-154, Registration RA85816</u>
Bashkirian Airlines Flight 2937
69 aboard, 69 killed

<u>Boeing 757, Registration A9C-DHL</u>
DHL Worldwide Express Flight 611
2 aboard, 2 killed

<u>Synopsis</u>: At night a Tupolev TU-154 airliner and a Boeing 757 cargo plane cruised at the same altitude on a collision course. The Tupolev pilots followed ATC instructions instead of TCAS Resolution Advisories. The two airplanes collided, killing all 71 people aboard.

The Tupolev TU-154 Flight: Several hours after darkness settled over Moscow, Russia, a Tupolev TU-154 tri-jet took off on a charter trip to Barcelona, Spain. Bashkirian Airlines (BTC) Flight 2937 carried a flight deck crew of five. In the cabin four flight attendants took charge of their 60 passengers. Eight passengers were adult chaperones, but the remaining 52 were Russian schoolchildren. The *Associated Press* later would explain:

> They were the chosen children, standout athletes or students with the best grades, selected to travel and represent their Muslim religion on a trip to the Spanish coast.

The captain was 40 years of age. He held a Russian Federation ATP certificate, and his logbook reflected 8,500 hours of flight time. The first officer, making his second trip to Barcelona, had 12,070 flying hours. Their TU-154 had position lights on the wingtips and tail plus anticollision lights on the tail and fuselage.

The Boeing 757 Flight: A twin-jet Boeing 757 freighter left Bahrain International Airport in the Kingdom of Bahrain on a routine flight to Brussels, Belgium. After a quick stop at Bergamo, Italy, the transport climbed north to its cruising altitude at flight level 360:

Boeing 757 (photo courtesy of U.S. NASA)

The Boeing 757 had been designed to replace the aging Boeing 727. The new passenger and cargo transport began commercial service in 1983. It flies both domestic and international routes. Two Rolls-Royce engines, mounted under the wings, power the narrow-body airliner. Depending upon the model and seating configuration, it can carry up to 280 passengers. Production ended in 2005 after 1,050 planes had been delivered.

The captain, age 47, held an ATP certificate issued by the United States and validated by the Kingdom of Bahrain. He had logged 11,942 flying hours. His first officer, similarly qualified, had 6,604 hours of flight experience. The certificates of airworthiness and registration for their aircraft had been issued in Bahrain. The plane had position lights on the wingtips and tail, plus strobe lights on the wingtips and on the top and bottom of the fuselage.

Trouble Ahead: At 489 knots the Tupolev flew west at flight level 360 on a heading of 274 degrees. At 506 knots the Boeing flew north at the same altitude on a heading of 004 degrees. Both planes were equipped with autonomous TCAS collision avoidance capability. For the non-flying layman a brief overview of TCAS is necessary:

High speed and high altitude flight can involve plane-to-plane closure rates approaching 1,000 knots. Engineers spent decades working on a reliable "last line of defense" for aerial collision avoidance. Traffic Alert & Collision Avoidance Systems (TCAS) were developed. This electronic system, built into an aircraft, can detect other planes that are a potential collision threat.

If conflicting traffic exceeds TCAS warning parameters the system will first give the pilots an automated cockpit display plus a synthetic voice Traffic Advisory: "Traffic, Traffic." At this point TCAS does not offer a suggested course of action.

However, if the conflicting traffic situation becomes critical, TCAS will give pilots a synthetic voice Resolution Advisory such as "Climb! Climb!" or "Descend! Descend!" Pilots must (1) follow the TCAS advisory, (2) ignore any contradictory instructions from ATC, and (3) inform ATC of the action taken. TCAS equipment in both planes will coordinate the advisories to make sure both aircraft are not given the same command.

Both planes were overflying Germany, but they were in airspace controlled by Zurich Center in Switzerland. The Tupolev ATC and CVR transcript, below, begins with the planes on converging courses at flight level 360. Verbiage from the Boeing CVR is omitted:

Capt: . . . . Captain, Bashkirian Airlines Flight 2937 (TU-154)
FO: . . . . . First Officer
Nav: . . . . Navigator
TCAS: . . . *TCAS synthetic voice*
Radio-: . . (prefix) Radio transmission, Flight 2937
Center: . . Zurich Center (ARTCC)

Radio-Capt: Aaahhh, Zurich, good evening, BTC two-nine-three-seven [is at flight level] three-six-zero. (2230:28)

Center: BTC two-nine-three-seven, squawk – aaahhh – seven-five-two-zero. (2230:33)

[at 2233:03 the Tupolev TCAS cockpit display shows another aircraft at the same altitude approaching from the left; the pilots try for over a minute to visually locate it; finally the Captain sees the lights on the approaching Boeing 757]

Capt: Here it *[meaning, the other plane]* is, in sight – look here, [the

TCAS altitude difference] indicates zero. (2234:36)

[the Tupolev TCAS issues a synthetic voice Traffic Advisory]

TCAS: Traffic, Traffic. (2234:42)

[at the same time the Boeing TCAS issues a Traffic Advisory]

[the Zurich Center controller suddenly notices the conflict]

Center: BTC two-nine-three-seven, aaahhh, descend [to] flight level three-five-zero, expedite, I have crossing traffic. (2234:49)

[the Tupolev pilots begin a descent]

[the Tupolev TCAS issues a synthetic Resolution Advisory]

TCAS: Climb! Climb! (2234:56)

[the Tupolev pilots ignore TCAS and continue their descent]

[at the same time the Boeing TCAS issues a Resolution Advisory, "Descend! Descend!" and the Boeing pilots begin a descent]

[both planes are now descending from the same altitude at the same rate of descent]

FO: It *[meaning, TCAS]* says climb? (2234:59)

Capt: [Don't worry], he *[the controller]* is guiding us down.

Center: BTC two-nine-three-seven, descend [to flight] level three-five-zero, expedite descent! We have traffic at your two o'clock position now at [flight level] three-six-zero! (2235:03)

Radio-Capt: [We are] expediting descent [to flight] level three-five-zero, [this is] BTC two-nine-three-seven. (2235:07)

Capt: Where is it *[meaning, where is the other plane]*? (2235:10)

FO: Here, on the left side!

Nav: It's going to pass beneath us!

[the Tupolev TCAS issues an urgent Resolution Advisory]

TCAS: Increase climb! Increase climb! (2235:24)

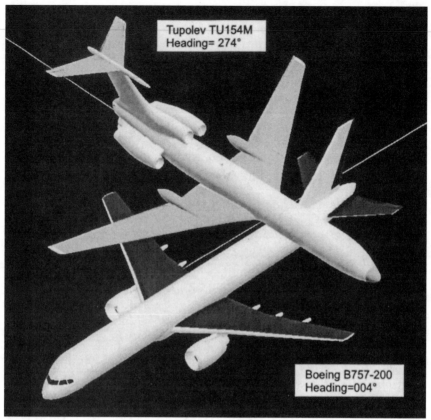

An illustration showing the position and heading of the two aircraft at the moment of aerial impact (illustration courtesy of German BFU)

FO: It *[meaning, TCAS]* says climb? (2235:26)

*[the Tupolev and the Boeing **collide** at 2235;32]*

Center: BTC two-nine-three-seven? (2236:01)

Center: BTC two-nine-three-seven? (2236:23)

[there is no response]

The Collision: At cruising speed the two aircraft rammed through each other at 34,890 feet. The Boeing vertical stabilizer sliced into the Tupolev cabin, cutting it in half:

The Boeing 757 lost most of its vertical tail in the collision and was destroyed by impact forces. The Tupolev TU-154 broke into several pieces [in flight]. . . . Forty occupants of the TU-154 fell out of the airplane [and into the night sky].

What was left of the two airliners fell to the ground. All 71 people on the two craft were killed as debris and bodies rained down near Ueberlingen, Germany. Remains of passengers and crew were later identified by using dental records and DNA protocols. German authorities conducted autopsies on the bodies of the seven flight deck crewmembers, and the pathological report stated:

> Death of all crewmembers was caused by extreme destruction of their bodies by the collision.

The Investigation: The German Federal Bureau of Aircraft Accident Investigation (BFU) dispatched teams to the accident site and to the ATC facility in Zurich, Switzerland. The Russian Federation, the Kingdom of Bahrain, Switzerland, and the United States assisted the German investigators. FDR data showed the two planes had torn through each other at a closure speed of 704 knots. There had been no aircraft systems anomalies. The investigation quickly focused on a host of abnormal conditions at Zurich Center:

1. The radar system was operating in a "fall-back mode" because of scheduled reconfiguration of the system, and the Short Term Conflict Alert (STCI) function was out-of-service.

2. The controller was not fully aware of the radar system fall-back mode limitations.

3. Neighboring ATC facilities had not been informed of the Zurich radar system limitations.

4. According to the duty schedule, two controllers had been responsible for Zurich Center airspace at the time of the accident. However, only one controller was actually on duty.

5. The lone controller on duty had to move back and forth between two radarscopes and monitor two radio frequencies.

6. Telephone lines to adjacent ATC facilities went out-of-service 12 minutes before the collision. Incoming calls were not switched to the bypass system, and they went unanswered.

7. The bypass telephone system was defective and – at the time of the collision – the controller was unsuccessfully trying to call Friedrichshafen ATC to coordinate an Airbus arrival.

8. From information on manual flight progress strips given to the controller, the conflict between the Boeing and the Tupolev could not be recognized without the radar display – and it was operating in fall-back mode.

Nonetheless, despite all the ATC problems the TCAS systems on the two planes should have allowed the pilots to avoid a collision. Both systems had functioned perfectly. What had gone wrong? An independent technical report explained it this way:

> In case of a conflict between TCAS Resolution Advisories and ATC instructions, the TCAS Resolution Advisories always take precedence. If one aircraft follows a TCAS Resolution Advisory, and the other follows ATC instructions, a collision can occur.

The Tupolev pilots had followed the controller's instructions instead of the TCAS Resolution Advisories. Their TCAS synthetic voice had commanded: *"Climb! Climb!"* and later *"Increase climb! Increase climb!"* The pilots had ignored TCAS and had relied on the controller to guide them out of danger:

> The initial Resolution Advisories would have ensured a safe vertical separation of both airplanes *if [emphasis added]* both crews had followed the TCAS instructions.

Probable Cause: In May 2004 the BFU finished its complex task and compiled an exhaustive Investigative Report, No. AX001-1-2/02. The horrible irony was that TCAS had given the pilots the ability to avoid the accident. If all pilots had obeyed their TCAS, (1) the Tupolev would have climbed, (2) the Boeing would have descended (as it did), and all would have been well.

The Boeing pilots had properly followed their TCAS advisories. Unfortunately the Tupolev pilots had ignored TCAS and had relied upon the errant ATC controller. The BFU found two *Immediate Causes* of the accident:

[First], the imminent separation infringement was not noticed by ATC in time.

[Second], the TU-154 crew followed ATC instructions to descend and continued to do so even after TCAS advised them to climb.

The BFU found three *Systemic Causes* of the accident:

[First], the integration of TCAS into the aviation system was insufficient. . . . The regulations concerning TCAS published by ICAO . . . were not standardized, [were] incomplete, and [were] partially contradictory.

[Second], management and quality assurance [at Zurich Center] did not ensure that during the night all open workstations were continuously staffed by controllers.

[Third], management and quality assurance [at Zurich Center] tolerated for years that, during times of low traffic flow at night, only one controller worked and the other one retired to rest.

Postscript: Seventy-one people had died in the accident, but 71 deaths would not be the final tally of victims. On 24 February 2004, twenty months after the accident, the lone Zurich controller who had been on duty at the time of the accident was murdered – stabbed to death in his home.

Collision at Flight Level 370
Mid-Air Collision over the Amazon Jungle, Brazil
29 September 2006

<u>Embraer EMB-135 Legacy 600, Registration N600XL</u>
New aircraft delivery flight
7 aboard, 0 killed

<u>Boeing 737, Registration PR-GTD</u>
Gol Lineas Aereas (Gol) Flight 1907
154 aboard, 154 killed

<u>Synopsis</u>: A Boeing 737 and an Embraer EMB-135 Legacy 600 flew toward each other on the same airway at the same altitude. They clipped wingtips at flight level 370. The Boeing airliner crashed, killing all 154 people on board. The Legacy made an emergency landing without injury to the seven persons aboard.

The Embraer EMB-135 Legacy 600 Flight: A brand-new Embraer EMB-135 Legacy 600 business jet took off from Runway 15 at Urbano Ernesto Stumpf Airport at San Jose dos Campos, Brazil. The new plane had recently rolled off the assembly line at the Embraer factory. Two corporate pilots were making this flight to deliver the plane to the purchaser, ExcelAire Service Inc., in New York. Five passengers rode in the cabin on this trip to the United States:

The Legacy is a business jet designed to carry 16 passengers in executive seating, or up to 37 passengers in an airline seating layout. The engines are mounted on the aft fuselage in front of the T-tail. The cockpit is equipped with a Honeywell integrated avionics system. Two computers interface with aircraft systems and manage data for five CRT displays. With fly-by-wire flight controls and a glass cockpit, the Brazilian twin-jet can cruise at 459 knots for over 3,000 miles.

The pilot-in-command, age 42, had emigrated to America when he was a child. He held an ATP certificate and had logged 9,375 hours of flight time. The copilot, age 34, also held an ATP certificate and had 6,400 hours of flight experience. They planned to fly to Manaus, Brazil, and refuel before continuing north.

Boeing T-43, a military variant of the civil Boeing 737 airliner (official U.S. Air Force photo)

The Boeing 737 Flight: Gol Lineas Aereas (Gol) Flight 1907 took off from Eduardo Gomes International Airport in Manaus, Brazil, at 1535 Hours on a scheduled flight to Presidente Juscelino Kubitschek International Airport in Brasilia. The Boeing 737 carried two pilots, four flight attendants, and 148 passengers. The airliner climbed to flight level 370 and cruised southeast on airway UZ-6.

The captain, age 44, had logged 14,900 hours of flight time. His first officer, age 29, had 3,850 hours of flight experience. Their Boeing 737 was a new short-field-performance model. The factory had delivered the new plane only 17 days before this final flight.

Trouble Ahead: After several clearances to intermediate altitudes, all of which were read back, Brasilia Center cleared the Legacy pilots to climb to flight level 370. An hour into the flight ATC told the pilots to change frequencies and contact a controller in a different sector. The pilots complied and radioed their new controller:

> November six-hundred XL at [flight level] three-seven-zero, good afternoon.

The new controller instructed the pilots:

> November six-hundred, squawk ident *[meaning, flash your trans-ponder]* – roger, maintain flight level three-seven-zero, [you are] under radar surveillance.

This verbal exchange would prove to be the last successful two-way contact between the Legacy pilots and ATC. Recorded radar data later would confirm the "ident" was displayed on radar.

Five minutes later the Legacy flew over Brasilia VOR. The pre-filed flight plan called for a descent to flight level 360. However, ATC recordings later would reveal the controller did not try to clear the pilots to descend. Consequently, the Legacy droned northwest along airway UZ-6 at flight level 370.

Meanwhile, the Boeing 737 cruised southeast on the same airway at the same altitude. The two aircraft were flying toward each other head-on at the speed of a rifle bullet. The following time-line, beginning at takeoff, pertains only to the Legacy and ATC:

1451 Hours: the control tower clears the Legacy for takeoff

1457 Hours: the pilots make radio contact with Brasilia Center

1511 Hours: the controller clears the pilots to flight level 370

1551 Hours: the last successful two-way radio contact

1556 Hours: the pilots overfly Brasilia VOR and turn northwest on airway UZ-6 at flight level 370

1602 Hours: the secondary (transponder) return, including altitude information, drops off Brasilia Center radar

1615 Hours: controller shift-change at Brasilia Center

1626 Hours: the new controller makes the first of seven radio calls in an unsuccessful effort to contact the pilots

1630 Hours: the primary return becomes intermittent on radar

1648 Hours: the pilots make the first of 12 radio calls in an unsuccessful effort to contact their controller

1653 Hours: the new Brasilia Center controller tries to hand-off the Legacy to Amazonia Center, but he still can not make radio contact with the pilots

1656 Hours: the Legacy **collides** with the Boeing 737

As of this writing, ATC and CVR transcripts have not been *officially* released. However, completed transcript content has been referenced by Brazilian investigators and the NTSB. Also, excerpts have been leaked to several technical aviation publications. The portion of the Legacy CVR transcript, printed below, contains parts of the leaked content. All verbiage comes from the two Legacy pilots, but they are not separately identified. Because of the *unofficial* status of this CVR content, readers are cautioned that verbiage may not be verbatim:

*[the CVR documents the **sound of a collision** at 1656]*

Pilot: Uh-oh! (1656)

Pilot: What the [expletive] was that? (1656)

Pilot: All right, just fly the airplane . . . .

Pilot: Where the [expletive] did he come from? All right, we're going down, declaring an emergency . . . .

Pilot: Did we hit somebody? Did you see something?

Pilot: We were on course. We're on [the right flight] level altitude. I don't know what the [expletive] we hit! (1703)

[the pilots make 5 unsuccessful attempts to radio ATC]

Pilot: Do we have a wingtip?

Pilot: No.

Pilot: It's all right . . . we'll make it, we'll make it . . . .

[the pilots fly toward Cachimbo Air Base]

Pilot: I don't know if we hit someone – (1710)

Pilot: We hit something, man! We hit another plane! I don't know where the [expletive] it came from!

Pilot: I never saw it, dude.

[24 minutes after the collision the Legacy touches down on the runway at Cachimbo Air Base]

Pilot: Good job, good job, good, we're good. (1720)

Pilot: We're alive! (1720)

[sound of passengers clapping in the cabin at 1720]

Pilot: If we hit someone – I mean – we were in the proper altitude, the [controllers] forgot about us . . . .

Pilot: But I'm worried about the other plane. What else could it have been?

Pilot: It was a hard hit, whatever it was.

Pilot: No kidding!

The Collision: The Boeing 737 and the Legacy *almost* missed each other. However, the Legacy left winglet had sliced into the Boeing outer left wing. The wing broke away, rendering the big airliner uncontrollable. It tumbled earthward, began to disintegrate at around 8,000 feet, and debris showered into the Amazon Jungle below:

> The Boeing 737 was destroyed by in-flight breakup and [ground] impact forces.

The Legacy pilots never saw what hit them. They had felt and heard a tremendous blow. The aerial impact had ripped the left winglet and part of the left horizontal stabilizer off the aircraft. One passenger was a journalist. His subsequent feature article in *The New York Times* would carry the following headline:

Colliding with Death at 37,000 Feet, and Living

In the newspaper article the journalist described the collision:

> I felt a terrible jolt and heard a loud bang, followed by an eerie silence . . . . at the end of the wing was a jagged ridge, perhaps a foot high, where the five-foot-tall winglet was supposed to be.

Awestruck passengers in the cabin had watched the leading edge of the left wing lose rivets and start to peel back. The pilots had begun an emergency descent and headed for the nearest airport, the Brazilian Air Force airfield at Cachimbo. At 2020 Hours, 24 minutes after the collision, the Legacy had touched down on the runway. No one aboard had been injured. Happy to be alive, the pilots crawled out and explained to Brazilian authorities that they had hit *something* at flight level 370.

ATC soon reported the Boeing airliner missing. The next day Brazilian military pilots spotted the crash site in the jungle. Helicopters hovered over the treetops, and potential rescuers rappelled down to seach for survivors. They found none. All 154 people on the airliner had been killed in the worst aviation accident in Brazilian history. Searchers began recovering bodies five days after the accident. They recovered the final human remains on 22 November, and victims were eventually identified through DNA protocols.

The Investigation: A team from the Brazilian Aeronautical Accident Prevention and Investigation Center recovered the FDR and CVR from the Boeing wreckage. At Cachimbo they accessed the FDR and CVR in the Legacy. In addition they gathered recorded ATC voice and radar data from Brasilia Center.

The United States was the country of registry of the Legacy and the country of manufacture of the Boeing 737. Accordingly, under the provisions of ICAO Annex 13 the NTSB provided an accredited representative and advisors to participate in the investigation.

The Boeing 737 had been flying southeast on airway UZ-6 at flight level 370 in compliance with its ATC clearance. Primary and secondary returns had been displayed on the controller's radarscope:

> There were no anomalies in communication with, or radar surveillance of, the Boeing 737 throughout the flight.

The Legacy had reached Brasilia VOR at flight level 370 and had turned northwest on airway UZ-6 toward Manaus. At that time the Boeing 737 had been flying southeast on the same airway at the same altitude. Under the control of Brasilia Center, the two aircraft had been cleared to fly straight toward each other.

The Legacy pre-filed flight plan had called for a descent to flight level 360 at the VOR. However, before reaching the VOR the pilots had made their last successful two-way radio contact with Brasilia Center. The controller had acknowledged the pilots' report that they were at flight level 370. Recorded radar data confirmed the transponder return, including altitude, had been displayed on the scope. The controller told the pilots they were "under radar surveillance." At no time, then or later, had he mentioned an altitude change.

At 1615 Hours there had been a shift change at Brasilia Center. The old controller briefed the new controller and *erroneously* told him the Legacy was at flight level 360. According to an independent technical report the new controller:

> . . . took no action to independently verify the Legacy's altitude, even when it was apparent from the [radarscope] target position symbol that the transponder signal was lost.

The new controller changed the electronic data strip to show the Legacy at flight level 360. With this error, in the absence of radio and radar contact, the last realistic chance to avoid the accident quietly slipped away. An independent technical report explains:

> Both aircraft were cleared by ATC to fly at the same assigned altitude. . . . This collision course was established more than an hour before the eventual accident; yet, because of a series of failures in the air traffic control system, the controllers never realized their error until it was too late.

Investigators questioned the cooperative Legacy pilots, who were detained in Brazil for over two months. However, 10 controllers had been placed on paid leaves of absence to get "psychiatric treatment." They were not available for interviews.

Probable Cause: On 16 November 2006 the Brazilian investigative team released a preliminary report. Based upon that report the NTSB promptly released its own interim report, File No. DCA06RA076A. Five months later the NTSB released a Safety Recommendation on 2 May 2007. A final report has not been issued as of this writing:

> There was no indication of a TCAS alert on board either airplane, no evidence of pre-collision visual acquisition by any flightcrew member on either aircraft, and no evidence of evasive action.

ATC had cleared the Boeing airliner and the Legacy to fly at flight level 370. The pre-filed Legacy flight plan had called for a descent to flight level 360 at Brasilia VOR. However, ATC never authorized the descent. An independent technical report explains:

> The filed flight plan is a *request [emphasis added]* for a clearance to operate under IFR, but a flight plan does not independently authorize such operation. . . . An ATC clearance overrides and supersedes the information contained in the filed flight plan.

Unable to contact the Legacy pilots by radio, ATC should have expected them to follow their clearance and remain at flight level 370. The controller could have instructed the Boeing pilots to climb or descend to eliminate the conflict. Yet, he never made an effort to do so, and the planes collided.

Both aircraft had been equipped with TCAS, the electronic "last line of defense" against mid-air collisions. For the system to function the TCAS unit and transponder in each aircraft must be operative. In this case TCAS did not generate a Traffic Advisory or Resolution Advisory in either cockpit. Had one system malfunctioned? Had one system been turned off? To date, no one knows for sure. Yet, the Legacy CVR recording has revealed:

> During the emergency descent to Cachimbo Air Base the [Legacy pilots] made a series of comments related to whether or not the TCAS was on. *[and also]* . . . for reasons yet to be determined, the collision avoidance system . . . was not functioning at the time of the accident.

In addition to the TCAS question there are other unknowns. For example, airspace over the vast Amazon Jungle is infamous for its radar blind spots and sporadic radio coverage. Yet, the accident can not be attributed to such shortcomings. Where the human element is concerned, people make honest – and often tragic – mistakes. One technical source has summed it up as follows:

> Brazilian authorities [have] confronted the truth – that their own air traffic controllers made a massive human error by placing the two jets at the same altitude in opposite directions along the same airway.

"Last Words" from Airline Cockpits

CVR technology has allowed verbatim documentation of sounds and conversation in the cockpit. CVR and ATC recordings reveal that, in times of dire peril, "last words" from the cockpit often are not actual words. Instead, they may be unintelligible exclamations.

Each entry, below, contains *intelligible* "last words" from an airline cockpit prior to an accident involving fatalities. The spoken words are followed by (1) the date, (2) the flight identification, (3) the type of aircraft, (4) the location, and (5) the cause of the crash. These fatal accidents have not been addressed previously in this book. The entries are arranged in chronological order.

Professional pilots rely upon experience and technical competence. Nonetheless, they know that "luck counts." Given a choice, they would never knowingly trade luck for skill. They know that, there, but for the grace of God . . . .

Something has hit us! Hang on!
> 30 June 1962; South African charter flight; DC-4; near Durban, South Africa – *mid-air collision*

Mayday! Mayday! Mayday! Clipper two-fourteen out of control!
> 8 December 1963; Pan Am Flight 214; Boeing 707; near Elkton, Maryland – *fuel tank explosion*

Skipper's shot! We've been shot! [I was] tryin' to help!
> 7 May 1964; Pacific Air Flight 773; Fairchild F-27; near San Ramon, California – *suicidal passenger shot both pilots*

We've lost all control! We don't have anything!
> 23 June 1967; Mohawk Airlines Flight 40; BAC-111; near Blossburg, Pennsylvania – *in-flight fire and loss of control*

Good God, I hit him! Sorry!
> 6 November 1967; TWA Flight 159; Boeing 707; at Cincinnati, Ohio – *runway collision*

What's that red light?
> 15 January 1969; Ariana Afghan Flight 701; Boeing 727; near London, England – *descent below MDA on approach*

We'll make a low pass and see if we can [see] the lights.
  20 March 1969; Travel Associates charter flight; DC-3; at New Orleans, Louisiana – *descent below MDA on approach*

Sorry! Oh, sorry, Pete! Sorry, Pete!
  15 July 1970; Air Canada Flight 621; DC-8; near Toronto, Canada – *inadvertent spoiler deployment on final approach*

Man, we ain't twenty feet [above] the water!
  17 June 1971; Allegheny Airlines Flight 485; CV-580; near New Haven, Connecticut – *in-flight collision with a two-story building*

I wish I knew where we were so we'd have some idea of the general [expletive] terrain around this [expletive] place.
  27 September 1973; Texas Intl. Flight 655; CV-600; Black Fork Mountain, Arkansas – *in-flight collision with a mountain*

All we got to do is find the airport.
  11 September 1974; Eastern Airlines Flight 212; DC-9; near Charlotte, North Carolina – *descent below MDA on approach*

We're gonna crash.
  11 February 1978; Pacific Western Flight 314; Boeing 737; Cranbrook, B.C., Canada – *aborted landing attempt*

We're not going to be able to make the airport.
  28 December 1978; United Airlines Flight 173; DC-8; near Portland, Oregon – *fuel exhaustion*

I don't like this.
  28 November 1979; Air New Zealand charter flight; DC-10; Antarctica – *collided with Mt. Erebus in white-out conditions*

Larry! We're going down, Larry!
  13 January 1982; Air Florida Flight 90; Boeing 737; at Washington, D.C. – *stall after takeoff due to ice on wings*

What's happened? What? What?
  1 September 1983; Korean Airlines Flight 007; Boeing 747; over the Okhotsk Sea – *shot down by air-to-air missile*

Raise nose! Raise nose! Power!
>12 August 1985; Japan Airlines Flight 123; Boeing 747; near Tokyo, Japan – *structural failure and loss of control*

We're dead!
>4 October 1986; Southern Air Transport Flight 15; L-100; at Kelly AFB, Texas – *locked controls, rolled inverted on takeoff*

Somethin's wrong . . . we're not gonna make it.
>31 August 1988; Delta Airlines Flight 1141; Boeing 727; near Dallas, Texas – *wrong takeoff configuration*

What the [expletive] was that?
>24 February 1989; United Airlines Flight 811; Boeing 747; over the Pacific Ocean – *in-flight structural failure*

That's it, I'm dead!
>7 June 1989; Surinam Airways Flight 764; DC-8; Paramaribo, Surinam – *descent below MDA on approach*

Keep calm!
>27 December 1991; SAS Flight 751; MD-81; near Gottrora, Sweden – *dual engine failure on takeoff*

We will turn back soon.
>31 July 1992; Thai International Flight 311; Airbus A-310; near Kathmandu, Nepal – *in-flight collision with a mountain*

What shall we do now?
>14 September 1993; Lufthansa Flight 2904; Airbus A-320; near Warsaw, Poland – *windshear induced runway overrun*

No, no, hold it . . . gimme flaps up . . . Whoa!
>7 January 1994; American Eagle Flight 3641; Jetstream 41; at Columbus, Ohio – *stall during final approach*

Aaahhh! We're finished!
>26 April 1994; China Airlines Flight 140; Airbus A-300; at Nagoya, Japan – *stall after inadvertent TOGA activation*

Hang on! What the [expletive] is this?
>8 September 1994; USAir Flight 427; Boeing 737; at Aliquippa, Pennsylvania – *flight control failure*

Go max power! Max power! Max power!
   16 February 1995; Air Transport International ferry flight; DC-8;
   at Kansas City, Missouri – *attempted three-engine takeoff*

Amy! I love you!
   21 August 1995; Atlantic Southeast Flight 529; EMB-120; near
   Carrollton, Georgia – *in-flight structural failure*

Oh, [expletive]! Pull up, baby!
   20 December 1995; American Airlines Flight 965; Boeing 757;
   near Cali, Columbia – *in-flight collision with a mountain*

Oh! What's happening?
   6 February 1996; Birgen Air Flight 301; Boeing 757; over the
   Caribbean Sea – *disorientation due to a blocked pitot port*

Mountains!
   29 August 1996; Vnokovo Flight 2601; Tupolev TU-154; near
   Spitsbergen, Norway – *in-flight collision with a mountain*

Speed's decreasing! Speed! Speed!
   8 October 1996; Aeroflot Flight 9981; Antonov AN-124; at San
   Francesco al Campo, Italy – *loss of control on missed approach*

[Runway] not in sight. Missed approach.
   6 August 1997; Korean Airlines Flight 801; Boeing 747; Nimitz
   Hill, Guam – *in-flight collision with a mountain on approach*

We are declaring an emergency now, [this is] SwissAir one-eleven.
   2 September 1998; SwissAir Flight 111; MD-11; near Halifax,
   Nova Scotia, Canada – *in-flight fire and loss of control*

This is a can of worms!
   1 June 1999; American Airlines Flight 1420; MD-82; at Little
   Rock, Arkansas – *attempted landing during a thunderstorm*

Turning left! Left! Left! Left! Left!
   10 January 2000; Cross Air Flight 498; Saab 340; at Nassenwil,
   Switzerland – *spatial disorientation*

Two engines! We lost two engines!
   13 January 2000; Sirte Oil Company charter flight; Shorts 360;
   over the Mediterranean Sea near Lybia – *dual engine failure*

Power. Power. Aaawww, [expletive].
> 16 February 2000; Emery Flight 17; DC-8; at Rancho Cordova,
> California – *mechanical failure, loss of pitch control*

I have no idea which way is up!
> 19 July 2000; Airwave Flight 9807; Gulfstream I; near Montreal,
> Quebec, Canada – *disorientation, loss of control*

The airspeed . . . the airspeed . . . Le Bourget . . . No!
> 25 July 2000; Air France Flight 4590; BAC Concorde; near Paris,
> France – *in-flight fire and dual engine failure on takeoff*

That's all, guys, [expletive]!
> 4 July 2001; Vladivostokavia Flight 352; Tupolev TU-154; near
> Burdakovka, Russia – *loss of control and stall on approach*

Hey, get out of here!
> 11 September 2001; United Airlines Flight 93; Boeing 757; over
> Somerset, Pennsylvania – *hijacked by terrorists*

Push the nose down!
> 8 January 2003; Air Midwest Flight 5481; Beechcraft 1900; at
> Charlotte, North Carolina – *improper elevator rigging*

Aircraft is turning right! Over bank! Over bank!
> 3 January 2004; Flash Air Flight 604; Boeing 737; near Sharm el
> Sheikh, Egypt – *spatial disorientation*

One-hundred knots – rotate – whoa!
> 27 August 2006; Comair Flight 5191; Bombardier CL-600; at
> Lexington, Kentucky – *attempted takeoff on wrong runway*

Glossary

*Like most technical pursuits, aviation has developed
a unique jargon. Much of it is unfamiliar to laymen.
This glossary of common terms, phenomena, and
acronyms will assist readers who are not versed in
aeronautical terminology:*

ADF: Automatic Direction Finder. A low-frequency radio receiver that provides *homing* direction toward ground stations.

AFB: Air Force Base.

Ailerons: On the trailing edge of wings, moveable surfaces that control the bank and roll of an aircraft.

Airspeed: Expressed in knots, the speed of an aircraft relative to the air mass through which it travels. (see "Knots")

Altitude: Unless otherwise specified, height above mean sea level, measured in feet. When used, the suffix "agl" means height "above ground level." (also, see "Flight Level")

Angle-of-Attack: Measured in degrees, the angle at which the wings meet the relative flow of air.

Approach Chart: Also called Approach Plate. A small chart that provides IFR electronic navigation instructions and altitude limits for a specified landing approach at a given airport.

Approach Control: (see "ATC")

ARSA: Airport Radar Service Area.

ATC: Air Traffic Control. The system for directing and separating aircraft in flight. It is subdivided into several domains; generally speaking: (1) the "Tower" is responsible for aircraft taking off and landing, (2) "Departure Control" and "Approach Control" are responsible for aircraft near the vicinity of the airport, and (3) the "Center" is responsible for the en route phase of flight.

ATIS: Automatic Terminal Information Service. A recorded radio message detailing weather and other concerns at an airport.

ATP: Airline Transport Pilot.

Attitude: The roll and pitch of an aircraft relative to the horizon.

Attitude Indicator: A gyroscopically controlled instrument, crucial for IFR flight, that displays the aircraft "attitude" relative to the horizon. For obvious reasons, also called an Artificial Horizon.

Black Box: (see "CVR" and "FDR")

Bug: An adjustable setting on the rim of aircraft instruments to highlight crucial airspeeds and altitudes.

CAB: The former U.S. Civil Aeronautics Board.

CAVU: Ceiling and Visibility Unlimited.

Center: Air Route Traffic Control Center. (see "ATC")

"Clean" Aircraft: An aircraft in cruise configuration with its high-lift devices and landing gear retracted.

Clearance: An approval by ATC for an aircraft to land or take off at an airport, or to fly specified routes and altitudes. Also called an ATC Clearance.

CVR: Cockpit Voice Recorder. Also called a Black Box, although it is painted bright orange. A crash resistant recording device that documents cockpit verbiage and other sounds.

Departure Control: (see "ATC")

DME: Distance Measuring Equipment. An electronic device that provides a constant cockpit display of the distance, in slant-range nautical miles, from an aircraft to a selected ground station.

Elevator: A movable surface, usually mounted on the horizontal stabilizer, that controls the vertical pitch of the aircraft.

FAA: U.S. Federal Aviation Administration.

FAR: U.S. Federal Air Regulations.

FDR: Flight Data Recorder (or Digital Flight Data Recorder). Also called a Black Box, although it is painted bright orange. A crash resistant recording device that documents operating parameters such as airspeed, altitude, rate of climb or descent, engine performance, acceleration, flight control movements, etc.

First Officer: An airline copilot, the second-in-command.

Flaps: Adjustable control surfaces on the wings. Flaps can be lowered to several incremental settings, usually up to a maximum of about 50 degrees, depending upon the type of aircraft. Lowering the flaps increases the curvature of the wing and increases aerodynamic lift (and drag), allowing the aircraft to land or take off at a lower airspeed. (see "Slats")

Flight Director: A "glass cockpit" computerized display of aircraft attitude plus electronic flight and navigation data.

Flight Engineer: A cockpit crewmember (on some aircraft) who is responsible for operation of aircraft systems.

Flight Level: Height in hundreds of feet, based upon a standard barometric setting (for example, 22,000 feet is *flight level 220*).

Flight Plan: A report filed with ATC and containing necessary details about a proposed VFR or IFR flight.

G: The force, positive or negative, acting upon an aircraft and its occupants in flight, measured in multiples of the gravitational force of the Earth.

GCA: Ground Controlled Approach. A military precision approach involving a ground controller who uses glideslope and glidepath radar to guide pilots to a safe landing in IMC weather.

GPWS: Ground Proximity Warning System. An electronic aircraft transceiver that can sense dangerous *ground proximity* conditions. If the GPWS detects such a condition it warns the pilots with an appropriate digitized verbal message such as (1) "*sink rate, sink rate*" or (2) "*whoop whoop, pull up, whoop whoop, pull up.*"

GMT: Greenwich Mean Time. Formerly called Zulu Time; more recently called Universal Coordinated Time.

GPS: Global Positioning System. The satellite-based navigation system funded, controlled, and operated by the U.S. Department of Defense. GPS was designed for the U.S. Armed Forces, but it can be used by anyone. With a GPS radio receiver one can determine his position anywhere on, or above, the Earth.

Ground Speed: Expressed in knots, the speed of an aircraft relative to the surface of the Earth. May be less or greater than airspeed, depending upon the wind direction.

Guard: Also called Guard Channel or Guard Frequency. The military UHF frequency (243.0 MHz), or the civil VHF frequency (121.5 MHz), reserved for use in an emergency.

Horizontal Stabilizer: At the rear of an aircraft, a horizontal airfoil that provides vertical pitch stability in flight.

Hours: Local time on a 24 hour clock (for example, 0915 Hours equates to 9:15 am, and 2215 Hours equates to 10:15 pm). If a colon and two digits follow the time (for example, 0915:51 Hours), the two digits represent *seconds*, not hundredths of a minute.

ICAO: International Civil Aviation Organization.

ICS: Internal Communications System. Also called Intercom.

IFR: Instrument Flight Rules. Rules and procedures for navigating by cockpit instruments alone under poor visibility conditions. Refers to either (1) the procedures, or to (2) the visibility condition that necessitates the procedures. (see "IMC")

ILS: Instrument Landing System. An electronic landing aid that provides (1) glideslope and (2) glidepath displays in the cockpit.

IMC: Instrument Meteorological Conditions. The poor visibility conditions that mandate IFR flight. (see "IFR")

Knot: One *nautical mile* per hour. A nautical mile is 6,076 feet in length (as opposed to 5,280 feet in a statute mile). Thus, 100 knots equals roughly 115 miles per hour.

Mach: Also called Mach Number. The numerical relationship between airspeed and the speed of sound.

Mayday: The international distress message, traditionally repeated three times. Mayday is derived from *m'aidez*, which means "help me" in French. It is the verbal equivalent of the former Morse Code

"SOS" distress message.

MCAF: Marine Corps Air Facility.

MCAS: Marine Corps Air Station.

MDA: Minimum Descent Altitude. During a landing approach on an IFR flight, the altitude below which an aircraft must not descend until the runway environment is in sight.

MEA: Minimum En Route Altitude. On an IFR flight, the en route altitude below which an aircraft must not descend. Flight at or above the MEA will ensure that the aircraft remains clear of high terrain and other obstructions to flight.

NAS: Naval Air Station.

NASA: U.S. National Aeronautics and Space Administration.

Nautical Mile: (see "Knot")

Navaid: The common term for an electronic Navigation Aid.

NDB: Non-Directional Beacon. A low-frequency transmitter of navigational use to aircraft with ADF equipment. (see "ADF")

NOTAM: Notice to Airmen. A cautionary written notice which details irregularities with airways, airports, and navigation aids.

NTSB: U.S. National Transportation Safety Board.

Pitot Static System: Interconnected aircraft tubes and sensors that allow display of airspeed, altitude, and rate of descent or climb.

Radar: An *acronym* (Radio Detection and Ranging) that, through extensive common usage, has evolved into a *word*.

Radial: A bearing from a VOR ground station. (see "VOR")

Rudder: A moveable surface, mounted on the vertical stabilizer, that controls the lateral yaw of an aircraft.

RVR: Runway Visual Range.

Sink Rate: The vertical descent rate, measured in feet per minute. It is crucial in the final moments as an aircraft prepares to land.

Slats: Enhanced lift devices on the leading edge of the wings. Slats can be incrementally extended to increase the curvature and the aerodynamic lift (and drag) of the wings, thereby lowering the landing and takeoff airspeeds of an aircraft. (see "Flaps")

SOB: Souls on Board. The number of persons aboard an aircraft.

Spatial Disorientation: A condition that exists when a pilot, because of poor visibility, can not detect the horizon and thus becomes disoriented because of vestibular illusions of the inner ear. Often called "vertigo" by laymen.

Speed Brakes: (see "Spoilers")

Spin: A perilous low airspeed condition which can occur at – or slightly above – stalling speed when an aircraft is turning left or right. The wing on the "inside" of a turn has a lower airspeed than the wing on the outside of a turn. Consequently, the *slower* inside wing can *stall* and lose lift, while the faster outside wing has not stalled. In such a condition (1) the *inside stalled wing* will drop, and (2) the fuselage will revolve – *spin* – around the stalled inside wing as the aircraft spirals toward the ground. (see "Stall")

Spoilers: Depending on their configuration and use, also called Speed Brakes or Dive Brakes. Flat panels which lie flush against the surface of the wings (and sometimes, the fuselage) in cruising flight. When extended they (1) *spoil* the smooth flow of air, (2) decrease unwanted lift, and (3) increase drag.

Spool: The increase or decrease (as in *spool up*, or *spool down*) in revolutions-per-minute of a turbine engine.

Stall: A phenomenon that occurs when the airspeed of an aircraft decreases below the minimum at which the wings can create sufficient lift for flight. (see "Spin")

Stick-Shaker: An electro-mechanical device (in some aircraft) that shakes the control column to warn of an approaching stall.

Tower: (see "ATC")

Transponder: An aircraft radio transceiver that enables Air Traffic Control radar to identify an aircraft and, usually, its altitude.

<u>Trim</u>: Adjusting control surfaces so that no pressure on the rudder or control column is required for a given flight regime.

<u>UHF</u>: Ultra High Frequency. An aviation radio frequency band used primarily by military aircraft. (see "VHF")

<u>V-1</u>: The calculated "go or no-go" *decision airspeed* on takeoff.

<u>V-2</u>: The calculated takeoff *safety airspeed*; the minimum control airspeed in case of engine failure, plus a safety margin.

<u>V-r</u>: The calculated airspeed at which to *rotate* for takeoff.

<u>VASI</u>: Visual Approach Slope Indicator.

<u>Vertical Stabilizer</u>: At the rear of an aircraft, the vertical airfoil that provides directional stability in flight.

<u>VFR</u>: Visual Flight Rules. The rules under which a pilot may fly and navigate by visual reference to the ground while remaining clear of clouds and other aircraft. (see "VMC")

<u>VFR-on-Top</u>: Flight in VFR conditions above a layer of clouds, smoke, fog, or haze.

<u>VHF</u>: Very High Frequency. An aviation radio frequency band used primarily by civil aircraft. (see "UHF")

<u>VMC</u>: Visual Meteorological Conditions. The conditions under which a pilot may fly under Visual Flight Rules. (see "VFR")

<u>VOR</u>: Very-high-frequency Omni-directional Range. Electronic navigation equipment for use by aircraft. Refers to either the (1) ground transmitting station, or to the (2) aircraft equipment that receives and displays the signal from such a ground station.

<u>WTS</u>: The former U.S. War Training Service, a civil program that provided basic flight instruction to prospective military pilots.

Military and Aviation
books by Marion Sturkey

BONNIE-SUE: A Marine Corps Helicopter Squadron in Vietnam

Murphy's Laws of Combat

Warrior Culture of the U.S. Marines (first edition)

Warrior Culture of the U.S. Marines (second edition)

MAYDAY: Accident Reports and Voice Transcripts from Airline Crash Investigations

MID-AIR: Accident Reports and Voice Transcripts from Military and Airline Mid-Air Collisions

Regional-Interest-Only
books by Marion Sturkey

GONE, BUT NOT FORGOTTEN: An Introduction

GONE, BUT NOT FORGOTTEN: Cemetery Plots and Verbatim Transcripts of Inscriptions and Epitaphs from Cemeteries of European Settlers, and their Descendants, in McCormick County, South Carolina

About the Author

Marion Sturkey began his flying career in the U.S. Marine Corps. He earned the Naval Aviator designation in 1965 and flew both fixed-wing aircraft and helicopters. His certifications include VIP Pilot, NATOPS Check Pilot, and Maintenance Test Pilot.

Marion became a Commercial Pilot after he left active duty. For two years he flew helicopters to support off-shore oil exploration and production operations in Louisiana and Texas. He then left flying behind and worked with the BellSouth Corporation for 25 years. During the last ten years of that career he served as guest instructor at Bell Communications Research in Illinois and New Jersey.

After retirement from the corporate world, Marion began writing and publishing books about military and aviation concerns. He is a frequent guest speaker at military bases and at functions for military veterans. He is a Life Member of the Marine Corps League and the Marine Corps Combat Helicopter Association. *MID-AIR* is his eighth book.

Aviation is proof that, given the will, we have the capacity to achieve the impossible.

[Edward V. "Eddie" Rickenbacker (1890-1973), World War I flying ace, and in his later years, an airline executive]

Dedication

I dedicate this book to all the entrepreneurs who have entered the arena with a dream of greatness and a will to make it happen. Without your spirit we'd all still be gathering berries and huddling around fires at night.

Acknowledgments

Thank you also to all the CISOs who have given us feedback and encouragement as we've published the two volumes of the CISO Desk Reference Guide. As we complete the CISO Desk Reference Guide Small Business Series, we hope this catalog we are building makes a difference. Thank you to Alan Watkins and Matt Stamper for their thorough reviews of this book. Thank you to my partners Gary Hayslip and Matt Stamper for their friendship and collaboration, this has been even more rewarding than I had hoped!

Thank you to Gwendoline Perez for creating original cover art and graphical illustrations. Nadine Bonney deserves special appreciation as well for painstakingly combing through my word usage disasters and under-appreciation for the Chicago Manual of Style. My most heartfelt thanks to Chris Lawrence for producing this book and helping us define the look and feel for our catalog.

Finally, I would like to thank the following local small business owners who were kind enough to provide details of their road through security so this book would be grounded in the real-world and not just the theoretical ramblings of someone who has lived cyber for decades. Their contributions have helped make Bring Your Own Cyber a reality:

Danielle Barger, Jennie Bradach, Tahra Doan, Chris Gluck, Ronn and Lauren Gruer, Jessica Harper, Marinda Neumann, Leslie Ordel, Dr. Mitchell Shulkin, and Tiffany Torgan.

Thank you as well to Adam Gladsden of Broad Street Labs for his counsel on small business cyber insurance.

Table of Contents

Bring your own cyber. As a small business owner, you are probably the president, the marketing VP, the sales VP, the finance VP and, if you use a computer or mobile phone to conduct business, like it or not, you're the security VP as well. Many small businesses use firms to provide IT services and security services. However, you still need to be aware and direct the work of those firms. You need to know the questions to ask, you need to know what your requirements are, and you need to sometimes push hard to make sure the necessary services are delivered to you and you aren't manipulated into buying what you do not need.

Though it takes years of training to become a cybersecurity professional, anyone can take prudent steps to make themselves and their company more secure. In the following nine chapters, I'm going to teach you the basics. I will outline what you need to know to protect your business, yourself, your employees, and your brand from most cybercrime.

There are three fundamental rules I'm going to live by throughout this book.

First, I promise no jargon. I'm going to use plain language to tell you how to become more secure. If I do use a technical term, I'll do so only because you hear it talked about and you need to know what it is. Then I will define it in plain language and tell you what it means to you.

Second, I'm going to be honest about the day-to-day work you need to do to be secure. It's not rocket science, but it does take discipline. That makes sense, right? If it was easy to be cybersecure by buying a product or a service, we'd all be secure by now.

Third, every chapter will come with prudent steps you can take right away to be more secure. I'll tell you what you have to do (the basics) and then I'll give you some more advanced advice (intermediate steps and pro steps) so you can take it to the next level. I promise I will not tell you to go do something and then not tell you how.

To say that a lot has been written about cybersecurity since 2011 when the SONY Play Station hack occurred is a flamboyant understatement. Cyber is both sexy and scary at the same time. And while tales of spooks and spies make for good reading, if you're so inclined, most of what we see is scary as hell. By now we all know that every single one of us has been the victim, directly or indirectly, of cybercrime. Our credit card numbers, our personal medical data, and other private information are being leaked or stolen all the time.

Thank you for picking up a copy of "Bring Your Own Cyber."

The following nine chapters are divided into two sections. In the first section we focus on securing your business and in the second section we focus on securing your brand.

We'll start in Chapter 1 with physical security. It's often overlooked when talking about cybersecurity, but it's critical for your physical protection as well as the first level of protection for cybersecurity.

After physical security, we'll move on to the cyber realm in Chapter 2 and start with awareness. Awareness is critical, both for yourself and to help your entire company prepare for and respond to cybersecurity issues you might encounter.

The next three chapters, protecting your networks (3), updates and backups (4), and access (5), address the mechanics of how to secure your company, whether you are the sole owner and employee or the leader of your crew.

In Chapter 6 we cover web and social media security and in Chapter 7 we deal with the very topical and very complex topic of data privacy. In Chapter 8 we dive into the complex world of cyber insurance. In every case, the goal is to make this very consumable. As stated in the preface, the information is presented jargon free. At the end of each chapter you'll find a series of actions you can take. These are divided into basic, intermediate, and pro steps. Chapter 7 also includes a set of controls and a checklist that you can use to address regulatory requirements for data privacy and, should you need to, qualifying for insurance coverage.

Finally, in Chapter 9, you'll find a summary of the key advice introduced throughout the book. After you have read, learned and taken key steps, Chapter 9 serves as a reminder to put your new knowledge to regular use. Use this as a catalyst to form the habits that will keep you safe. And when the inevitable breach occurs, look to appendix A for who to call and what to do to recover.

With that as preamble, let's begin.

Section 1

Securing Your Business

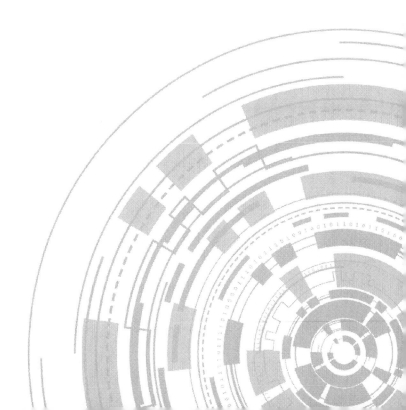

Introduction

You've probably heard the term "cyber hygiene" and might wonder what it means. It refers to the basic steps we take to keep our systems operating securely. The term cyber hygiene probably originates from drawing the analogy to personal hygiene, habits we've learned as children that are necessary to keep us clean and healthy.

Cyber hygiene should become the habits that we learn as entrepreneurs in the digital age. These are straightforward steps that we take to protect our systems and our data from harm.

In Section 1, we're going to go through these straightforward steps, and I will tell you what you need to do to form the habits you'll need to stay digitally safe and secure.

You can think of this section as teaching you the tools to set up the security layers to protect your valuable digital assets. In Chapter 1, this begins with the physical layer – all security starts with a locked door. You also need to be aware of the cyber threats and teach awareness to your employees so you and your employees can better defend your business from cybercrime. Chapter 2 addresses cyber awareness.

Chapter 3 expands the security that started with the physical to your network and your computers. Chapter 4 covers the critical topics of updates and backups. In Chapter 5, we add another layer of security…controlling who has access to logon to your systems.

At the end of each chapter, you'll find a list of straightforward steps you can take to make your business more secure.

Chapter 1

Lock the Doors

Yes, we are talking about cybersecurity, but fundamentally, it is still security. And security starts with simple physical things, like doors and locks and safes. One of the most essential actions any small business owner can take is to quite simply lock the doors. Of course, I mean more than just locking the front door.

Locking the front door only locks out the least determined strangers. The front door doesn't stop determined thieves, and it locks employees, contractors, and service people in. Parts of this first chapter may seem very obvious to many of you. I include this information for three reasons.

First, one of the easiest cybercrimes to commit is one where the thief already has the account number, or even worse, the password. A thief can get those easily if they (or a buddy) are unsupervised in a room with bank statements and computers freely lying around. You may have heard the term "insider crime." If a disgruntled employee or the cleaning service (insiders) have access to the computer or the accountant's reports, you are open to becoming the victim of insider crime.

Second, walking through this will make access control and other cybersecurity techniques easier to understand, as you'll have something familiar to relate it to. Cybersecurity is still security, and many of the methods we've used for decades to be more physically secure can help us be more cyber-secure if we learn how to apply those techniques in our digital lives.

And third, most people don't know how to commit cybercrime. It typically takes specialized training or access to malicious computer code. Yes, there are high school dropouts who commit cybercrime, but most of them receive training from someone who knows how to do it. But anyone can riffle a desk or walk out of the office with a laptop computer in a delivery box. The two are therefore related. One can lead to the other, and weak physical security increases the chances a cybercrime will occur.

Layers of Security

Security pros talk about layers of security (what we call defense in depth). Each layer of security is another door or another lock or another password. Anything that makes someone stop and take action or go around a barrier is another layer.

For most businesses, there are multiple layers of physical barriers that can and should be used to add protection for the business' most important possessions. Important possessions include obvious things like cash registers and computers, but also include documents such as bank statements, invoices, bills, receipts, payroll reports, and reports from the accountant.

A simple example of layers of security for a store is to have an office with a door inside the store and a desk with a lockable drawer inside the office. This provides three layers of security. You have a lock on the front door, a lock on the office door, and a lock on the desk drawer. If you consistently put private or sensitive documents in the locked drawer, the confidential materials that should be protected are behind those three layers of security.

Now let's say we have a computer server – the kind of computer that "runs the office." Perhaps it has the "shared disk" that everyone uses as a file server, maybe it has the printer attached, perhaps it's the computer that stores customer data. Having the customer data makes the computer server a precious possession, both to you and to thieves. It might make sense, then, to put this computer in a closet with a door that locks or lock it in a cabinet. Once again, you are putting a valuable asset, the computer server, behind three layers of physical security.

Be Careful with the Keys

To make this strategy of using multiple layers of security to protect critical possessions work, you also need to focus on who has which keys. Only the most trusted people who need to have keys to the desk drawer or server closet get them. As much as we'd like to implicitly trust the people we've hired, especially those who have been with us for years or are friends of the family, the unfortunate reality is that small businesses (those with less than 150 employees) experience more than half of all embezzlements.[1] Why? The biggest reason is that small businesses often lack the same kinds of safeguards that are commonplace at larger companies.

This lack of safeguards makes small businesses targets. Sometimes small businesses are just targets of opportunity. If someone experiences a negative life-altering event and knows no one is watching the till, the temptation may be too high for that person to resist. This type of motivation certainly predates the age of cybercrime.

As I said above, almost anyone can pull off in-person crime; it just takes an opportunity. One reason people don't commit more crimes

---

[1] Hiscox embezzlement study:
https://www.hiscox.com/newsroom/press/hiscox-embezzlement-study

is that they would be in physical peril when the offense is committed in person. They can get caught, be arrested, and put in jail.

The alter-ego of in-person crime is cybercrime, which can only be done by those who know how. But cybercrime is usually done remotely, with little to no personal jeopardy. Instead of robbing the till our unlucky soul might be tempted to write down some account numbers and passwords and pass these along. This unfortunate soul may be someone you don't even know. Perhaps they work for the cleaning company, or ironically, the security firm that checks the doors at night.

Lock the doors when you aren't there, put important papers and the server under lock and key, and pay attention to whom you give the keys.

Prudent Security

Physical barriers don't have to be perfect or impenetrable. It is not necessary to stop everyone from ever getting unauthorized physical access to a server or sensitive documents. What we want to do is prevent most people from getting unauthorized physical access and slow down the more determined thieves so that it is not worth it for them to try, or so that you or the police have time to catch them before they complete the crime.

It isn't always necessary that every layer of security is in good working order. Firms install video cameras to record who is present in a particular place and what they do, but they also install fake cameras to fool any would-be thieves into thinking they are being watched. Who wants to take the chance? The camera, working or not, acts as an additional layer of security (in this case as a deterrent). A working camera has the additional benefit of helping the police catch the thieves if they do commit a crime. Any casual viewing of the local news shows a pretty remarkable record of thieves caught on camera.

Besides keeping your possessions safe, multiple layers of security can also help you lower your insurance rates. Just as being near a fire hydrant and installing a smoke detector can help you reduce your homeowner insurance premium and having a car alarm can lower your car insurance premium, having a physically secure location makes a loss (theft) less likely and may cause your insurance company to charge less for your business insurance policy. You'll see that this same technique, applied digitally, might allow you to qualify for cyber insurance if you need it, which we discuss in Chapter 8.

Shore Up Your Virtual Defenses as Well

In the virtual world, the layered defenses are just as important, and the basics are often referred to as "good cyber hygiene." We'll go into each of these steps in the chapters that follow, but here is a summary of the key steps to good cyber hygiene that you can take as a small business owner.

Step 1: Stay current with updates. We talked about multiple layers of security earlier in this chapter. In the world of cybersecurity, one of the most important layers of security is to make sure that all of the systems you use – the office computer (especially servers), your laptop, the smartphones and tablet computers, and all the software that turns these pieces of metal and plastic into indispensable tools – are up to date with the latest versions. All of these devices and applications are constantly being updated by the manufacturers to fix the bugs that they discover, that their customers discover, and that the cybercriminals discover. So, the first rule of good cyber hygiene is to keep your systems up to date. This especially applies to anything you've just installed because the manufacturing and packaging dates could be months before the ship date. Your first order of business once you install it is to update it. Almost every device and application has an easy way in the settings or on the help menu to check for updates. Start there.

Step 2: Install protection against malware. In Chapter 3 we're going to cover the key tools you need to keep your company safe. In the past all you needed to do was run an anti-virus program on your PC, but those days are long gone. Part of your basic hygiene is to select, install, and maintain protection against malware. Malware protection won't stop every single attack against you, but it will knock out a good number of them and give you a fighting chance.

Step 3: Implement routine backups. In Chapter 4 we're going to go into detail on backups. At a high level, an essential part of your computer hygiene program is to make sure you are taking routine backups, and that these backups are stored someplace safe and where you can get access to them in an emergency. Having one set in your office and another set in the cloud or in an offsite storage facility (one that specializes in storing backups) is a best practice you should consider. Your backups also need to be regularly tested. It is astonishing how many backups are never tested and then fail when they are needed.

Step 4: Control access. In Chapter 5 we will cover access control. Making sure that the right people have access to your key systems is essential to protecting yourself. Another key to good computer hygiene is to limit who has access, just as you would determine who should have the keys to which door, which closet, and which drawer.

Step 5: Set strong, unique passwords and change them regularly. I know you have heard all the warnings about passwords, and we're certainly not saying that managing your passwords is easy. The security industry is working on ways to eliminate passwords. As of this writing, marginal progress is being made and some companies are using smartphone verification and other means of eliminating the dreaded 13-character mixed case password. In the meantime, good password management is fundamental to good hygiene. I'll show you how in Chapter 5.

Step 6: Manage your handheld devices. BYOD policies, short for Bring Your Own Device, assume that you and your employees will use your personal handheld devices while on the job. We've become too dependent on them as a society to leave them home or off all day. In Chapters 3 and 4 I'll go over some key steps for good handheld device hygiene.

Step 7: Train your employees. You have probably heard it said that your employees are your first line of defense, and your Achilles' heel, often in the same breath. What is true in the physical world is equally true in the digital world. Your employees need to know what to do with SPAM and phishing emails (defined and covered in the next chapter), when to alert you about unusual activity, and how to handle sensitive customer information to both protect the customer's privacy and keep you out of regulatory trouble. An employee training program is even mandated in some industries.

Take It to the Bank

At the beginning of this chapter, I mentioned doors, locks, and safes. I would be remiss if we didn't talk about the ultimate safe…the bank. We're going to go more in-depth on preventing these crimes in Chapters 2, 7, and 9 on Cyber Awareness, Cyber Insurance and Be Ready, respectively. But for now, there are two general forms of cybercrime involving banks that I'd like to cover.

In the past, what we read about most in the news was credit card fraud. At the dawn of Internet commerce, back in the 1990s, the consumer's fear of credit card fraud was a real barrier to people being willing to shop online. As a result, banks and merchants started to guarantee that no consumer would be held liable if fraudulent transactions were made on their cards. The result of this policy, along with the convenience of shopping online, was a huge increase in the number of online merchants. The downside was that with all the shopping that moved online, there was a great incentive to commit credit card fraud. There was just too much money to ignore. And instead of having to rob individual stores or steal credit card info from the local shop, credit card fraud could be done online, often from foreign countries.

To fight back, the payment card industry (or PCI) eventually adopted credit cards with embedded chips. Credit cards with chips are now widely used throughout the United States and Europe, with just a small percentage of merchants relying on the magnetic strip (notably gas stations, which still present a significant risk of skimming, or surreptitiously reading, the magnetic strip on a credit card).

The result is that credit card theft is not as lucrative as it once was. Also, it takes time and effort to get money back from credit cards. The thief has to go through a series of actions, including, for example, buying merchandise and reselling it or fraudulently returning it for refunds, taking cash advances, or purchasing gift cards and selling them.

These two factors, the difficulty in monetizing the card (turning the card into money) and the fact that the embedded chips make it harder to commit credit card fraud in the first place, have pushed cyberthieves to branch off and master stealing directly from your bank account. Direct bank fraud is the second major form of cybercrime involving banks I want to cover. And you can pretty much guess who the favorite target is. Yup, small businesses. The simple reasons why are that there is typically much more money in a business bank account than a personal bank account, and there is more movement between bank accounts, for example, to pay suppliers or to make payroll. This larger pot of gold and the greater amount of activity act as cover makes for an inviting target.

There are two major forms of bank fraud that affect small businesses. The first is wire transfer fraud and the second is what is known as "business email compromise" (also called BEC). Let's talk about wire fraud first. Cyber thieves work hard to get your bank "credentials," by which we typically mean user-IDs and passwords. They use these credentials to impersonate you online, and with access to your online bank account, can transfer money to another bank using a wire transfer. Typically, this money is transferred outside of the country, usually in several "hops" or steps, and often to a destination that makes it easy to make a quick withdrawal. The money is then "laundered" or turned into something that has value, is quickly redeemable for cash, and is not directly related to the account. In many cases, these accounts are closed as soon as the money is withdrawn or when it's evident that authorities somewhere are sniffing around the account.

There are several ways to protect yourself from this happening, even assuming that cyberthieves can get your credentials. First, if you do not need to do wire transfers, have that feature turned off on your account. Second, make sure that you are at least notified or, even better, that a separate dialog is required with you, perhaps a phone call or a text message, if any features, including wire transfer, are turned off and on. Third, consider a dollar limit for wire transfers from your account. Fourth, require a confirming action, again either a call or text, for any wire transfer. And finally, consider having a second "signer" on any check or transfer over a specified dollar limit. These steps taken together provide multiple layers of security and should help you limit the damage that any cyberthieves can do.

The second type of bank fraud that typically hits small businesses is business email compromise or BEC. BEC happens when the cyberthieves impersonate you in front of your employees. How this happens is first, the cyberthieves "case you" – they conduct enough reconnaissance to know, for instance, how you typically write emails and what your signature file looks like. They might also watch and learn about what is happening with your business at the moment. Armed with this information, they send an email to an employee designed to fool them into wiring money to an account they control.

They might send an email to your accountant saying, "Hey Jocko, I finally got the Ajax contract, can you wire $24,995 in good faith money to this account so we can close the deal." They might know that only you call your accountant John "Jocko" and that you are working on the Ajax deal. With these details, the email looks so convincing that there is a good chance John will fall for it.

Assuming Jocko was enticed by the email, the same safeguards we put in place to combat wire fraud work here too. With no wire transfer feature turned on, Jocko can't wire the money. If there is a dollar limit below $24,995, he can't send all of the cash. If you are notified or, better yet, your approval is sought first, that acts as a circuit breaker as well.

In the next chapter on Cyber Awareness, we'll help you make these kinds of crimes even less likely, but for now, here is a summary of the key steps we've covered to help you physically secure your business.

Basic Steps
- Lock the doors when you're not around
- Lock cash registers and offices
- Put important papers in locked drawers

Intermediate Steps
- Make sure everything valuable is behind three layers of security
- Be careful who you give keys to and make sure only the most trusted people have all the keys

Pro Steps
- Secure your bank accounts by limiting or removing wire transfer capabilities
- Make sure all money movement requires a second verification step
- Turn on notification whenever changes are made to the features of your bank accounts

Chapter 2

Cyber Awareness

When we talk about cyber awareness, we're referring to three basic things. The first is general awareness about the topic – what is going on in the world of cybercrime that you need to be aware of to keep your people and your company safe. The second is how to train your employees so that they can help you keep your company and themselves safe. The third is an understanding of industry-specific rules, such as merchants needing to protect credit card data and doctors needing to protect medical data.

Let's take them in order.

Cybercrime Awareness

At the present time, the biggest cybercrime threats that small business owners face fall into two categories. The first is untargeted, completely random crimes. You just happen to be the person walking down the alley when the mugger picks the next person they see to rob. The second is targeted crimes, which are the result of identifying and casing specific victims that are likely to have something of particular value to the criminal.

The random crime usually starts with some kind of mass email or computer virus campaign. This often takes the form of SPAM email – which is email addressed to millions or hundreds of millions of accounts using databases of stolen email addresses. These SPAM emails are used to plant malware (usually computer viruses but can be any kind of malicious computer code) on the devices of anyone unlucky enough to click on or respond to the message. These viruses then begin some kind of malicious activity or focus on spreading themselves to other computers, or both.

Once a computer is infected or "pwned" (which means completely owned by the criminal), the goal is to extract value. Typically, that involves installing ransomware to force the computer's owner to pay a ransom to get their files back, stealing data from the computer, or using the computer for other nefarious activity, unbeknownst to the owner. This could be something like launching a DDoS (Distributed Denial of Service) attack or doing "crypto jacking." More and more often these days, the attacks combine two or more of these. Why not, once the thief have access, they want to bleed the owner as much as possible.

DDoS attacks are used to vandalize other sites by flooding them with an overwhelming number of simultaneous requests from thousands to millions of "pwned" computers. This activity can bring these sites to a halt by interfering with normal computer transactions, or the flood of unwanted attention can serve as a distraction to mask other forms of more serious cybercrime.

Crypto jacking means hijacking a victim's computer and using it to perform calculations that help the criminal mine crypto currencies (such as Bitcoin, Litecoin, and Ethereum). While these might seem like crimes that don't involve you, think again. If your computer is busy doing the criminal's bidding, it is not printing your invoice, or worse than that, the slowdown is so severe in many cases that you are unable to use the infected computer for anything. And while they've got you doing their crypto mining, they'll use you for sending SPAM and attacking other computers.

And you can be sure that if criminals have control of your computer, they will also help themselves to your bank account and personal information while they are on your device. There are no victimless crimes in the cybercrime world. To recover from having your computer pwned often requires a lot of time and money. Some viruses are extremely difficult to remove, and most non-experts armed only with commercial anti-virus software don't fully succeed. Little scraps of the virus remain, causing ongoing grief or the re-infestation of the computer. Even if you are successful at restoring your computer, you may have to defend yourself against retribution sought by the ultimate victim of the attack launched from your device.

The Scourge of Ransomware

Infecting computers with ransomware is another common tactic of cybercriminals. Ransomware is a form of malicious computer code that surreptitiously encrypts (scrambles) all the files on a victim's computer with hidden keys and then holds access to those files for ransom. In theory, if you pay the ransom, you get the key and that allows you to get your files back.

But, as pirates of the high seas, kidnappers, and various other scoundrels have demonstrated throughout history, paying ransom is not as reliable a business model as charging ransom. Some cybercriminals have no intention of giving you back the files. Sometimes they can't because they don't really know how to use the ransomware. Sometimes they deploy badly (or dastardly) written ransomware that only focuses on encrypting and not decrypting (unscrambling), and sometimes they are only interested in robbing you so giving you back the files is of no real interest to them.

In most cases, once ransomware takes hold, the only reliable way to get your files back is to recreate them or restore them from backup.

Beginning in 2019 and becoming more prevalent in 2020, ransomware attacks started coming with a new extortion angle. Instead of simply holding the victim's files for ransom, a new twist was added. It went something like "if you don't pay us, we'll release all this information we stole from you." There is very little you can do to combat this issue once your files have been stolen. The temptation to pay certainly goes up, and each decision is unique. But consider these two factors. First, choosing to encrypt all your sensitive data makes this attack less successful, and second, if you aren't storing data that can damage you when released, you have little to worry about. These are both preventive measures, but combined with a good backup scheme, they can make you mostly immune to ransomware and ransomware extortion. In Chapter 3 on protecting your network, we discuss encrypting files for an extra layer of protection.

So, what about targeted cybercrime, the kind where the criminal cases specific victims? It is tempting to assume that because you run a small business, no one is interested in robbing you specifically. Nothing could be further from the truth.

Just as thieves case small businesses in the real world all the time – when is the safe emptied, are expensive tools left lying around, is the door open and the cash register unattended at certain times – so are cybercriminals interested in profiling small businesses to go after payroll, commit wire transfer fraud, steal customer credit card information, medical and dental records, and anything else of value. Small businesses that have large clients are also cased for access to those larger clients. Ever since Target was breached through their HVAC vendor in 2013, this approach is in every cybercriminal's arsenal.

These crimes all start in much the same way. It is relatively easy to estimate the size of the payroll or business turnover from publicly available information, such as the Dunn & Bradstreet (D&B) record for your firm. In addition to formal records, there are other methods to profile your firm. If you have ever put a company team picture on your website, you have informed a cybercriminal of your relative size and the size of your payroll. The thieves can also make easy guesses. For instance, real estate offices deal with clients who are processing large transactions for home purchases, medical and dental offices have healthcare records, as does a branch office of a physical therapy provider. Of course, any shop keeper likely has a merchant account with a bank.

Just as businesses specialize in the services they deliver, cybergangs also specialize. There are gangs that specialize, for instance, in payroll fraud, credit card theft, medical records theft, and wire transfer fraud.

Gone Phishing

Armed with information about who they want to target, there are many ways of getting into the victim's systems. One technique, called phishing (which means casting a net in a specific region or at a particular industry) uses email messages more specific than mere SPAM, to fool people who might be expecting certain types of messages. Perhaps a message that appears to be from a big bank that runs payroll for lots of customers in a specific industry. How about a message that appears to come from an insurance provider that many doctors or dentists accept? These targeted messages work on a surprising number of businesses.

Two other forms of targeted messaging are spear-phishing and whaling.

Spear-phishing uses messages targeted with a specific person in mind (perhaps a personal assistant) and might be so finely tuned that the temptation to click on the link is overwhelming. The message might be as specific as "That was a great play your son Bobby made last night, here's a short clip from the game video I shot, enjoy. See you next week." Clicking on that attachment usually results in the installation of malware.

Whaling means going after a high-profile target, like a key executive, high-priced lawyer or prominent surgeon. Whaling is more of an objective than a technique, and spear-phishing is often the technique used in whaling. The messages are more official, seeming to come from a business colleague, and typically usually address business concerns.

As I mentioned in Chapter 1, another category of fraud is "business email compromise," also know by the acronym BEC. BEC has been used broadly to describe all manner of fraud that involves email. However, BEC is a specific kind of fraud that is especially important to small businesses. It usually involves a wire transfer to get the money out of your account and often, out of the country. And it is usually accompanied by a sense of urgency. BEC works because of the pressure techniques applied. As I said in Chapter 1, put the extra safeguards on your bank account and you can shut BEC down.

We can't go into the entire world of cybercrime in this book. But one final thought on this topic is that cyber thieves will try to blend in with their surroundings. They will try to take advantage of events, good or bad, scheduled or out of the blue. Just like phone scams for fake charities ramp up when natural disasters occur, so too do cyber scams ramp up around any event in the news. So be especially vigilant around current emergencies, events, and trends.

For those who are interested in learning more, search for the above terms in your favorite search engine and enjoy the reading.

Now that we know the general threat landscape for small businesses, it's time to start thinking about what to do about it.

Stay Informed

A small business owner has a lot to keep up on, and time taken from growing your business must not be wasted. Focus only on what you have to know to survive. The unfortunate reality is that in today's connected world, that includes keeping up on cybersecurity threats. I am not recommending that you become a cybersecurity expert, but to be your own CISO (Chief Information Security Officer), you should keep up on the threats that affect small businesses in general as well as those that specifically affect your industry. I recommend that you add a small number of tasks to your daily alerts and your weekly routine.

There are several excellent publications that can help you stay up on the current threats. I recommend the following:

General and global cybersecurity news:

- Wired (wired.com)
- TechCrunch (techcrunch.com)
- The CyberWire (thecyberwire.com)

Specific threats and cybercrime activity:

- SC Magazine (scmagazine.com)
- Security Magazine (securitymagazine.com)
- Krebs on Security (krebsonsecurity.com)

Technical information for the cyber enthusiast:

- Threatpost (threatpost.com)
- The Hacker News (thehackernews.com)

In addition to these targeted publications, most major dailies, including the New York Times, The Wall Street Journal, The Los Angeles Times, The Atlantic, and The Washington Post, as well as news services such as Bloomberg, Business Journal, Reuters, The Associated Press, and National Public Radio all cover current cybersecurity events and cybersecurity awareness features. The foregoing is not an exhaustive list. The point is that cybersecurity is an important topic of coverage for nearly all publications and to be up on this topic, you should start reading articles about cybersecurity that apply to small business or to your industry. Focus on those articles in your favorite publications and add one or two of the targeted publications I listed to keep up on cyber threats. Add to that seeking out and focusing on articles about cybersecurity from your favorite industry publication.

Employee Training

Security professionals are fond of saying human hardware (our brain) is the most susceptible to hacking. You can think of social engineering (used to cleverly target phishing and spear-phishing attacks) as a branch of hacking that targets the human brain. The truth is we are susceptible and will always be susceptible. We evolved from loose collections of hominids and humanoids that learned to behave in specific ways to survive. Some behaviors are meant to ward off predators, some to prepare for common defense, others to quickly decide who to trust.

We can't unlearn these behaviors overnight when it took us millions of years to learn them in the first place. We are curious, we want to be helpful, we are confident when we should not be and fearful when it is unwarranted. What we need to do to combat our human nature is to automate and train.

Unfortunately, corporate security training has the reputation of being among the most dreaded activities for any corporate citizen. It's lumped together with workplace harassment training and anti-bribery training and the corporate new-hire classic: "how to log onto your new expense reporting system in 20 easy steps."

Well, fortunately, you don't run a large multi-national corporation and you don't have 10,000 employees to train. You don't have to sit your employees in front of a one-hour video and watch them find inventive ways of not watching it. But you still need to train your employees and keep cyber hygiene on the front of their minds.

Focus Your Training

With your employees, insist that they take care to protect both customer data and your sensitive data. Focus on training that targets the following topics:

- Phishing
- Passwords
- Protecting sensitive material
- Cyber-safe Behavior

Encourage employees to adopt cyber-safe behaviors, including:

- Apply maximum privacy settings on their social media accounts such as LinkedIn, Facebook, Twitter, and Instagram.
- Allow only their contacts to see their personal information such as birth date and location.
- Use a password manager.
- Always choose two-factor authentication when it is available.
- Lock devices when not in use.
- Inform you immediately if they lose a device, or one is stolen.
- Do not pass around USB (thumb) drives.

By limiting the amount of personal information that is available online, vulnerability to spear phishing attacks as well as identity theft can be reduced.

There are several good training solutions available, some for free. U.S. anti-malware maker ESET offers a free program which is gamified, meaning users get badges and certifications as they advance through the program. For many people that's just enough to make the training palatable. How you deliver it is less important than whether or not your employees are taking an active role in protecting the company.

Our suggestion is to make it personal. Help your employees understand how they can be more secure online. Keep in mind that if one of your employees is victimized by identify theft, she will be focused on recovering from that and not on your customers. Your employees are much more likely to take the training to heart if it makes them feel more personally secure at the same time. Whichever way you decide to deliver the training, make sure it is focused on their security along with yours.

If your employees aren't likely to benefit from an online training program, how about treating everyone to lunch once a quarter and having a round table discussion? If you can't lead the discussion, find someone who can. We'll list some excellent resources at the end of the book, including options for connecting with security professionals who are dedicated to helping make the digital world safer.

An excellent infographic to post can be found at ready.gov at this address:
https://www.ready.gov/sites/default/files/2020-03/cyberattack_information-sheet.pdf

This full-color PDF can be printed out and posted in the lunchroom and in other conspicuous places.

Industry Specific Regulations

You have heard over and over that being compliant isn't the same thing as being secure. Still, it is imperative that companies comply with the security and privacy regulations that apply to them. The most common for small companies include handling credit cards and safeguarding protected health information.

If your firm handles credit cards, as the owner you have signed an agreement with your bank mandated by the Payment Card Industry (PCI). You should bookmark the PCI security standards website, which can be found at www.pcisecuritystandards.org. On this site you will find helpful guides for how to comply with the terms of the contract you signed. You will also find helpful suggestions for awareness training for your employees. Just type "awareness training" in the search box.

An Internet search for this phrase or similar might yield results for:

- Webroot.com – a company that provides security for small businesses
- Knowbe4.com – a company that provides privacy training
- pcisecuritystandards.org – which is the PCI site
- ESET.com – anti-malware firm with free employee training

Healthcare providers also need to comply with the Health Information Portability and Accountability Act (HIPAA). You'll want to explore and bookmark the following site:

- www.hhs.gov/hipaa – HIPAA compliance guidelines and training program

Above all, keep it easy, focus on the minimum number of behaviors you want to change. It's about what your people do, not what they know.

In the next chapter, Protecting Your Network, we'll help you use the tools of the trade that you'll need to secure your business, but for now, here is a summary of the key steps we've covered to help you address cyber awareness and employee training.

Basic Steps

- Add a weekly review of cybersecurity-related articles from your favorite news sources
- Bookmark sites with specific rules for your industry
- Train your employees to recognize and resist phishing attacks

Intermediate Steps

- Implement a cybersecurity training program for all employees
- Work with your employees to help them update their online settings to protect their privacy and decrease the likelihood they will be phishing victims
- Add a weekly task to review one of the recommended publications from the general and global cybersecurity news category

Pro Steps

- Add a weekly task to review one of the recommended publications from the specific threats and cybercrime activity
- Host a monthly or quarterly lunch and learn with your employees to talk about how they can be safer online

Glossary

The Dark Web – A subset of the deep web that uses the traditional Internet infrastructure but requires special software for access (the most famous of which is TOR). Some believe (incorrectly) that the dark web is entirely for cyber criminals. However, many cyber criminals do use the dark web for their activity. Nefarious code and services are typically traded on the dark web.

Identity theft – Typically means using someone else's identity to open financial accounts, for instance bank accounts, credit cards, or personal loans, or impersonating someone to create a false tax return or submit fraudulent reimbursement requests for healthcare-related expenses.

Waterholes – Poisoning websites you use with malware – for small businesses the danger is in websites your industry uses.

Credential stuffing – Using credentials harvested from various breaches to commit account takeover (ATO) fraud.

Pretexting – Engaging in a back and forth email conversation to get you to lower your guard – over time, as they gain your trust, you are more susceptible to being asked to do something you would not have done for a stranger.

Vishing – Voice phishing – mass voice mail messages meant to get you to act.

Tailgating – Mostly a physical security issue – following someone in the door.

Quid-pro-quo – Exactly what it sounds like – someone does you a favor and you are compelled to do a favor back – it's surprising how small a favor can yield a big piece of the puzzle for a criminal.

Baiting – Leaving a thumb drive or SD drive or some other external storage device where you can find it – often loaded with malware – baiting can also refer to dangling something many people want, like a drawing for a drone or something similar.

Trolling – Making random unsolicited (often controversial) comments on various internet forums with the intent to provoke a reaction and engage the reader.

Cat phishing – Creating a fictitious persona on social media to engage, befriend, and (possibly) defraud another.

Chapter 3

Protecting Your Network

Talk of firewalls and routers and VPNs is very confusing if you don't work with these tools every day. In this chapter, we're going provide the information you need to know to make good, confident choices about which types of tools you need for your business. If you are working with a systems or security service provider, you will know the questions to ask and requests to make.

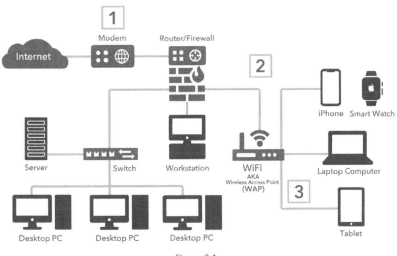

Figure 3.1

We've included a diagram here that depicts a simple SOHO network. SOHO means Small Office, Home Office. It is a term that has been around for some time and while the concept hasn't changed, the components of the network have. For an office, in the past you'd usually see a modem, a router, one or more laptops, perhaps a server and a couple of printers. At home, you'd probably have only one printer, and no server. Today, you can add a few smartphones, perhaps one or more tablet computers, a WiFi hotspot, networked TVs (even at the office), drones, wireless printer(s), and a host of smart devices (home or office) including lighting, thermostat, an Echo or other voice-controlled personal assistant, and perhaps some wearable technology such as a Fitbit or Apple Watch.

With this as a backdrop, we're going to go through each of these items and describe the protection to have for each. As usual, at the end of the chapter we'll have basic, intermediate, and pro steps.

Your Perimeter

The portion of the diagram labeled "1" consists or your modem and your firewall, and together, they form a perimeter for your network.

Your modem, as shown, connects your network to the Internet. The job of the modem is to connect with a device maintained by your Internet service provider (ISP), often a phone company or cable company (the carrier). Any data transmitted to the modem by the carrier will be passed to your network and any data transmitted to the modem from your network will be passed to the carrier.

Your firewall, as shown, acts as a barrier from your modem to your network. It does this by applying rules such as what data types to permit (e.g., email or web traffic) or who (people and programs) to permit into your network.

To keep your perimeter devices safe, do the following:

Physical security – It is advisable to keep your modem and firewall out of sight and physically protected. Depending on your facility, this means at a minimum keep them up off the floor, so people don't trip on them and either unplug or damage them. Try to keep them dry and relatively cool. Limit access by placing them behind a door or behind a locked door, such as in a closet.

<u>Device security</u> – These devices often come with rudimentary security turned on. There might be a login (for instance, the "admin" account) that is required to configure the device, but the password might be set to a factory default. Make sure you change the password for or disable all built-in logins, so the defaults don't work. Many of these devices come with default logging capabilities. Consider turning this feature on and monitoring the activity for potential issues, such as configuration changes you didn't make.

<u>Electrical power strategy</u> – Do not overload electrical outlets and use power strips and power cords that are appropriately rated. Consider battery backups (such as a UPS – uninterrupted power supply) to maintain network connections during brief brownouts or to protect against power surges. This can keep you productive and help you avoid replacing damaged equipment.

Make sure the UPS you buy can handle the power draw from your devices. At the end of this chapter we'll give a little detail for the D-I-Y network admins out there. A good rule of thumb is one 450VA UPS for your modem, routers, firewalls and WiFi routers and then a 450VA or 600VA UPS for every two to three "systems." Include the combo of laptop or server with monitor for each system.

Your Network

The portion of the diagram labeled "2" is your network. This includes your router and your wireless router (or WiFi router). Sometimes engineers refer to all the devices that you connect together as your network, some include the wires you use to connect the devices together, and some refer to your routers, wireless routers, and wires as your network infrastructure. For simplicity, we'll refer to your router, wireless router, and switch as your network.

- Your router connects your devices together with cables and routes network traffic to the right part of your network.

- Your wireless router connects your devices together using wireless transmission.

Most modern networks consist of both routers and wireless routers. For small networks, they may be in the same device.

To keep your network devices safe, do the following:

<u>Physical security</u> – As with the devices that make up your perimeter, it is advisable to keep your router and wireless router out of sight and physically protected. Again, depending on your facility, this means at a minimum keep them up off the floor, keep them dry and relatively cool, and limit access by placing them behind a door or behind a locked door, such as in a closet.

<u>Device security</u> – These devices also often come with rudimentary security turned on. As with modems and firewalls, there might be a login required to configure the device, and again the password might be set to a factory default. Make sure you change the password for or disable all built-in logins, so the defaults don't work on these devices as well.

Your WiFi Routers

You should change your network name to something non-descript. For instance, if installed by a cable company, you'll get a default name that looks like every other network name from that cable company, but with a different number (e.g., ABCD5534, ABCD4512). Change that name to something that does not make it obvious which cable provider you use and does not identify your company name. This makes it harder for the thieves to formulate an attack that will work on your device and makes it hard for them to be sure it's even your router.

Set broadcast to "No." This prohibits the network from advertising itself.

Turn on MAC address authentication, which requires you to add each device on your network rather that allowing devices in range to auto-configure themselves.

If you are going to provide WiFi for your customers or guests, buy, install, and configure a separate WiFi router and learn how to restrict access from that router to the rest of your network. Ask a professional for help if needed.

Choose WPA2 Personal (AES) for your security setting. If that is not available as a choice, use WPA/WPA2 Mode and if that is not available, use WPA or consider getting a newer model. These describe encryption protocols. The newer devices support better encryption and you should use it.

If the WiFi router is your only router, enable both DHCP and NAT. DHCP stands for Dynamic Host Configuration Protocol and NAT stands for Network Address Translation. These two services manage the assignment of IP addresses to your devices and the translation of those addresses to the Internet, respectively. If it is not your only router, your first choice is to have the non-WiFi router do both NAT and DHCP. If it is not obvious what you have and what should take precedence, ask a professional.

Many of these devices also come with default logging capabilities. Consider turning the logging on and monitoring the activity for potential issues, such as configuration changes you didn't make or high amounts of network traffic when no one should be using the network.

Electrical power strategy – As with all systems, avoid overloading electrical outlets, and consider battery backups to keep you productive and help you avoid replacing damaged equipment. Some say that laptops and phones already have battery backups and these batteries act as surge protectors. While that is true, these are often your most expensive devices and your most critical devices for your ongoing operations. Plugging them into a UPS is a prudent extra step. They don't take much of a draw from your UPS.

Your Computers

The portion of the diagram labeled "3" includes all of your computers, or more broadly, computing devices. Until the last few years, the term computer referred to micro, mini, and mainframe computers. These were loosely differentiated by size and whether they were intended for one user or many users. With the introduction of smartphones and tablets beginning in the early 2000s, we've come to acknowledge that "computing devices" is a much more inclusive term.

When phones were focused on a few related tasks (sending and receiving calls, perhaps sending a text message), they had limited computing power and simple operating systems. Now, phones, Fitbits, body cameras, drones, and smart appliances have robust operating systems, always-on wireless connections, and computing power to spare. Any of these computing devices can be on your network and need to be protected.

With that background, we'll refer to your computers as:

- Your systems, including your server(s), workstations, and laptops
- Your devices, including your smartphones and tablets, connected cash registers, connected medical devices, any smart devices such as personal assistants, wearable technology, and home automation such as lighting, heating and air, security devices including cameras, appliances, printers, and entertainment devices

To keep your computers safe, do the following:

<u>Physical security</u> – With the obvious exception of your server(s), it seems impractical to lock your computers away. However, you should still consider practical physical precautions with your equipment. All devices should be stored out of harm's way and kept dry and relatively cool. For special-purpose devices, such as cash registers or connected medical devices, storing under lock and key when the device is not in use is certainly appropriate. These are valuable systems and, where possible, consider securing them with a cable lock. Also, do not leave laptops unattended in public places.

<u>Device security</u> – Initial device setup includes some basic security settings, such as setting the boot (power on) password, and network settings to allow the device to join your network or your WiFi network. Allocate the proper amount of time for this process; rushing may lead to mistakes in your configuration. Read the installation manual first and acquire the information you will need for setup, including network information, to avoid accepting default settings that are not right for you.

Your Smartphones and Tablets

For smartphones and tablet computers, always lock the device when not in use. When in public settings, never leave your smartphone or tablet computer unattended or unlocked. Besides the obvious threat that small portable devices are easy to steal, unlocked devices can provide a trove of information in just seconds or minutes. Set a reasonable timeout for automatic locking of the device. Lock the device manually when not in use. Consider activating facial recognition, fingerprint recognition, or security patterns for added security. Do activate the security setting for deleting all data at the tenth consecutive incorrect password so that brute force attacks cannot be applied to breaking your security if you lose the device.

Manage your account with your carrier as well. All major carriers offer two-factor authentication for your mobile account. This is a good idea for two reasons. First, it secures your account, which protects you from someone making changes you don't want. Second, and the major reason mobile carriers offer (often require) two-factor authentication, it protects you from having someone transfer your phone number to a phone they control (called SIM fraud – SIM stands for subscriber identity module) which allows them to intercept your onetime passwords for bank accounts, email, social media sites, etc.

In addition to keeping your device up to date, you should also reboot (power off, power on) your devices regularly. Rebooting helps keep your device running smoothly. Review the privacy settings for your apps, remove apps and files (photos, videos, etc.) you don't need, and remember to delete the files from your deleted photos album. Having at least 20% of your memory available at all times helps your device function properly. Besides taking up space you need, each photo comes embedded by default with GPS coordinates and a time stamp. That's how your social media accounts know when and where a photo was taken. You don't want this information just sitting around for photos you don't need.

With the millions of attacks that occur against mobile devices every day, I recommend running security software on your smartphone and tablet computer. At a minimum, the software you select should include 24/7 scanning, protection against trojans, malware, adware, spyware, ransomware, and phishing scams. For added protection, consider using a firewall on your mobile device.

A firewall on a mobile device performs different functions than your network firewall or a firewall running on your server or laptop. On mobile devices, it helps you control the apps you run and allows you to approve, app by app, what has access to your data. While this is supposed to be configurable either within the app or in your system preferences, not all apps give as much control to the user for configuration as they should and malicious apps work to circumvent the control you should have.

This is not meant to be an endorsement of any product, but you might start your research by looking at Bitdefender, Panda, Scanguard, TotalAV, and Webroot for full suites and Mobiwol, Netguard, or NoRoot for a standalone firewall.

Servers, Workstations, and Laptops

For servers, workstations, and laptops, set the inactive screen timeout to no more than 15 minutes in case you forget to lock the screen when you step away. Set a reasonably complex boot password. By reasonably complex, I mean a minimum of 15 characters with some complexity. This is the one case where you will need to come up with a password that you can remember. A phrase is better than a random string of characters. Make your phrase more than 20 characters and you've got a great password. "I won my first game of risk when I was 20" is an example of something very easy to remember. At 41 characters it might even be too long.

As with devices, there are specific categories of security software that I recommend small business owners install. At a minimum, all systems, including servers, should run anti-virus software. Today's anti-virus is very different than that from even a few years ago. Besides protecting against viruses, modern anti-virus packages, now often referred to as anti-malware software (a broad category of software with malicious intent) or security suites or packages, go well beyond viruses to include trojans, adware, spyware, ransomware, and phishing scams. Modern anti-malware also often provides firewalls, password vaults, and VPNs. Whichever anti-malware solution you choose should explicitly reference protection against the malware mentioned above. If the security package you choose does not include a VPN or firewall, you should consider installing those in addition. Again, without endorsing any specific product, you might start your research by looking at Bitdefender, ESET, TotalAV, Trend Micro, and Webroot for full suites; ZoneAlarm and Comodo for standalone firewalls; and NordVPN, CyberGhost, IPVanish, and ProtonVPN for VPN capabilities.

If you must store sensitive data, meaning financial information, protected health information, trade secrets, or any files you would not like other people to see, consider using file or folder encryption software. Again, not endorsing any particular software, searching for nCrypted Cloud, Boxcryptor, Cryptomator, and UBIQ will show you options for easily encrypting files or folders that contain sensitive data. Using applications such as these, you'll be able to drag and drop files into a file manager that encrypts files before they are stored in the cloud or even while they are on your hard drive.

Internet of Things (IoT) Devices

For personal assistants, wearable technology, and home automation such as lighting, heating and air, security devices including cameras, appliances, printers, drones, and entertainment devices, it is tempting to ignore the security concerns because we either don't know much about them or they serve a singular purpose and require little attention to succeed in the mission we have for them.

In addition to these attributes, a lack of imagination about what these devices might be made to do and a single-minded goal to get cheap, single-purpose devices to market quickly, have led to security that is often quite lax. In every case, do your research and reward the manufacturers with the devices rated as more secure with your business. Choose the most secure settings in setup, keep the devices updated, and pay attention to the news. This is a market segment where many customers are price sensitive and that is the dangerous part of the market. Devices are hijacked every day for nefarious use. As I said in Chapter 2, there are no victimless crimes in the cybercrime world.

It may be difficult to see the harm, but we're not talking about penny ante network bandwidth theft. Unsecured security cameras and microphone-enabled entertainment devices can be used to spy on their owners and unpatched networked appliances are routinely commandeered to create botnets that are then used to attack the service providers we rely on every day, such as banks, power grids, hospitals, and emergency services.

Take some prudent steps to secure these devices and especially any virtual assistant you use to control your home/office automation:

- As with the earlier advice, don't keep default passwords, plus turn on the strongest encryption option your equipment will support.
- Consider using a touchpad unconnected to the rest of your smart home to control your security systems to separate the commands from your lighting, heating, air and entertainment devices.
- Watch what you say. Voice-controlled systems use cloud computing to listen to and interpret your commands. Don't speak passwords and answers to security questions in range of your devices' microphones.
- Go a step further and turn off your mic when you don't need it.
- Enable voice recognition. This sounds counter-intuitive given that I just said mind what you say, but by turning on voice recognition you train your devices to your voice and that decreases the likelihood another person's voice can seize control of your automation.
- If you're technically savvy and are willing to take it a step further, configure a guest network for your IoT devices which includes all the televisions, speakers, lights, wearables, and home/office automation.

There are also security devices that you can install that provide a secure guest WiFi network for your IoT devices and protect the rest of your network if any of these devices are compromised. Makers of these devices (again, not an endorsement) include Norton Core, Bitdefender, Cujo, TP-Link Archer, Netgear Armor, and Fingbox.

Electrical power strategy – At the risk of being repetitive, clean and reliable power are essential. The good news for devices in this category is that other than servers and printers, everything in this category runs on batteries. That means you have two added protections: the battery takes over for brownouts and short-lived blackouts, and because the battery delivers all the power to the device, even when you are plugged into an outlet you still get surge protection. But don't let it go to your head. These devices are indispensable to your everyday life and in this context, more importantly, to your business. Make sure you are fully charged, routinely inspect battery health, and keep UPS units fully charged. For bonus points, have an extra UPS charging with no devices plugged in. In the event of an emergency, you can charge up all your devices while waiting for the power to be restored. For extra bonus points, have a generator available to recharge the UPS for up to seven days.

One brief caveat, though. Feel free to plug an inkjet printer into a UPS, but don't plug your laser printer into a UPS. A laser printer draws tremendous power as it cycles power to the fuser and this cyclic draw will severely damage a UPS. That fuser has to be hot enough to melt dry ink so it can be applied to a page. Inkjets don't have the same problem, they use microjets to spray liquid ink. However, you can plug your laser printer into a surge suppressor to help protect it from damage from power spikes.

In the next chapter, Updates and Backups, we'll detail what you should be doing to keep your systems up to date and appropriately backed up, but for now, here is a summary of the key steps we've covered to help you protect your network.

Basic Steps

- Always keep your systems in dry, cool, secure places. Keep them up off the floor and out of the line of foot traffic.
- Lock all devices
- Train your employees to recognize and resist phishing attacks
- Install and use a firewall and anti-malware software

Intermediate Steps

- Protect your critical systems with a UPS
- Configure your routers and WiFi routers with advanced settings to restrict use outside your network

Pro Steps

- Separate WiFi for personal and entertainment devices
- Size your UPS according to a maximum calculated amperage load

How to Calculate Your Maximum UPS Load

Add up the amps listed on each of your devices (see the labels usually on the back or underside of the device). If you have to convert from amps to watts, remember that volts times amps equal watts. So, a 1.25-amp draw on a 120-volt circuit is equivalent to 150 watts or VA (volt-amps) of power.

Your UPS is going to clean the power delivered to the devices plugged into it using either Power Factor Correction or Capacitors. Capacitors are more common for the size UPS you'll most likely be using, and, unfortunately, watts and volt-amps are not 1:1 with capacitor-driven power supplies. The factor ranges from .55 to .75. OK, that all means you should not load your UPS beyond about 60% of the stated capacity. So, if you need 900 volt-amps of capacity, a good rule of thumb is to get a 1.5KVA UPS. A 1.5KVA UPS should be the only thing plugged into a 15-amp circuit.

Glossary

Modem – Modem stands for "modulator-demodulator," which originally meant that it modulated a digital signal from a device to be transmitted over an analog phone line and then demodulated from the phone line back to a digital signal. The modern modem modulates and demodulates between signals of different types.

Firewall – Taken from the physical wall that is often used in construction to provide a barrier to contain potential fire threats, a network firewall limits the data that is allowed from outside to inside a computer network and vice-versa.

Router – A router directs data, also known as network traffic, from one device to another, either across network segments within a network or from one network to another.

Network – A collection of systems that form a cooperative unit within which various resources, such as storage, printers, and data, are shared for a common objective of the users.

Server – Typically a computer that houses or drives shared resources, such as disk drives, printers, tape backup systems and sometimes software, such as accounting, payroll, or customer relationship management (CRM) software.

Workstation – A powerful, single-user computer that is custom-built for specific functions such as CAD (Computer Aided Design), media production, or software development.

WiFi Router – A wireless router that connects to your devices via radio frequency (RF) signals rather than through cables.

Cable Lock – A flexible, heavy-duty, durable cable, often reinforced with metal fibers, that can be used to secure laptops and other portable or movable high-value devices to a building or other immovable structure.

UPS – Uninterruptable power supply. For our purposes these are battery systems ranging from 250 volt-amps (VA) to 600 volt-amps, 900 volt-amps, or as much as 1.5KVA (1,500 volt-amps). In the smaller range, there are probably just the battery and outlets, and in the larger sizes, you'll typically have gauges for load and health and maybe software for configuring alarm signals.

Chapter 4

Updates and Backups

A great first step before we dive into the mechanics for performing updates and taking backups is to create an inventory of all the systems (devices, computers, etc.) and applications (email, accounting, inventory control, etc.) that you use to run your business. With this inventory in hand, you'll be better able to validate that your backups are covering all bases and all systems are accounted for when performing updates.

Looking again at the three categories of devices defined in Chapter 3 – your perimeter, your network, and your computers – we're now going to cover updating and backing up your devices. For computer professionals, updating and backups are typically referred to as "basic hygiene" and these activities are considered critical to protecting your systems.

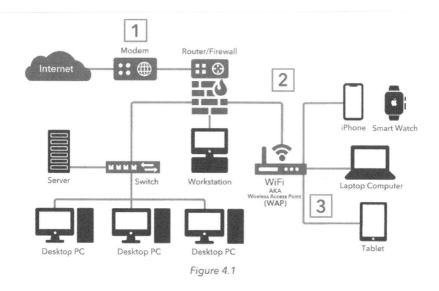

Figure 4.1

Update Strategy for Your Perimeter

Make sure that all recommended firmware and software updates are applied in a timely manner. When potential vulnerabilities are discovered for specific devices, cybercriminals search for all susceptible devices to target them. The sooner you apply firmware and software updates, the less susceptible you become. Subscribe to the mailing list from your device's manufacturer to inform you of updates. A common function of the device registration process is to add you to the mailing list for the device you've bought. This is a great reason to register your devices. The process also activates your warranty. If the manufacturer doesn't offer automated notification, make yourself a note (perhaps with a calendar invite or a recurring task in a task tracking application) to check for updates on the manufacturer's website at least every 90 days.

Backup Strategy for Your Perimeter

Most SoHo modems and firewalls have the ability to save a copy of the configuration file, sometimes in the form of an email to an address of your choice. Save this copy of the configuration file somewhere safe, such as a thumb drive you put in a safe or a Google drive which is a loud application. If you have to recover your firewall, restoring the configuration from a backup copy can save you a lot of time. If you can, save a text version of the configuration settings as well. If you need to replace a damaged firewall, it could prove very helpful to have a text version you can use as a reference as you configure your replacement firewall. If you hire an consultant, it will help them understand your configuration.

Update Strategy for Your Network

Make sure that all recommended firmware and software updates are applied in a timely manner. These updates come out frequently (at least quarterly) and deploy fixes to errors in code as well as patches to vulnerabilities discovered since the last update. Do not make the mistake of letting your patches get out of date, and do not draw quick conclusions about quality because of frequent patches. Sometimes the volume of patches is driven by the disproportionally higher number of attacks on particularly popular devices. Vulnerabilities that were unknown at the time the code was originally written are discovered often and staying diligent with patches is an essential security practice.

This strategy applies to other devices on your network as well. Office equipment such as printers occasionally have updates you should apply, but home and office automation, including lighting systems, security, heating and air, and streaming devices can require updates as well. A good rule of thumb is if the system was professionally installed, buy a service plan that includes maintenance and updates. If you installed it yourself, follow the plan for routers, firewalls, and modems and stay informed on the manufacturer's recommendations. Keep in mind that everything is connected and therefore any device not updated is vulnerable to attack and makes your network more vulnerable.

Backup Strategy for Your Network

Most routers and wireless routers also have the ability to save a copy of the configuration file, sometimes in the form of an email. As with the modem and firewall, restoring the configuration from backup can save you a lot of time. As with the firewalls and modems, also try to save a plain text copy of your configuration file that you can consult as a reference. Having this information readily available can mean the difference between being down for a few hours or a few days. That's time during which you are not running your business.

Update Strategy for Your Computers

The update strategy for your laptops and workstations needs to be much more comprehensive. For devices such as phones, tablets, fitness devices, and watches, make sure the operating system is updated whenever new updates come out for the version you have. Your device should give you a notification, usually as a badge or banner on the device. Also, keep your apps up to date as new versions are provided. Watch for new features and options and make sure you understand what these new features do for you or to you. Do not be shy about turning off new features until you understand how they add value to your daily routine. Every feature represents a security risk. In Chapter 7 I dive deeper into data privacy and the key take-away is that every piece of data you leak provides an opportunity for someone to exploit you. So be mindful about what you share and make sure the value you receive in return is worth it.

You'll need a similar level of diligence for your laptops, desktops, and servers. If you are using only general purpose applications, such as web browsers, a word processor, a spreadsheet tool, and email, then the likelihood is high that updates will be frequent enough and comprehensive enough to stay up with threats, and come with a relatively low risk of causing disruption to your other operations. However, though the risk is low, there is risk that any update could corrupt key software or data. It is always prudent to complete a full backup before applying any update.

If you are using specialized software, such as an accounting package, point-of-sale software, or software that is specific to your industry, you'll want to stay in touch with the vendor's support staff and follow their instructions for applying updates. Also keep in mind the suggestions provided below for backups in support of your updates.

Backup Strategy for Your Computers

As with your update strategy, the backup strategy for your computing devices needs to be comprehensive. Starting with your personal devices, you should have both local and remote backups for any device that holds your data. By remote backups, I mean backups that are either saved directly to a cloud service or made locally but then sent offsite for safekeeping. Personal fitness devices capture data but typically send it to a cloud service, or a phone or PC you own. These devices typically don't need to be backed up as they can easily be restored. Phones and tablets typically store apps, music, photos, videos, documents, and other data you purchase or generate. These devices should be backed up as the data is specific to you and cannot be restored by reinstalling. Local backups can be achieved by synching your device on a personal computer. Remote backups can be achieved by subscribing to a cloud service. These services are often available from your vendor, as for instance the iCloud is available to Apple device owners or OneDrive for those who use Microsoft.

For your laptops, desktops, and servers, the best practice is to have full and incremental backups scheduled over a monthly, weekly and daily basis, following the 3-2-1 backup rule. The 3-2-1 backup rule says to keep three backup copies, using at least two different types of media (CD, tape, thumb drive) and storing at least one offsite.

Unless you have the time, discipline, and expertise to faithfully execute such a plan, I suggest you add a backup service to your list of critical purchases. Software packages to consider (again, not an endorsement) include Acronis, Paragon, Backblaze, and IDrive. The best backup strategy is one that works, so test your backups periodically and make sure essential files are being backed up and can be restored. An added bonus for a successful backup strategy is that a good backup, as we discussed in Chapter 2 on cyber-awareness, is the one of the most effective strategies to combat ransomware attacks.

A point I just made bears repeating – the best backup strategy is one that works. And it has probably become apparent by now that good systems hygiene is about diligence and caring about the details. As a small business owner, we know you don't have time to obsess over the details, and we know you can't spend any more money on your systems than you have to. Spending the time to make sure that your devices are updated, your systems are being routinely backed up, and you can restore your files from backup are among the most important steps you can take to be secure.

In the next chapter, Access and Passwords, I'll show you how to appropriately grant and limit access to your critical systems, but for now, here is a summary of the key steps we've covered to help you stay on top of updates and backups as a key step in securing your systems.

Basic Steps

- Register your devices with the manufacturer to trigger your warranty and be added to routine notifications for updates
- Apply updates for perimeter and network devices as soon as they are available
- Backup all personal devices both locally and using a cloud service

Intermediate Steps

- Make a backup copy of configuration files for perimeter and network devices
- Periodically check the manufacturer's website for any updates in cases where automatic notification is not available
- Implement a 3-2-1 backup strategy using a backup solution that automates daily, weekly, and monthly backup processes

Pro Steps

- Make a backup copy before applying updates
- Routinely test your backup strategy by successfully restoring files from the backup copy

Chapter 5

Access and Passwords

In this chapter, we're first going to talk about who has access to your systems and how you protect yourself and your employees when accessing your essential technology. We call this process "Access Management." We'll describe how to create and manage strong passwords. As with Chapter 1, when we talked about locking the doors, access in the digital world has its analogies to access in the physical world. We double lock critical doors, with deadbolts or bars. We are careful about who we give keys to, and which keys. Expensive cars have valet keys that can turn on the ignition but not open the trunk or glove compartment. If we have a security system on our home, we are used to having a unique code to change the programming, beyond just enabling or disabling the alarm. If we have children, especially "tweens," we might have parental controls on TVs and phones. The same concepts apply to cybersecurity for our businesses.

In Chapter 1 we talked about deciding who needed to have keys to the shop, the office, closets, and drawers, and how to put critical systems behind more than one layer of protection. In this chapter, we're concerned with what users have access to what files and systems and how access is controlled and made safe. There are four main questions to answer about access management:

1. Who do you give access to your systems?
2. What extra steps should you take to secure access?
3. How do you share passwords (when you must) and transfer files securely?

4. What precautions should you take outside your home or office?

Who Do You Trust?

Let's take these in order. Question 1 is "Who do you give access to your systems?"

First, let's talk briefly about the types of accounts. Basically, there are four different types of accounts:

1. **User accounts** are the regular accounts that typically belong to employees not maintaining the systems for you.
2. **Privileged accounts** are accounts that you use to configure applications and some server functionality.
3. **Systems accounts** are accounts that run for you, to drive printers, execute backups, and perform other functions.
4. **Administrator accounts** are the keys to the kingdom that allow you unfettered access to every aspect of the server or any system or device for which you have an administrator password. A common administrator account is called the "local admin."

It's surprising how many large businesses give too many people access to administrator passwords. The same is true for small businesses. The combination of feeling overwhelmed by having to make decisions about who gets what access, believing that not having a needed password will slow down employees who are critical to supporting the business, and not really knowing how to set up access correctly causes us to want to throw in the towel and just give the administrator passwords to everyone who might seem to need them.

So, who needs the administrator passwords? Well, as the owner, you need them. If there is a very trusted employee who performs most of the systems administration tasks, that person needs them. And if you have hired an IT firm, that firm needs them. That's it.

Who needs privileged accounts? This is also a small list. Don't just assume the accountant needs all the passwords for all the systems that have anything to do with accounting. Perhaps he or she needs none of them. It could be they just need you to run a report from your accounting system. Maybe your IT people can configure the accounting software so that the accountant can have a regular account. Even software for small businesses (like QuickBooks) and for slightly larger firms (like Peachtree) have support for different user roles and accounts. You can give employees access to only enter their time worked or run specific reports, and many other similarly restrictive account types. You might hear this referenced as "least privilege access" meaning each account is given the least amount of privilege to get the job done.

As a small business, you aren't likely to have a lot of systems accounts, but the ones you do have are critical and often function out of sight. Make sure you or your IT firm disables all system accounts that are not in use and changes any default passwords on system accounts you do use. That's an essential step to cut down on the number of ways cyberthieves can surreptitiously attack you.

It is difficult for a small business owner to know what system accounts come with any system they might buy, especially industry-specific software or even general-purpose systems like payroll and accounting. One way around this, for most of the software you will need, is to subscribe to a service rather than buy and install a package yourself. This gives you two advantages: first, you aren't installing something that you then have to configure, secure and maintain, and second, there is a dedicated team of security professionals protecting your system just as they protect everyone else. Don't be scared off by the never-ending news coverage of cyber-attacks. The least common form of attack is against the category of "software as a service." We'll teach you steps you can take to protect yourself against those attacks later in this chapter when we talk about password security.

One final recommendation about accounts is to make sure everyone has a unique account. Don't let people share group accounts unless those accounts are so limited in functionality that they could never be used for anything else. If something goes wrong, you want to be able to trace bad activity (purposeful or mistaken) back to a single person in order to hold them accountable or give them extra training.

Here is a simple plan for how to assign accounts:

1. Assign everyone who needs access to computer systems for your company their own, unique user account.
2. Disable or have your IT firm (or the service you subscribe to for your software) disable all system accounts you are not using. There should be very few of them and disabling the ones you don't need should be part of your process for setting up a system, such as accounting or payroll.
3. Only give out access to the special, privileged accounts to the very few employees who need them.
4. Only you and your IT team should ever have access to the administrator accounts.
5. If you do use any shared accounts, and we all know it happens, make sure you change the password when someone leaves. Also see below for a more in-depth conversation about sharing passwords and files.

Who Are You?

Question 2 is "What extra steps should you take to secure accounts?" Before we tackle that, let's think about why this is important. When we talk about hacking systems, we're talking about gaining access to the system physically or virtually (logging on at the keyboard or remotely) and then impersonating some account. The holy grail for hackers is to impersonate the administrator but hacking usually starts by impersonating a user. That is why you need to care about the accounts and why you need to be sure to give people only the access they need.

Now to answer the question, the two most critical extra steps to take to secure accounts are to use strong, unique passwords and to use two-factor authentication (2FA) whenever it is available. We're going to cover strong, unique passwords later in this chapter. Right now, let's talk about and demystify two-factor authentication.

Two-factor authentication means having a second factor, like a one-time password (OTP) such as an SMS code or a code from a key fob, smartcard, or some biometric verification (like a fingerprint, face scan, or voice print) to verify who you are. Besides your password, you provide a second verification method or "factor." That's all it means. The first attempts at this are familiar to a lot of folks – those tokens or key fobs that people often still carry around that display a code that changes every minute or two.

Now we have a wide range of options, including having a text code sent to your phone, getting a code from an app on your phone (not the same thing and more secure in some cases), and unlocking an app with your fingerprint, voice command, or facial scan. The two-factor technique builds on the idea that someone can easily steal your password, but it's more difficult to steal a physical thing from you (like a phone or key fob) and even more difficult to fake who you are. There is also activity around moving away from passwords altogether and using only fingerprints, voice, and other forms of identity proofing because there is nothing to remember. Other more esoteric techniques include measuring your walking gait or monitoring your keyboard typing profile. These last two are not intuitive but are amazingly accurate. Who would have thought that no two people walk precisely the same way or type in precisely the same manner?

Initially, 2FA was limited to banks, but now many of the systems and services we use online employ 2FA, and our advice is to use it whenever you can. A common misconception is that 2FA is only necessary for your most critical accounts, such as a bank account. But common access such as your email account now functions as a utility and is often used as the first step to identify you. Treat your email, your financial accounts, and all software that you rely on, including social media, as critical and use 2FA when it is offered. Once you get used to 2FA, you'll barely notice the extra step and it will not encumber your life.

> *Sharing is sometimes more demanding than giving.*
>
> - Mary Catherine Bateson

Question 3 is "How do you share passwords and transfer files securely?" You have heard it said that you should never share passwords, and we just told you above to give everyone in your company a unique account. But as we also noted, it's sometimes required to share passwords and files. There are safe ways and not-so-safe ways to do it.

Let's deal with sharing passwords first.

Rule 1: Do not do it for mere convenience. The easiest way to lose control of your accounts and passwords is to share them freely, as we noted above. Only do it when the cost of not doing so is significant lost productivity, in emergency situations that truly put life, health, or your business at risk, for systems that simply cannot be configured for multiple users, or for low-risk access with many users where the burden of administering the passwords significantly outweighs the security risk.

Rule 2: Change the password whenever any of the following conditions are met:

- You shared the password for a one-time use of the system in question, and that use is concluded.

- Individual(s) within the group that share a password to an asset you protect have left your employment.
- You believe the password has been compromised or see unusual activity.
- You can't remember the last time you changed the password.

In this case, assets are broadly defined. Many of us have lived in a housing complex with a numeric password entered on a keypad to access the trash collection cage. Use of the disposal service is an asset which the complex pays for, and former residents who know the passcode often use the service for free trash disposal. Hence, housing complexes periodically change their passwords to cut down on the freeloaders. Your assets are worth more than trash collection, so change the password.

Now let's look at sharing files. Again, there are safe ways and not-so-safe ways to share files, and safety should go up in proportion to the sensitivity or value of the data being shared. Documents with unmasked sensitive information, in other words, documents that don't X out items such as a social security number, a bank account number, salary figures, or personal information about employees or customers, should never be sent in clear text.

But what does that mean? It means don't attach a Word document or an Excel spreadsheet or a PDF file to an email message without encrypting the attachment first. But how do you safely do that? If you are a sophisticated user, and you are sending your data to a sophisticated user, you can use any number of encryption tools you both agree on and share the key securely, perhaps using a public, private key exchange. That is way too complex for the average user and even if you know how to do it, the chances are good that not all of your clients or vendors know how.

The easiest way to exchange files is with a secure file storage/transfer service. If you are communicating with a bank, financial services firm, or healthcare provider, they should provide a service for you. In most cases, these are safe to use for the exchange of sensitive information.

But what if your accountant or lawyer or business colleague doesn't have a service they provide or subscribe to? In that case, there are several very good low-cost general-purpose services available. Among these are drop.net, Dropbox, Box, Google Drive, and others. For specific types of data exchange, other services such as DocuSign facilitate secure exchanges. Again, this is not an endorsement of any particular product or service.

With all the stories of insecure file exchange and misuse, what precautions should you take? Glad you asked. Here are a series of steps that in some cases echo best practices we've already hit upon:

- Turn on two-factor authentication
- Enable email notifications, so you are informed when files are added or accessed
- Enable logging when available
- Use a VPN for access (see the next section for VPN information)
- Use strong passwords
- Delist linked devices you don't need for file sharing
- Confirm receipt of files you are sharing and delete them once the other party has them

Don't Leave Home Without It

Question 4 is "What precautions should you take outside your home or office?" Recurring themes of the constant news we hear about hacking are the stories involving travel. We've heard about how unsafe airport WiFi is and how dangerous it is to connect to the WiFi in the hotel lobby or the local coffee shop. We've heard about hacked cars and hacked planes, and now we're hearing about Airbnb guests reporting stealth video cameras. Is it safe to travel and use